AF827993

PROCESS ADVANCEMENT IN CHEMISTRY AND CHEMICAL ENGINEERING RESEARCH

PROCESS ADVANCEMENT IN CHEMISTRY AND CHEMICAL ENGINEERING RESEARCH

Edited by

Gennady E. Zaikov, DSc

Vladimir A. Babkin, DSc

APPLE ACADEMIC PRESS

Apple Academic Press Inc.
3333 Mistwell Crescent
Oakville, ON L6L 0A2
Canada

Apple Academic Press Inc.
9 Spinnaker Way
Waretown, NJ 08758
USA

©2016 by Apple Academic Press, Inc.

Exclusive worldwide distribution by CRC Press, a member of Taylor & Francis Group

No claim to original U.S. Government works

Printed in the United States of America on acid-free paper

International Standard Book Number-13: 978-1-77188-105-0 (Hardcover)

International Standard Book Number-13: 978-1-4987-1931-5 (eBook)

All rights reserved. No part of this work may be reprinted or reproduced or utilized in any form or by any electric, mechanical or other means, now known or hereafter invented, including photocopying and recording, or in any information storage or retrieval system, without permission in writing from the publisher or its distributor, except in the case of brief excerpts or quotations for use in reviews or critical articles.

This book contains information obtained from authentic and highly regarded sources. Reprinted material is quoted with permission and sources are indicated. Copyright for individual articles remains with the authors as indicated. A wide variety of references are listed. Reasonable efforts have been made to publish reliable data and information, but the authors, editors, and the publisher cannot assume responsibility for the validity of all materials or the consequences of their use. The authors, editors, and the publisher have attempted to trace the copyright holders of all material reproduced in this publication and apologize to copyright holders if permission to publish in this form has not been obtained. If any copyright material has not been acknowledged, please write and let us know so we may rectify in any future reprint.

Trademark Notice: Registered trademark of products or corporate names are used only for explanation and identification without intent to infringe.

Library and Archives Canada Cataloguing in Publication

Process advancement in chemistry and chemical engineering research / edited by Gennady E. Zaikov, DSc, Vladimir A. Babkin, DSc.

Includes bibliographical references and index.
Issued in print and electronic formats.
ISBN 978-1-77188-105-0 (hardcover).--ISBN 978-1-4987-1931-5 (pdf)
1. Chemical engineering. 2. Chemistry. I. Zaikov, G. E. (Gennadi™i Efremovich), 1935-, author, editor II. Babkin, V. A. (Vladimir Aleksandrovich), 1952-, author, editor

TP155.7.P76 2015 660 C2015-906504-6 C2015-906505-4

Library of Congress Cataloging-in-Publication Data

Process advancement in chemistry and chemical engineering research/[edited by] Gennady E. Zaikov, DSc, Vladimir A. Babkin, DSc.

pages cm
Includes bibliographical references and index.
ISBN 978-1-77188-105-0 (alk. paper)
1. Chemical engineering. 2. Chemical industry. I. Zaikov, G. E. (Gennadii Efremovich), 1935-
II. Babkin, V. A. (Vladimir Aleksandrovich), 1952-

TP155.P767 2015 660--dc23 2015036085

Apple Academic Press also publishes its books in a variety of electronic formats. Some content that appears in print may not be available in electronic format. For information about Apple Academic Press products, visit our website at **www.appleacademicpress.com** and the CRC Press website at **www.crcpress.com**

ABOUT THE EDITORS

Gennady E. Zaikov, DSc

Head of the Polymer Division, N. M. Emanuel Institute of Biochemical Physics, Russian Academy of Sciences, Moscow, Russia

Professor, Moscow State Academy of Fine Chemical Technology, Russia Professor, Kazan National Research Technological University, Kazan, Russia

Gennady E. Zaikov, DSc, is Head of the Polymer Division at the N. M. Emanuel Institute of Biochemical Physics, Russian Academy of Sciences, Moscow, Russia, and professor at Moscow State Academy of Fine Chemical Technology, Russia, as well as Professor at Kazan National Research Technological University, Kazan, Russia. He is also a prolific author, researcher, and lecturer. He has received several awards for his work, including the the Russian Federation Scholarship for Outstanding Scientists. He has been a member of many professional organizations and on the editorial boards of many international science journals.

Vladimir A. Babkin, DSc

Professor and Head, Research Department, Volgograd State University of Architecture and Engineering, Sebryakovsky Branch, Volgograd, Russia

Vladimir Aleksandrovich Babkin, DSc, is Professor and Head of the Research Department at Volgograd State University of Architecture and Engineering, Sebryakovsky Branch in Volgograd, Russia. Professor Babkin graduated from Bashkir State University in 1976 (Ufa, Russia) as a physicist, specializing in the field of applied quantum chemistry. He is the author of more than 1,200 scientific papers (including 14 monographs.)

CONTENTS

LIST OF CONTRIBUTORS

M. I. Abdullin
Bashkir State University, Ufa 450077, Russia.E-mail: ProfAMI@yandex.ru

M. Anachkov
Institute of Catalysis, Bulgarian Academy of Sciences, Bonchev St. #11, Sofia 1113, Bulgaria

D. S. Andreev
Sebrykov Department, Volgograd State Architect-build University, Akademicheskaya ul., 1, Volgograd 400074, Volgograd Oblast, Russia

L. Arabuli
Faculty of Exact and Natural Sciences, Institute of Inorganic–Organic Hybrid Compounds and Non-traditional Materials, Ivane Javakhishvili Tbilisi State University, 1 Ilia Chavchavadze Avenue, Tbilisi 0179, Georgia

M. I. Artsis
N.M.Emanuel Institute of Biochemical Physics, Russian Academy of Sciences,4 Kosygin str., Moscow 119334, Russia. E-mail:Chembio@sky.chph.ras.ru

B. Arziani
Tbilisi State Medical University, Tbilisi, Georgia

V. A. Babkin
Sebrykov Department, Volgograd State Architect-build University, Akademicheskaya ul., 1, Volgograd 400074, Volgograd Oblast, Russia

Kh. Barbakadze
Faculty of Exact and Natural Sciences, Institute of Inorganic–Organic Hybrid Compounds and Non-traditional Materials, Ivane Javakhishvili Tbilisi State University, 1 Ilia Chavchavadze Avenue, Tbilisi 0179, Georgia

A. A. Basyrov
Bashkir State University, Ufa 450077, Russia

T. Batakliev
Institute of Catalysis, Bulgarian Academy of Sciences, Bonchev St. #11, Sofia 1113, Bulgaria. E-mail: todor@ic.bas.bg

A. Berlin
N.N. Semenov Institute of Chemical Physics, Russian Academy of Sciences, 4 Kosygin str., Moscow 119991, Russia. E-mail:Berlin@chph.ras.ru

W. Brostow
Laboratory of Advanced Polymers & Optimized Materials (LAPOM), Department of Materials Science and Engineering, University of North Texas, 1150 Union Circle No.305310, Denton, TX 76203-5017, USA.E-mail: wbrostow@yahoo.com

A. M. T. D. P. V. Cabral
Faculty of Pharmacy, University of Coimbra, 3000-295 Coimbra, Portugal. E-mail:acabral@ff.uc.pt

T. Datashvili
Laboratory of Advanced Polymers & Optimized Materials (LAPOM), Department of Materials Science and Engineering, University of North Texas, 1150 Union Circle No.305310, Denton, TX 76203-5017, USA

I. Didbaridze
Kutaisi Akaki Cereteli State University, 59, Tamar Mephe st., 59, Kutaisi 4600, Georgia

V. Yu. Dmitriev
Sebrykov Department, Volgograd State Architect-build University, Akademicheskaya ul., 1, Volgograd 400074, Volgograd Oblast, Russia

A. Fainleib
Institute of Macromolecular Chemistry, National Academy of Science of Ukraine, Kiev, Ukraine. E-mail:fainleib@i.kiev.ua

V. Georgiev
Institute of Catalysis, Bulgarian Academy of Sciences, Bonchev St. #11, Sofia 1113, Bulgaria

K. Giorgadze
Faculty of Exact and Natural Sciences, Institute of Inorganic–Organic Hybrid Compounds and Non-traditional Materials, Ivane Javakhishvili Tbilisi State University, 1 Ilia Chavchavadze Avenue, Tbilisi 0179, Georgia

A. B. Glazyrin
Bashkir State University, Ufa 450077, Russia

M. D. Goldfein
Saratov State University, Saratov, Russia.E-mail: goldfeinmd@mail.ru

N. G. Grigor'eva
Institute of Petrochemistry and Catalysis of RAS, 141 pr. Oktyabria,450075 Ufa, Russia

O. Grigorieva
Institute of Macromolecular Chemistry, National Academy of Science of Ukraine, Kiev, Ukraine

D. Horák
Institute of Macromolecular Chemistry, Academy of Sciences of the Czech Republic, Heyrovskeho Sq. 2, 162 06 Prague 6, Czech Republic

A.V. Ignatov
Sebrykov Department, Volgograd State Architect-build University, Akademicheskaya ul., 1, Volgograd 400074, Volgograd Oblast, Russia

N. P. Ivanova
Belarusian State Technological University, Sverdlova Str. 13a, Minsk, Republic of Belarus

G. Jioshvili
Faculty of Exact and Natural Sciences, Institute of Inorganic–Organic Hybrid Compounds and Non-traditional Materials, Ivane Javakhishvili Tbilisi State University, 1 Ilia Chavchavadze Avenue, Tbilisi 0179, Georgia

P. Jurkovič
VIPO, Partizánske, Slovakia

S. G. Karpova
Institute of the Russian Academy of Sciences, N.M. Emanuel Institute of Biochemical Physics, Russian Academy of Sciences, Moscow, Russia

List of Contributors

A. P. Knyazev
Sebrykov Department, Volgograd State Architect-build University, Akademicheskaya ul., 1, Volgograd 400074, Volgograd Oblast, Russia

N. N. Kolesnikova
Institute of the Russian Academy of Sciences, N.M. Emanuel Institute of Biochemical Physics, Russian Academy of Sciences, Moscow, Russia

G. G. Komissarov
N.N.Semenov Institute for Chemical Physics, Russian Academy of Sciences, Kosygin St. 4, Moscow 119991, Russia.E-mail: komiss@chph.ras.ru; gkomiss@yandex.ru

G. V. Kozlov
Kh.M. Berbekov Kabardino-Balkarian State University, Chernyshevsky st., 173, Nal'chik 360004, Russian Federation

E. T. Krut'ko
Belarusian State Technological University, Sverdlova Str. 13a, Minsk, Republic of Belarus

O. S. Kukovinets
Bashkir State University, Ufa 450077, Russia

B. I. Kutepov
Institute of Petrochemistry and Catalysis of RAS, 141 pr. Oktyabria,450075 Ufa, Russia

G. Lekishvili
Tbilisi State Medical University, Tbilisi, Georgia. E-mail: gleki@gmail.com

N. Lekishvili
Faculty of Exact and Natural Sciences, Institute of Inorganic–Organic Hybrid Compounds and Non-traditional Materials, Ivane Javakhishvili Tbilisi State University, 1 Ilia Chavchavadze Avenue, Tbilisi 0179, Georgia.E-mail: nodar@lekishvili.info; gleki@gmail.com

V. M. M. Lobo
Department of Chemistry, University of Coimbra, 3004–535 Coimbra, Portugal. E-mail: vlobo@ci.uc.pt

M. Marônek
Slovak Academy of Sciences, Polymer Institute of the Slovak Academy of Sciences, 845 41 Bratislava, Slovakia

J. Matyašovský
VIPO, Partizánske, Slovakia. E-mail: upolnovi@savba.sk

I. Michalec
Slovak Academy of Sciences, Polymer Institute of the Slovak Academy of Sciences, 845 41 Bratislava, Slovakia

V. M. Misin
N. M. Emanuel Institute of Biochemical Physics, Russian Academy of Sciences, Moscow, Russia

A. K. Mikitaev
Kh.M. Berbekov Kabardino-Balkarian State University, Chernyshevsky st., 173, Nal'chik 360004, Russian Federation

T. V. Monakhova
Institute of the Russian Academy of Sciences, N.M. Emanuel Institute of Biochemical Physics, Russian Academy of Sciences, Moscow, Russia

R. R. Nabiev
Kazan National Research Technological University, 68 Karl Marx Street, 420015 Kazan, Republic of Tatarstan, Russian Federation

I. I. Nasyrov
Kazan National Research Technological University, 68 Karl Marx Street, 420015 Kazan, Republic of Tatarstan, Russian Federation

S. S. Nikulin
Voronezh State University of the Engineering Technologies, Voronezh, Russia

F. F. Niyazi
Fundamental chemistry and chemical technology, South-West State University, 305040 Kursk, street October 50, 94, Russia. E-mail: farukhniyazi@yandex.com

I. Novák
Department of Welding and Foundry, Faculty of Materials Science and Technology in Trnava, 917 24 Trnava, Slovakia

S. V. Novikova
Kazan National Research Technological University, 65 Karl Marx str., Kazan 420015, Tatarstan, Russia

N. K. Nuriev
Kazan National Research Technological University, 68 Karl Marx Street, 420015 Kazan, Republic of Tatarstan, Russian Federation

V. Patsula
Institute of Macromolecular Chemistry, Academy of Sciences of the Czech Republic, Heyrovskeho Sq. 2, 162 06 Prague 6, Czech Republic

A. A. Popov
Institute of the Russian Academy of Sciences, N.M. Emanuel Institute of Biochemical Physics, Russian Academy of Sciences, Moscow, Russia

K. Y. Prochukhan
Bashkir State University, Kommunisticheskaya ul., 19, Ufa 450076, Republic of Bashkortostan, Russia

N. R. Prokopchuk
Belarusian State Technological University, Sverdlova Str. 13a, Minsk, Republic of Belarus

Yu. A. Prochukhan
Bashkir State University, Kommunisticheskaya ul., 19, Ufa 450076, Republic of Bashkortostan, Russia

I. N. Pugacheva
Voronezh State University of the Engineering Technologies, Voronezh, Russia

S. Rakovsky
Institute of Catalysis, Bulgarian Academy of Sciences, Bonchev St. #11, Sofia 1113, Bulgaria

Z. Raskildina
Ufa State Petroleum Technological University, 1 Kosmonavtov Str.,450062 Ufa, Russia

A. C. F. Ribeiro
Department of Chemistry, University of Coimbra, 3004–535 Coimbra, Portugal.
E-mail: anacfrib@ci.uc.pt

E. G. Rozantsev
Saratov State University,Saratov, Russia

M. Rusia
Faculty of Exact and Natural Sciences, Institute of Inorganic–Organic Hybrid Compounds and Non-traditional Materials, Ivane Javakhishvili Tbilisi State University, 1 Ilia Chavchavadze Avenue, Tbilisi 0179, Georgia

N. Sagaradze
Faculty of Exact and Natural Sciences, Institute of Inorganic–Organic Hybrid Compounds and Non-traditional Materials, Ivane Javakhishvili Tbilisi State University, 1 Ilia Chavchavadze Avenue, Tbilisi 0179, Georgia

M. Samkharadze
Kutaisi Akaki Cereteli State University, 59, Tamar Mephe st., 59, Kutaisi 4600, Georgia

A. V. Sazonova
Fundamental chemistry and chemical technology, South-West State University, 305040 Kursk, street October 50, 94, Russia. E-mail:ginger313@mail.ru

A. I. Sergeev
Institute of the Russian Academy of Sciences, N.N. Semenov Institute of Chemical Physics, Russian Academy of Sciences, Moscow, Russia

R. A. Shagidullina
Kazan National Research Technological University, 65 Karl Marx str., Kazan 420015, Tatarstan, Russia

L. S. Shibryaeva
Institute of the Russian Academy of Sciences, N.M. Emanuel Institute of Biochemical Physics, Russian Academy of Sciences, Moscow, Russia

D. A. Shiyan
Kazan National Research Technological University, 68 Karl Marx Street, 420015 Kazan, Republic of Tatarstan, Russian Federation

L. Šoltés
Institute of Experimental Pharmacology of the Slovak Academy of Sciences, 845 41 Bratislava, Slovakia

R. Stoika
Institute of Cell Biology, National Academy of Science of Ukraine, Drahomanov St. 14/16, 79005 Lviv, Ukraine

O. V. Stoyanov
Kazan State Technological University, Kazan, Tatarstan, Russia

E. S. Titova
Volgograd State Technical University, Volgograd, Russia

V. V. Trifonov
Sebrykov Department, Volgograd State Architect-build University, Akademicheskaya ul., 1, Volgograd 400074, Volgograd Oblast, Russia

J. A. Tunakova
Kazan National Research Technological University, 65 Karl Marx str., Kazan 420015, Tatarstan, Russia. E-mail: juliaprof@mail.ru

N. V. Ulitin
Kazan National Research Technological University, 68 Karl Marx Street, 420015 Kazan, Republic of Tatarstan, Russian Federation. E-mail: n.v.ulitin@mail.ru

A. J. M. Valente
Department of Chemistry, University of Coimbra, 3004–535 Coimbra, Portugal.
E-mail: avalente@ci.uc.pt

M. Valentin
Department of Welding and Foundry, Faculty of Materials Science and Technology in Trnava, 917 24 Trnava, Slovakia

L. M. P. Veríssimo
Department of Chemistry, University of Coimbra, 3004–535 Coimbra, Portugal.
E-mail: luisve@gmail.com

G. E. Zaikov
Bashkir State University, Ufa 450076, Russia

G. E. Zaikov
N.M. Emanuel Institute of Biochemical Physics, Russian Academy of Sciences, 4, Kosygin St., Moscow 119334, Russian Federation. E-mail: Chembio@sky.chph.ras.ru

B. A. Zasonska
Institute of Macromolecular Chemistry, Academy of Sciences of the Czech Republic, Heyrovskeho Sq. 2, 162 06 Prague 6, Czech Republic

T. A. Zharskaya
Belarusian State Technological University, Sverdlova Str. 13a, Minsk, Republic of Belarus

M. V. Zhuravleva
Belarusian State Technological University, Sverdlova Str. 13a, Minsk, Republic of Belarus

S. S. Zlotsky
Ufa State Petroleum Technological University, 1 Kosmonavtov Str.,450062 Ufa, Russia

D. Zurabishvili
Faculty of Exact and Natural Sciences, Institute of Inorganic–Organic Hybrid Compounds and Non-traditional Materials, Ivane Javakhishvili Tbilisi State University, 0179 Tbilisi, Georgia

LIST OF ABBREVIATION

AAS	Atomic absorption spectroscopy
AMPA	2,2'-Azobis(2-methylpropionamidine)
APC	Acid powder-like cellulose
APTES	Aminopropyl triethoxysilane
ATR–FTIR	Attenuated total reflectance–Fourier transform infrared spectroscopy
BD	Beginning of the decay
BR	Butyl rubber
CMC	Critical micelle concentration
CPMG	Carr–Purcell–Meiboom–Gill
DLS	Dynamic light scattering
DMEM	Dulbecco's modified Eagle's medium
DGEBA	Diglycidyl ether of bisphenol-A
DSC	Differential scanning calorimetry
DTA	Differential-thermal analysis
DVM	Digital voltmeter
ED	End of the decay
EPC	Epoxy polymer coating
EPR	Electron paramagnetic resonance
ETP	Electron transport particles
FID	Free induction decay
GLC	Gas–liquid chromatography
GNC	Globular nanocarbon
HA	Hexylamine
HA-mPEG	Hydroxamic acid methoxy polyethylene glycol
HMDA	Hexamethylenediamine
LDPE	Low-density polyethylene
MCC	Microcrystalline cellulose
MCPBA	Meta-chloroperbenzoic acid
MFI	Melt flow index
MP	Melting points
mPEG	Methoxy polyethylene glycol
MRA	Mechanical-rubber articles
MRI	Magnetic resonance imaging
MST	Micro-scratch tester
NPC	Neutral powder-like cellulose
NS	Number of scans
PA	Polyamide

PA-mPEG	Phosphonicmethoxy polyethylene glycol
pAPh	p-aminophenol
PB	Polybutadiene
PCA	Principal components analysis
PDMAAm	Poly(N,N-dimethylacrylamide)
PE	Polyethylene
PEG	Polyethylene glycol
PLS	Partial least squares
PP	Polypropylene
PU	Polyester urethane
PUS	Sulfur-containing polyester urethane
PUSI	Ionomer of sulfur-containing polyester urethane
QSRR/QSAR	Quantitative structure–property relationship
RD	Recycle delay
RDF	Radial distribution functions
SEM	Scanning electron microscopy
TEM	Transmission electron microscopy
TEOS	Tetraethyl orthosilicate
TFPA	Trifluoroperacetic acid
TGA	Thermogravimetricanalysis
TMOS	Tetramethylorthosilicate
XRD	X-ray diffraction

PREFACE

This volume contains peer-reviewed chapters and is devoted to the publication of original research works from chemistry and their broad range of applications in chemical engineering. It covers theoretical and practical application of modern chemistry.

This book deals with different aspects of chemistry and chemical engineering. The book includes the most significant new research papers and other original contributions in the form of reviews and reports on new concepts and research being done in the world, thus ensuring its scientific priority and significance.

This is a multidisciplinary book dealing with many structural aspects of modern chemistry and chemical engineering. This volume will be of interest to researchers because it explores the principles of chemical bonding and matter organization, the impact of structural aspects on a chemical property or transformation, and the application of the newest physical methods in chemical structure research. The volume covers studies on the structure of single molecules and radicals, molecular assemblies, gases, liquids (including water and solutions), amorphous and crystalline solids, surfaces, films and nanoparticles (including inorganic, organic and organometallic compounds), molecular and polymeric materials, and single crystals and minerals.

PREFACE

CHAPTER 1

BIOACTIVE NITROGEN-CONTAINING COMPOUNDS WITH SPATIAL CARBOCYCLIC GROUPS: SYNTHESIS, MODELING OF PHYSICAL PROPERTIES, AND USE FOR CREATION OF INORGANIC–ORGANIC HYBRID MATERIALS WITH SPECIFIC PROPERTIES

N. LEKISHVILI[1], KH. BARBAKADZE[1], D. ZURABISHVILI[1], G. LEKISHVILI[2], B. ARZIANI[2], A. FAINLEIB[3], O. GRIGORIEVA[3], W. BROSTOW[4], and T. DATASHVILI[4]

[1]Faculty of Exact and Natural Sciences, Institute of Inorganic–Organic Hybrid Compounds and Non-traditional Materials, Ivane Javakhishvili Tbilisi State University, 0179, Tbilisi, Georgia, nodar@lekishvili.info

[2]Tbilisi State Medical University, gleki@gmail.com

[3]Institute of Macromolecular Chemistry, National Academy of Science of Ukraine, Kiev, Ukraine, fainleib@i.kiev.ua

[4]Laboratory of Advanced Polymers & Optimized Materials (LAPOM), Department of Materials Science and Engineering, University of North Texas, 1150 Union Circle No.305310, Denton, TX 76203-5017, USA, wbrostow@yahoo.com

CONTENTS

ABSTRACT

Quantitative "structure–properties" relationships (QSRR/QSAR) based on experimental data for the construction of models of dependences of the physical properties of bioactive nitrogen-containing compounds with bioactive spatial carbocyclic groups on molecular structures were studied. Several sets of molecular descriptors were used. The presence of the dataset outliers was controlled by using Principal Components Analysis (PCA); to ascertain the quality of models cross-validation was used. Virtual bioscreening and activity toward various microorganisms of the obtained compounds were established. Inorganic–organic hybrid materials with specific properties based on organic heterochain polymers and coordination compounds of some transition metals based on bioactive nitrogen-containing compounds were obtained and studied. Tribological properties and stability toward various factors (e.g., O_2, CO_2, and moisture complex action), photochemical stability (toward, e.g., ultraviolet and visible light), and isothermal aging of the created composite materials were studied. From preliminary investigations, it was established that the created bioactive composites could be used for prevention of materials from biodeterioration and noncontrolled biodegradation, inhibition of growth and expansion of microorganisms, which are causal factors of infectious diseases, and for prophylaxis and treatment of the above-mentioned diseases.

1.1 INTRODUCTION

Biodegradation of synthetic and natural materials by various microorganisms affects a wide range of industries and techniques. According to the existing statistical data, more than several hundred kinds of such aggressive microorganisms are known,[1–3] which damage especially carbon-containing polymers and materials. The actions of microorganisms on polymers are influenced by two different processes: (a) deterioration and degradation of polymers, which serve as a native substance for growth of the microorganisms (direct action), and (b) the influence of metabolic products of the microorganisms (indirect action). Losses caused by destruction of natural and synthetic materials with micromycets reach enormous amounts and constitute annually milliards of dollars.[4, 5]

Historical buildings, archeological artifacts (made of metals and their alloys, leather, and/or wood), museum exhibits, and collections of artwork all need protection from the influence of various aggressive microorganisms. So the protection of cultural heritage and various synthetic and natural materials is a global problem.[6]

One of the ways to protect synthetic materials from the action of microorganisms is the creation of novel polymer coatings with high bioactivity by modification of various polyfunctional adhesive polymer matrices with biologically active compounds.[7] Therefore, synthesis of compounds for bioactive composites and created antibiocorrosive coatings based on them for various natural, synthetic, and artificial materials is extremely significant and requires further developments.[8]

For the time being in the various spheres of science and technique, scientists are focusing special attention on asymmetric carbocyclic compounds (such as adamantane and its derivatives). Modification of various bioactive compounds by immunotropic and membranotropic adamantane groups has a great potential because of improvement in their hydrolytic stability and the increase in their biological activity.[9–12]

1.2 EXPERIMENTAL DETAILS

Nitrogen-containing compounds with spatial (adamantane) carbocyclic groups and adamantane-containing hydrazide coordination compounds of transition metals were synthesized by us earlier.[13]

Polyester urethane "PU"

(1) The mixture of 50 g (0.1 mole) polyoxypropylenglycole, 2.6 g (0.05 mole) diethylenglicole, and 26.1 g (0.3 mole) toluylendiisocyanate (the mixture of 2,4- and 2,6-isomers with ratio 65:35 wt.%) was stirred for 1 h at 80–90°C until formation of hexamethylendiisocyanate. The monitoring of the reaction was carried out by the content of the NCO group (the optimum content of the NCO group for the above-mentioned polyester urethane is ~7.6%).

(2) To hexamethylendiisocyanate, 0.05 mole hydrazine hydrate (the 1% solution of hydrazine hydrate in dimethyl formamide) was added for the purpose of chain lengthening. The mixture was stirred for 24 h. The monitoring of the reaction was carried out by infrared spectroscopy, in particular, by disappearance of the characteristic absorption bands of the NCO group in the IR spectra of the research sample (2273 cm^{-1}). It was established that for polyester urethane $M_w \approx 50{,}300$.

Sulfur-containing polyester urethane "PUS"

For fuctionalization of hexamethylendiisocyanate, its sulfurization was carried out with 98% sulfuric acid (5 wt.% of hexamethylendiisocyanate). To the mixture dropwise sulfuric acid was added for 20 min until 6.5–6.8% content of the NCO group. The reaction was carried out under the stirring condition at 80–90°C for 5 h. The degree of sulfurization of the obtained polyester urethane was ~3 wt.% (the degree of sulfurization was determined by titration of sulfo-groups). It was established that for sulfur-containing polyester urethane $M_w \approx 97{,}400$.

Ionomer of sulfur-containing polyester urethane "PUSI"

The sulfurized hexamethylendiisocyanate ($M_w \approx 118{,}000$) was cooled at 40–45°C and then the transfer of the SO_3H group was carried out from acid to salt form by the addition of equimolar (in relation to sulfuric acid) triethylamine. To the obtained ion-containing hexamethylendiisocyanate was added 0.05 mole 1% water solution of hydrazine hydrate under the stirring condition for 24 h for the purpose of chain lengthening.

The antibiocorrosive coatings were prepared in the following way: to the cyclohexanone solution of the polymeric matrix definite quantity of modifier (3 wt.%) and bioactive coordination compound (3–5 wt.%) was gradually added under the stirring condition until the formation of light color homogeneous mass. Later, the obtained composition was laid in the form of a thin layer on the surface of the selected various materials (wood, plastic, Teflon, etc.) for protection and was delayed in the air for 24–48 h at room temperature. After hardening, a homogeneous, smooth, mechanically stable protective layer was produced.

Methods of analysis

Standard methods for obtaining and purifying bioactive compounds were used.

The tribological properties of the polymeric matrices were determined using a Spanish Nanovea pin-on-disk tribometer (Micro Photonics Inc., 4972 Medical Center Cir # 4, Allentown, PA 18106, United States), micro-scratch tester (MST) (CSM, Neuchatel, Switzerland), and a Nicon Eclipse ME 600 Microscope. Standard microbiological methods were used for the study of the bactericidal and fungicidal properties of synthesized compounds.

1.3 RESULTS AND DISCUSSION

To establish the correlation between the structure of the possible bioactive compound and some of its fundamental properties, we synthesized and studied some anilides and nitroanilides[14] containing spatial carbocyclic groups (adamantane) with various organic radicals in the benzene ring (Scheme 1.1, Table 1.1). In order to select these compounds, we considered the availability of their synthesis and possibility of their perspective wide commercialization.

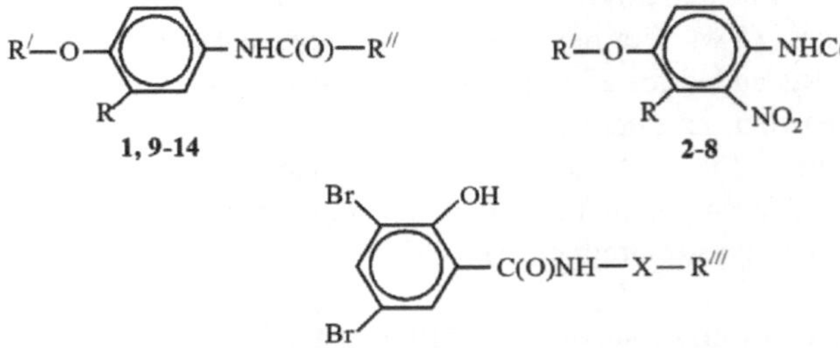

1, 9-14

2-8

15-16 R'''=Ad; X = - (15), C_6H_4 (16)

SCHEME 1.1 Nitrogen-containing derivatives of adamantane.

TABLE 1.1 Experimental Components Studied.

Compound[a]	R	R^1	R^2	Compound	R	R^1	R^2
1	H	CH_3	Ad[b]	8	Cl	4-ClC_6H_4	Ad
2	H	CH_3	Ad	9	H	C_2H_5	Ad
3	H	C_2H_5	Ad	10	H	Ad	CH_3
4	H	Ad	CH_3	11	H	Ad	$CH_2C_6H_5$
5	H	Ad	C_6H_5	12	H	Ad	C_6H_5
6	H	Ad	$CH_2C_6H_5$	13	H	Ad	Ad
7	H	Ad	Ad	14	Cl	4-ClC_6H_4	Ad

[a]Scheme 1.1; [b]Ad: adamantyl.

The techniques of the quantitative structure–property relationship (QSRR/QSAR) are used for establishing reliable models of biological activities and the physical–chemical properties of organic and element-organic molecules. The aforementioned approach is based on the representation of molecular structures with numeric quantities. They are calculated via straightforward algorithms and are known as molecular descriptors. Among the latter, considerable attention is granted to the autocorrelation-based descriptors.[14]

Let us give a brief outline of the approach. Suppose s_1, s_2, ..., s_n are variables, which present various molecular substructures. Most often, the paths of the corresponding molecular graphs are considered. As is clear, n, that is, the number of the paths depends on the number of atoms, or, in the graph-theoretical context, on the number of vertices. However, to build statistically reasonable models, one needs to represent molecules with vectors (i.e., series of molecular descriptors) of the same length. Therefore, we introduce a set of the template variables, t_1, t_2, ... t_m, where m remains constant, that is, is independent of the size of molecules of the dataset in question. The aforementioned substructures are characterized by special functions. An example is the product of numeric values of a physical property of the atoms of the substructure. That is, if we consider only paths as substructures, we have

$$f(s_i) = p_i^0 p_i^1 .$$

Here, p^0 is a physical property of the origin of the ith path and p^1 is that of the terminal vertex (atom) of the path. Examples of the physical property are sigma- and pi-charges, electro negativities, etc. Therefore, we arrive at a vector of the products of the physical properties of the paths, f, which has dimensionality equal to n.

Autocorrelation vectors are defined as linear transforms of the f vectors of the dimensionality n to the vectors a of the dimensionality m. The transform is given by kernel matrices (K):

$a = fK$, or in the functional form: $a(t) = \int K(t,s) f(s) ds$

The kernel matrix (K) has dimensionality (n,m). The vector a will always be of size m, and therefore, independent of the size of the molecule. The kernels are given by various functions. The kernel of the 3D-MoRSE descriptor, for example, is given as follows:

$$K(t,s) = \frac{\sin tr(s)}{tr(s)}$$

Here, $r(s)$ is the Euclidean distance.

QSPR calculations: We used MDL Isis Draw 2.5 SP4 to build molecular models. Afterward, we concatenated the models into the dataset by use of EdiSDF 5.02. The textual format of the dataset was SDF. We used the VCC-Lab e-Dragon web application for the calculation of molecular descriptors. Statistic 6.0 was a tool of our choice for building PCA and Partial Least Squares (PLS) models.

Among many available molecular descriptors at our disposal, we selected Radial Distribution Functions (RDF),[15] Crystal Structure Codes (MoRSE),[16] WHIM,[17] GETAWAY,[18] and traditional topological indices.[19] Our dataset[20] contained 16 compounds.

In order to detect outliers, that is, the compounds, which did not belong to the modeling population, we performed PCA.[21] However, unlike our previous contribution,[22] we did not identify any of the investigating compounds as outliers.

As one can see (Fig. 1.1), compounds 7, 8, 13–16[12] are situated somehow farther than the main group. This alone does not allow for their removal. For example, the Burden eigenvalues reveal that only compound 8 is an outlier (Fig. 1.2). When we used Randič-type invariants, none of the compounds left the main group (Fig. 1.3). Therefore, we decided to keep all of the compounds in the training dataset. It is noteworthy that the Randič-type invariants clearly output several clusters of compounds.

Our final step was establishment of relationships between these descriptors and the retention factors measured experimentally. Our studies show that the best model was achieved by employing, again, the GETAWAY descriptors. We used PLS[23] as the number of predictors was much higher than that of cases. We used cross-validation to define the optimal number of latent variables. In our study, we used 13 compounds in the training set and 3 for the cross-validation tests. Of course, the prediction power was lower in the case of cross-validation. We modeled both melting points (mp) and retention factors (R_f) within the same model, which, therefore, had two responses.

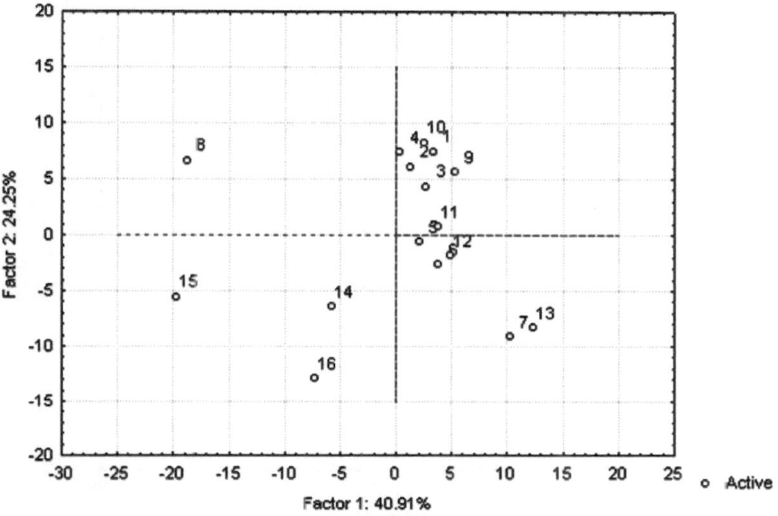

FIGURE 1.1 The outlier detection by means of PCA. The descriptor used is GETAWAY.

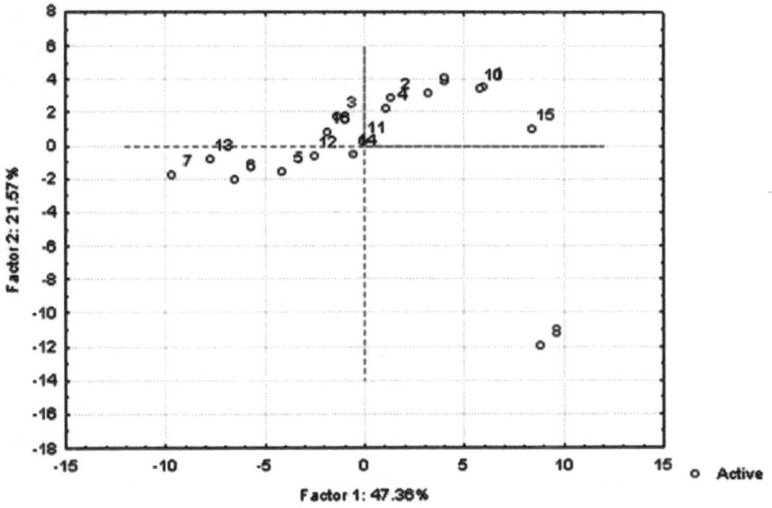

FIGURE 1.2 The outlier detection by means of PCA. The descriptors used are the Burden eigenvalues.

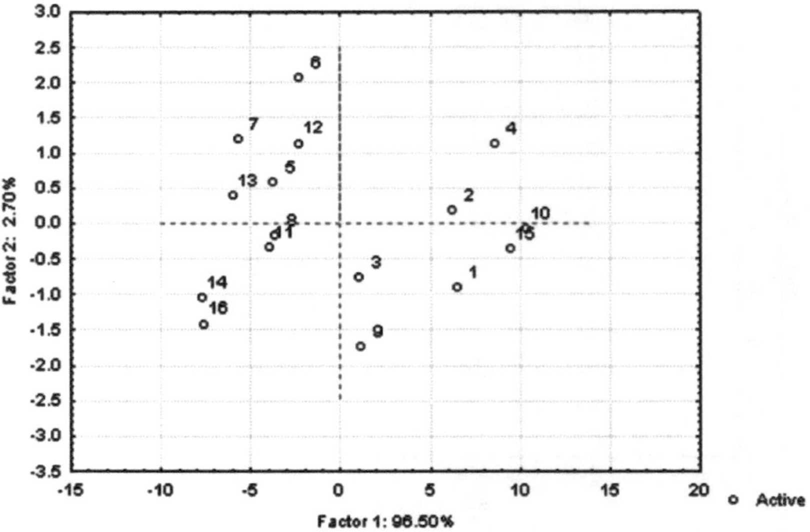

FIGURE 1.3 The outlier detection by means of PCA. The descriptors used are the Randič-type invariants.

The results of modeling look impressive as the square of the average correlation coefficient was as high as 0.92. The PRESS was also good enough (Table 1.2). One can examine the experimental and calculated values (Table 1.3). A reader should take into account that compounds 4, 13, 16 produced the test (validation) set.

TABLE 1.2 The Statistical Parameters of the Model.

Latent Var No.	Increase R^2 on Y	Average R^2 on Y	Increase R^2 on X	Average R^2 on X	R^2 for mp	R^2 for R_f	Sc. Press, mp	Sc. Press, R_f	Average Sc. Press
1	0.391833	0.391833	0.286528	0.286528	0.577891	0.205776	2.254355	1.727976	1.991166
2	0.286881	0.678714	0.275862	0.562390	0.611797	0.745632	2.094057	1.922039	2.008048
3	0.033169	0.711883	0.206076	0.768466	0.666069	0.757697	1.780881	1.833227	1.807054
4	0.041860	0.753743	0.109336	0.877802	0.670454	0.837033	1.763578	1.837925	1.800751
5	0.121957	0.875700	0.023599	0.901401	0.886302	0.865099	1.268438	1.445234	1.356836
6	0.028227	0.903927	0.048229	0.949629	0.910597	0.897258	1.211559	1.240578	1.226068
7	0.025273	0.929201	0.030173	0.979802	0.912324	0.946077	1.198989	1.180627	1.189808
8	0.023788	0.952989	0.004540	0.984342	0.939416	0.966562	1.389769	1.137673	1.263721

TABLE 1.3 Experimental vs. Calculated R/s.

Compound	mp, calc	R_f, calc	mp, exp	R_f, exp
1	184.9834	0.518455	178.5	0.60
2	126.5095	0.704182	134.5	0.65
3	142.8738	0.705937	134.5	0.73
4	102.3756	0.591046	121.5	0.42
5	154.5519	0.776873	152	0.80
6	128.6926	0.633454	128.5	0.65
7	171.7885	0.867685	173	0.86
8	167.9010	0.917788	169	0.91
9	193.2231	0.447994	206.5	0.40
10	166.1550	0.486902	166.5	0.50
11	178.3934	0.473237	160	0.41
12	170.2936	0.525245	181.5	0.53
13	198.3374	0.549713	240.5	0.77
14	175.8135	0.426382	176.5	0.45
15	226.3155	0.555867	226.5	0.55
16	183.8285	0.146642	194.5	0.17

1.3.1 STUDY OF THE BIOACTIVE PROPERTIES OF ADAMANTANE-CONTAINING ANILIDES AND NITROANILIDES

The modeling of the dependence of the structures of used nitrogen-containing adamantane derivatives on their fundamental properties (melting point and R_f) allows carrying out the theoretical evaluation of the possible bioactivity of the same structures. We carried out the preliminary virtual bioscreening of the obtained compounds by using the internet-system program PASS C&T.[24] The estimation of the probable bioactivity of chosen compounds was carried out via parameters P_a (active) and P_i (inactive); when P_a > 05, the compound could also show bioactivity experimentally. From the above-mentioned virtual bioscreening, based on the analysis of the obtained results, the synthesized compounds with experimentally high probability (Table 1.4) (P_a > 0.5) possibly will show the following bioactivity: antibacterial, antibacterial activity enhancer, anthelminthic, antiviral (*Arbovirus, Influenza, Picornavirus*), lipid metabolism regulator, and urologic disorders treatment.

Herewith, in the halogenated nitrogen-containing adamantane (15, 16) derivatives, an increase in antiviral (Influenza) activity was observed. The synthesized compounds with experimentally high probability (Table 1.5) possibly will show the following bioactivity: anti-infective, dopamine release stimulant, anthelminthic, antiparasitic.

TABLE 1.4 Relative Bioactivity of Some Nitrogen-Containing Adamantane Derivatives (1–14).

Compound	P_a/P_i								
	Lipid metabolism regulator	**5 Hydroxytryptamine** release inhibitor	**Urologic disorders** treatment	**Antiviral (influenza)**	**Antiviral (picornavirus)**	**Antiviral (arbovirus)**	**Membrane integrity agonist**	**Dependence treatment**	**Antibacterial Activity** enhancer
1	–	**0.900/0.004**	**0.872/0.010**	**0.791/0.004**	**0.705/0.004**	–	**0.703/0.101**	**0.741/0.014**	–
2	–	**0.834/0.004**	**0.813/0.022**	**0.782/0.004**	**0.673/0.013**	**0.869/0.003**	–	**0.545/0.056**	**0.436/0.239**
3	–	**0.823/0.004**	**0.797/0.027**	**0.816/0.004**	**0.689/0.010**	**0.899/0.003**	**0.312/0.170**	**0.389/0.174**	**0.371/0.311**
5	–	**0.371/0.087**	–	**0.450/0.052**	**0.572/0.053**	**0.762/0.010**	**0.548/0.155**	**0.350/0.218**	**0.550/0.113**
6	**0.388/0.213**	**0.256/0.179**	**0.327/0.315**	**0.269/0.205**	**0.509/0.097**	**0.613/0.108**	–	–	**0.663/0.022**
7	**0.344/0.241**	**0.778/0.005**	**0.745/0.049**	**0.638/0.011**	**0.679/0.012**	**0.803/0.005**	–	**0.404/0.160**	**0.498/0.172**
8	–	**0.746/0.005**	**0.755/0.044**	**0.549/0.023**	**0.724/0.005**	**0.820/0.004**	**0.549/0.154**	**0.543/0.057**	**0.593/0.070**
9	–	**0.891/0.004**	**0.855/0.013**	**0.781/0.004**	–	–	**0.752/0.083**	–	–
10	**0.973/0.003**	–	–	–	–	–	**0.770/0.075**	–	–
11	**0.966/0.004**	–	–	–	–	–	–	–	–
12	**0.970/0.003**	–	–	–	–	–	**0.838/0.043**	–	–
13	**0.971/0.003**	**0.876/0.005**	**0.818/0.021**	**0.775/0.005**	**0.705/0.004**	–	–	–	–
14	–	**0.843/0.005**	**0.788/0.030**	**0.752/0.005**	–	–	**0.792/0.065**	**0.720/0.017**	–

We tested the bactericidal and fungicidal activity of synthesized compounds according to the method described in Ref. 25. To this target we applied the test microorganisms—*Pectobacterium aroideae, Fusarium arenaceum, Autinomyces Griseus*, and *Fusarium proliferate*. The test results showed that synthesized compounds have revealed selected bactericidal properties and have suppressed the development of research cultures. The microbiological study of the investigated compounds confirmed the evaluated virtual concepts.

1.3.2 INORGANIC–ORGANIC HYBRID MATERIALS WITH SPECIFIC PROPERTIES BASED ON SYNTHESIZED COMPOUNDS

One of the modern ways to protect the synthetic and natural materials from the action of microorganisms is a creation of novel antibiocorrosive coatings with high bioactivity and multivectorial and directional action based on inorganic–organic hybrid composites.[8] Antibiocorrosive coatings mainly contain two components—biologically active compound and polymer matrix, where the bioactive compound dropped. Some polyfunctional hetero-chained organic polymers, such as polyurethane elastomers, polyurethane acrylates and ionomers, have been successfully used as a matrix for creation of antibiocorrosive coatings. In this regard, special interest has been developed in many respects in bioactive, including bactericide and fungicide polyurethanes.[26, 27]

At the first stage of the research, for the preparation of *"short-term"* action polymeric composites, the following polyurethanes were selected:

1. Polyester urethane "PU":

$$\cdot\!\left[(OR)_n-O\overset{\overset{O}{\|}}{C}NH-Ar-NH\overset{\overset{O}{\|}}{C}\right]_2\!\cdot OR'-O\overset{\overset{O}{\|}}{C}NH-Ar-NH\overset{\overset{O}{\|}}{C}\cdot\!\left[NHNH\overset{\overset{O}{\|}}{C}NH-Ar-NH\overset{\overset{O}{\|}}{C}\right]_3\!\cdot,$$

where $R = -CH_2CH(CH_3)-$; $R' = -(CH_2)_2O(CH_2)_2-$.

2. Sulfur-containing polyester urethane "PUS":

$$\cdot\!\left[(OR)_n-O\overset{\overset{O}{\|}}{C}NH-Ar-NH\overset{\overset{O}{\|}}{C}\right]_2\!\cdot OR'-O\overset{\overset{O}{\|}}{C}NH-Ar-NH\overset{\overset{O}{\|}}{C}\cdot\!\left[NHNH\overset{\overset{O}{\|}}{C}NH-Ar-NH\overset{\overset{O}{\|}}{C}\right]_3\!\cdot,$$

where Ar =

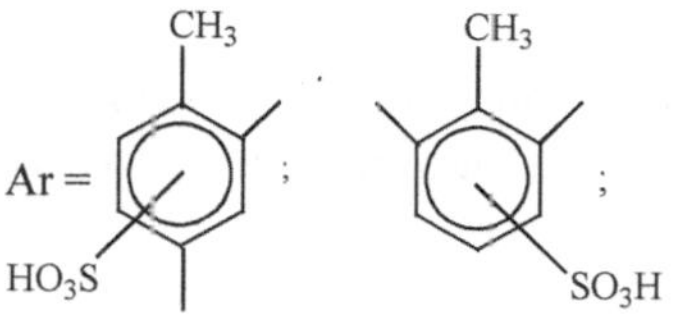

$R = -CH_2CH(CH_3)-$; $R' = -(CH_2)_2O(CH_2)_2-$.

3. Ionomer of sulfur-containing polyester urethane "PUSI":

$$\cdot\!\left[(OR)_n-O\overset{\overset{O}{\|}}{C}NH-Ar-NH\overset{\overset{O}{\|}}{C}\right]_2\!\cdot OR'-O\overset{\overset{O}{\|}}{C}NH-Ar-NH\overset{\overset{O}{\|}}{C}\cdot\!\left[NHNH\overset{\overset{O}{\|}}{C}NH-Ar-NH\overset{\overset{O}{\|}}{C}\right]_3\!\cdot,$$

where Ar = 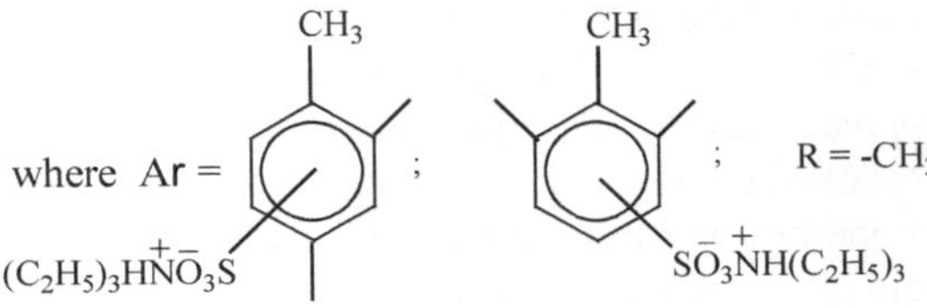 ; R = $-CH_2CH(CH_3)-$; R' = $-(CH_2)_2O(CH_2)_2-$.

4. Polyester urethane "BUTYP T-261" based on 4,4-dimethylmethanediisocianate and oligobuthyleneglicoladipinate.

Some adverse effects, in particular, low scratch and wear resistance and environmental degradation, have hindered many important applications of polymers.[28, 29] Thus, for physical and/or chemical modification of the above-mentioned polyurethanes for the purpose of improving tribological properties, silicon-organic oligomers were used, which will be able to improve several properties of the initial polymer systems including mechanical (elastic, adhesive strength) and thermophysical (frost- and thermal-resistant) and hydrophobicity:

(a) bis(hydroxyalkyl)polydimethylsiloxane:

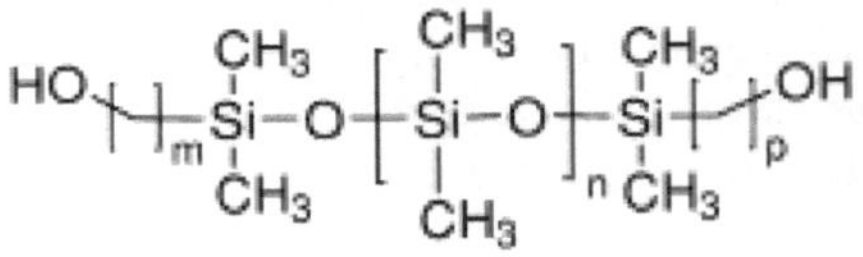

(b) α,ω-dihydroxymethylvinyloligoorganosiloxane:

$$HO\{[Si(CH_3)_2O]_{98.5}[Si(CH_3)(CH{=}CH_2)O]_{1.5}\}_xH, \text{ where } x = 3–5.$$

As bioactive compounds for antibiocorrosive coatings we used other nitrogen-containing adamantane derivatives[13, 30–32] such as transition metals coordination compounds of adamantane-1-carboxylic acid hydrazide synthesized by us[30,13] considering their stability and ability to form dipole–dipole and hydride bonds with the polymeric matrix. It was expected that hydrazide metal complexes will show various applications in the fields such as fungicidal, bactericidal, antiparasitic, antioxidative, and cytotoxic studies.[31, 32, 33, 34]

The most probable structures of the obtained coordination compounds[13] are given in Scheme 1.2.

By using obtained polymeric matrices and coordination compounds (3–5 wt.%) new effective and stable inorganic–organic antibiocorrosive coatings were obtained (Table 1.6).

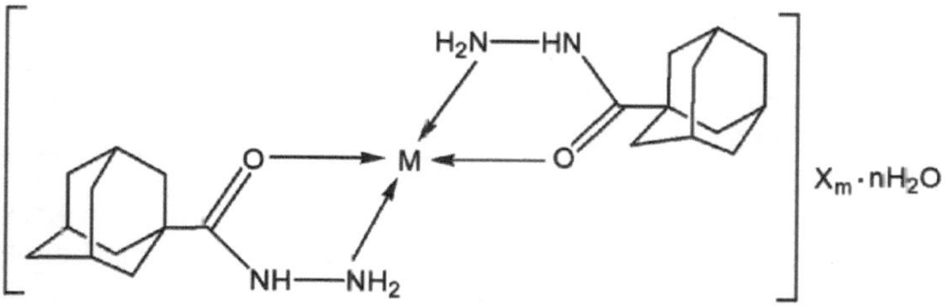

M = Cu, Co, Ni; X = NO_3^-, CH_3COO^-; m = 1, 2; n = 1, 2.

SCHEME 1.2 Adamantane-containing hydrazide coordination compounds of some transition metals.

TABLE 1.5 Relative Bioactivity of 1-(N-3,5-Dibromobenzoyl)aminoadamantane (15) and 4¢-(1-Adamantyl)-2-hydroxy-3,5-dibromobenzoylanilide (16).

Compound	P_a/P_i							
	Antiviral (influenza)	Antiviral (arbovirus)	Antiviral denovirus)	Antiviral (picornavirus)	ntiinfective	Dopamine release stimu-lant	Anthelmintic (nematodes)	Antiparasitic
15	0.857/ 0.003	–	0.548/ 0.005	–	0.778/ 0.005	0.848/ 0.009	0.478/ 0.057	0.422/ 0.035
16	0.903/ 0.002	0.731/ 0.020	0.617/ 0.004	0.520/ 0.088	0.807/ 0.005	0.749/ 0.029	0.544/ 0.030	0.512/ 0.017

TABLE 1.6 Polymeric Composites and Antibiocorrosive Coatings Based on Polyurethanes and Coordination Compounds.

1	Polyester urethane "PU"
2	Sulfur-containing polyester urethane "PUS"
3	Sulfur-containing polyester urethane "PUS"/3% α,ω-dihydroxymethylvinyloligoor ganosiloxane
4	Ionomer of sulfur-containing polyester urethane "PUSI"
5	Polyester urethane "BUTYP T-261"
6	Polyester urethane "BUTYP T-261"/3% coordination compound
7	Polyester urethane "BUTYP T-261"/5% coordination compound
8	Polyester urethane "BUTYP T-261"/3% bis(hydroxyalkyl)polydimethylsiloxane

9	Polyester urethane "BUTYP T-261"/3% α,ω-dihydroxymethyl-vinyloligoorganosiloxane
10	Polyester urethane "BUTYP T-261"/3% α,ω-dihydroxymethylvinyl oligoorganosiloxane/3% coordination compounds

The tribological properties of the polymeric matrices and antibiocorrosive coatings were studied by using a Nanovea pin-on-disk tribometer from Micro Photonics Inc. The tribometer provides dynamic friction results under rotation conditions.[35] A total of 440 steel balls each with a diameter of 3.2 mm made by Salem Specialty Balls were used. The tests were performed under the following conditions: temperature 20 ± 2°C, speed 100 rpm, radius 2.0 mm, load 5.0 N. The number of revolutions was 3000.

As shown in Figures 1.4 and 1.5, the dynamic friction for polyurethane matrices mainly depends on their structure and the structure of the modifier. By the analysis of curves of dynamic friction variation for nonmodified and modified polyurethanes and antibiocorrosive coatings based on them (Figs. 1.4 and 1.5), it was established that the value of dynamic friction is high for nonmodified polyurethane matrices and coatings based on them. The addition of silicon-organic oligomers (3–5 wt.%) to polyurethane matrices leads to a decrease in the above-mentioned parameter (Fig. 1.6).

By the doping of bioactive coordination compounds in the polyester urethane matrix based on 4,4-dimethylmethanediisocyanate and olygobuthylenglycole adipinate, the value the coefficient of dynamic friction is increased. This fact could be explained by the influence of their spatial structure. It was shown that the modification of polyester urethanes by silicon-organic oligomers the dynamical friction was reduced (Fig. 1.6), and is conditioned by high plasticized ability of flexible siloxane oligomers.

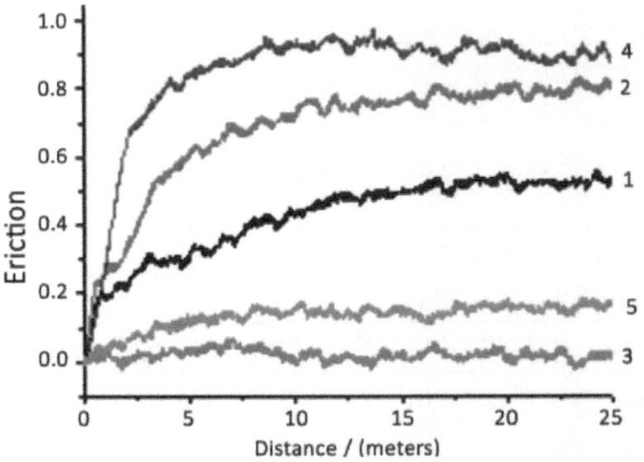

FIGURE 1.4 Dependence of the dynamic friction on the distance of sliding for polyurethane matrices and antibiocorrosive coating films based on them (1–5, Table 1.6).

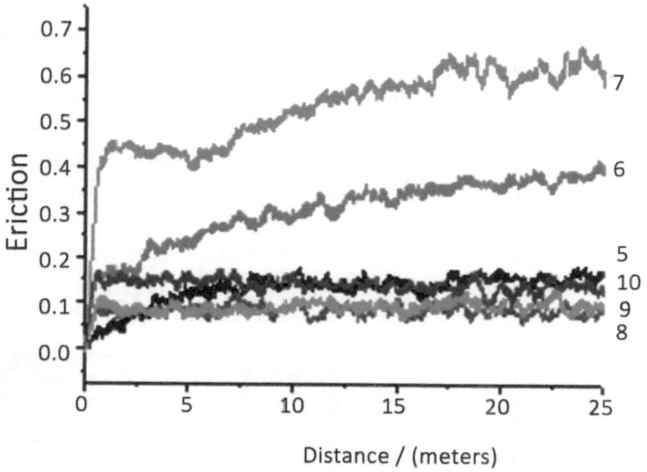

FIGURE 1.5 Dependence of the dynamic friction on the distance of sliding for the matrix based on polyester urethane ("BUTYP T-261") and corresponding antibiocorrosive coatings (5–10, Table 1.6), nonmodified and modified with silicon–organic oligomers.

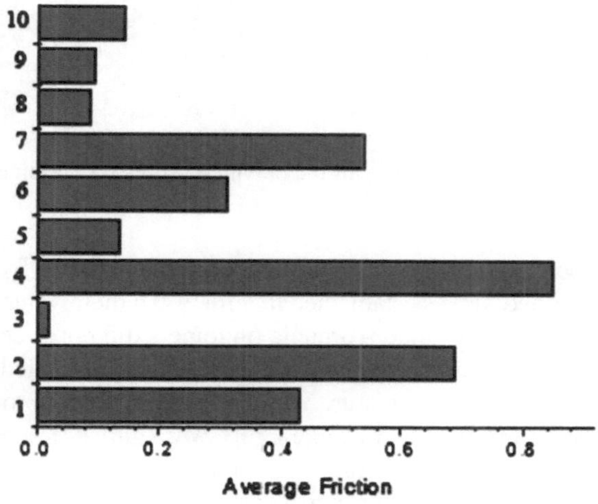

FIGURE 1.6 The comparison of the values of average dynamic friction for polymer matrices and antibiocorrosive coatings based on them (Table 1.6).

The obtained results were also confirmed by studying surface morphology (scanning electron microscopy) of the films from modified and nonmodified polymeric matrices for the purpose of creating antibiocorrosive coatings (Fig. 1.7). Plastic deformation of the corresponding films under the constant load equal to 5 N showed that split behavior of sulfur-containing polyester urethane "PUS," modified by α,ω-dihydroxymethylvinyloligoorganosiloxane, characterized with less crack nucleation in comparison with non-modified "PUS" (Table 1.6). It must be noted that polyester urethane "BUTYP T-261," modified by bis(hydroxyalkyl)polydimethylsiloxane (Table 1.6), demonstrated least crack nucleation among the tested polymeric matrices.

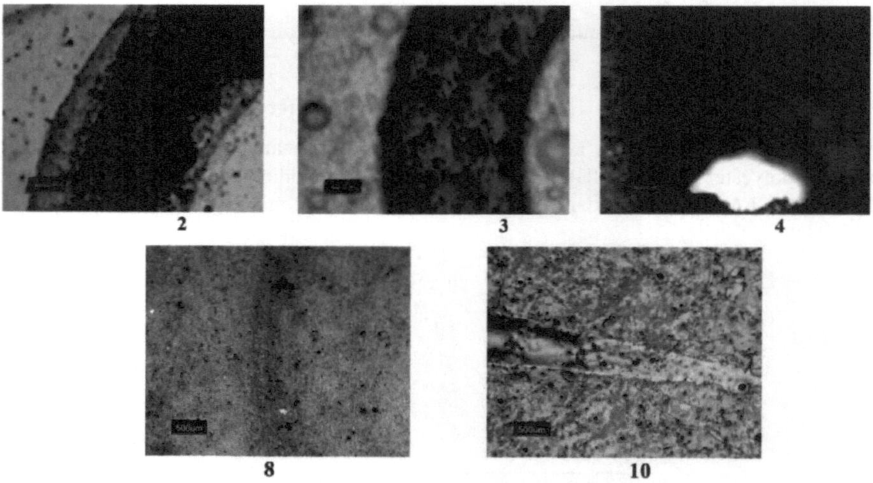

FIGURE 1.7 The surface morphology of the polymeric films for antibiocorrosive coatings based on nonmodified and modified polyurethanes (Table 1.6).

The hydrophobility of the obtained antibiocorrosive coatings by the gravimetric method was determined. It was established that for 300 h their water absorption ability for modified matrices by silicon–organic oligomers did not exceed 0.1%.

The influence of isothermal aging (40 and 60°C), the so-called weather ability (stability toward O_2, CO_2, and moisture complex action) and photochemical stability (stability toward ultraviolet and visible light), was studied. It was shown that for more than 3 months (at room temperature) the initial appearance (state), color, optical transparency, and mechanical properties (surface homogeneity without of split formation) of antibiocorrosive coatings (Table 1.6, 8–10) did not worsen.

By preliminary investigations, it was established that the created antibiocorrosive composites could be used for the prevention of materials from biodeterioration and noncontrolled biodegradation; the inhibition of growth and expansion of microorganisms, which are causal factors of infectious diseases; prophylaxis; and treatment of above-mentioned diseases.

The created composites can also be recommended for the preparation of protective covers with multivectorial application (film materials and impregnating compositions) stable for biocorrosion; materials with antimycotic properties for prophylaxis and treatment of mycosis; biologically active polymer composite materials for the protection of museum exhibits; and for human protection during their contact with microorganisms.[36]

1.4 CONCLUSION

1. Quantitative "structure–properties" relationships (QSRR/QSAR) based on experimental data for the construction of models of dependences of the physical properties of biologically active 16 nitrogen-containing compounds with bioactive spatial carbocyclic (adamantane) groups, synthesized by us, on molecular structures were studied.
2. Several sets of molecular descriptors were used. The presence of the dataset outliers was controlled by using PCA; to ascertain the quality of models, cross-validation was used.
3. Based on the performed research, one concludes that the best models have been acquired by the use of the GETWAY description of the obtained compounds.
4. Virtual bioscreening and activity toward various microorganisms of the obtained compounds were studied. The area of their application was established. The microbiological study of the investigated compounds confirmed the evaluated virtual concepts.
5. Polyurethanes with various structures as a polymeric matrix for antibiocorrosive coatings were chosen and used. This is mainly due to the tribological properties improvement in silicon–organic oligomers used, photochemical stable and "*Short-time*" active inorganic–organic antibiocorrosive coatings were obtained. The tribological properties of the obtained polymeric composites and antibiocorrosive coatings were determined. It was established that the modification process could used to improve the tribological properties of polyurethanes.

ACKNOWLEDGMENT

The authors thank Shota Rustaveli National Science Foundation for financial support.

KEYWORDS

- **QSRR/QSAR**
- **models**
- **descriptor**
- **calculation**
- **bioactive**
- **bioscreening**
- **antibiocorrosive**
- **tribology**
- **composite**

REFERENCES

1. Lekishvili, N.; Barbakadze, Kh.; Zurabishvili, D.; Lekishvili, G.; Arziani, B.; Fainleib, A.; Grigorieva, O.; Brostow, W.; Datashvili, T. *Oxid. Commun.* (*Int. J.*) **2012,** *35* (4), 746–761.
2. Gu Ji-Dong. *Int. Bioterior. Biodegr.* **2003,** *52* (1), 69–91.
3. Kurdina, M. I.; Malikov, V. E.; Zarikova, N. E.; et al. *Bull. Dermatol. Venerol.* **2002,** *5*, 49–52.
4. Lekishvili, N.; Barbakadze, Kh.; Zurabishvili, D.; Lobzhanidze, T.; Samakashvili, Sh.; Pachulia, Z.; Lomtatidze, Z. Antibiocorrosive Covers and Conservators Based On New Carbofunctional Oligosiloxanes and Biologically Active Compounds. *Oxid. Commun.* (*Int. J.*) **2010,** *33* (1), 104–124.
5. Howard G. T. *Int. Bioterior. Biodegr.* **2002,** *49* (4), 242–252.
6. Nakaya T. *Prog. Org. Coat.* **1996,** *27*, 173–180.
7. Hazziza-Laskar J., et al. *J. Appl. Polym. Sci.* **1995,** *58* (1), 77–84.
8. Sydnes, L. K. 41st IUPAC World Chemistry Congress. Chemistry Protection Health, Natural Environment and Cultural Heritage. Programme and Abstracts: Turin, Italy, August 5–11, 2007.
9. Morozov, I. S.; Petrov, V. I.; Sergeeva, S. A. In Pharmacology of Adamantanes. Volgograd Medical Academy: Volgograd, 2001.
10. Arcimanovich, N. G.; Galushina, T. S.; Fadeeva, T. A. Adamantanes—Medicines of the XXI Century. *Int. J. Immunorehab.* **2000,** *2* (1), 55–60.
11. Kovtun, V. Y.; Plakhotnik, V. M. Using of Adamantane Carbonilic Acids for Modification of Drugs Properties and Biologically Active Compounds. *Chem. Pharm. J.* **1987,** *8*, 931–940.
12. Barbakadze, Kh.; Lekishvili N., Pachulia, Z. Adamantane-Containing Biological Active Compounds: Synthesis, Properties and Use. *Asian J. Chem.* **2009,** *21* (9), 7012–7024.
13. Lekishvili N. and Kh. Barbakadze. *Asian J. Chem.,* **2012,** *24* (6), 2637–2642.
14. Gasteiger, J.; Engel, Th., Eds. In *Chemoinformatics, a Textbook*; Wiley-VCH: Weinheim, 2003.
15. Hemmer, M. C.; Steinhauer, V.; *J. Gasteiger. Vibrat. Spectr.* **1999,** *19*, 151–164.
16. J. Schuur, *Gasteiger. Anal. Chem.,* **1997,** *83*, 2398–2405.
17. R. Todeschini, M.; Lesagni, E.; Marengo. *J. Chemom.* **1994,** *8*, 263–273.
18. V. Consonni, R.; Todeschini, L.; Pavan. *J. Chem. Inf. Comput. Sci.* **2002,** 42, 682–692.

19. Todeschini, R.; Consonni V. In *Handbook of Molecular Descriptors*; Wiley-VCH: Weinheim, 2001.
20. L. I. Denisova, V. M.; Kosareva, K. E.; Lopukhova, I. G.; Solonenko. *Kh. F. Zh.* **1975**, *9*, 18–21.
21. Otto, M. In *Chemometrics*; Wiley-VCH: Weinheim, 1999.
22. Lekishvili, G.; Asatiani, L.; Zurabishvili, D. *Georg. Chem. J.* **2004**, *4* (3), 253–256.
23. Geladi, P.; Kowalski, B. R. *Anal. Chim. Acta* **1986**, *185*, 1–17.
24. Sadim, A. B.; Lagunin. A. A.; Filiminov, D. A.; Poroikov, V. V. Internet-System of Prognoses of the Spectrum of Bioactivity of Chemical Compounds. *Chem.-Farm. J.* **2002**, *36*, 21–26.
25. Barbakadze, Kh.; Zurabishvili, D.; Lomidze, M.; Sadaterashvili, I.; Lobzhanidze, T. and Lekishvili, N. *Proc. NAS Georgia, Chem.* **2008**, *34*, 45–52.
26. Savelyev, Yu. V.; Levchenko, N. I.; Rudenko, A. V. et al. Ukraine Patent 63092 (Jan. 15, 2004).
27. Savelyev, Yu. V. In *Handbook of Condensation Thermoplastic Elastomers*; Fakirov, S., Ed.; Willey-VCH, 2005, pp. 355–380.
28. Chen, Y. K.; Kukureka, S. N.; Hooke, C. J.; Rao, M. *J. Mater. Sci.* **2000**, *35*, 1269.
29. Yamamoto, Y.; Takashima, T. *Wear* **2002**, *253*, 820.
30. Yan, W.; Zheng, Y. Y.; Bao, D. *Synth. React. Inorg. Met. Org. Nano-Met. Chem.* **2005**, *35*, 237.
31. Deepa, K.; Aravindakshan, K. K. *Synth. React. Inorg. Met. Org . Nano-Met. Chem.* 2005, *35*, 409.
32. Holm, R. H. *Coord. Chem. Rev. 100*, 1990, 183.
33. Rehman, S. U.; Mazhar, M.; Amin, M.; Sadiq, B.; Xuequing, A.; George, S.; Khalid, E.; M. K. *Synth. React. Inorg. Met. Org. Nano-Met. Chem.* **2004**, *34*, 1379.
34. Adeoye, I. O.; Adelowo, O. O.; Oladipo, M. A.; Odunola, O. A. Comparison of Bactericidal and Fungicidal Activities of Cu(II) and Ni(II) Complexes of *para*-methoxy and *para*-hydroxy Benzoic Acid Hydrazide. *Res. J. Appl. Sci.* 2007, *2* (5), 590–594.
35. Brostow, W.; Datashvili, T.; Huang, B. *Polym. Eng. Sci.* 2008, *48,* 292–296.
36. Barbakadze, Kh.; Lekshvili, N.; Zurabishvili, D.; Pachulia, Z.; Gurjia, Zh.; Giorgadze, K.; Tsintsadze, G. 2nd International Caucasian Symposium on Polymers and Advanced Materials, Tbilisi, Georgia, 2010, p. 20.

CHAPTER 2

RELATIONSHIP BETWEEN THE STRUCTURE, MOLECULAR DYNAMICS, AND THERMO-OXIDATIVE STABILITY OF LOW-DENSITY POLYETHYLENE AND BUTYL RUBBER BLENDS

T. V. MONAKHOVA[1], L. S. SHIBRYAEVA[1], N. N. KOLESNIKOVA[1], A. I. SERGEEV[2], S. G. KARPOVA[1], and A. A. POPOV[1]

[1]Institute of the Russian Academy of Sciences, N.M. Emanuel Institute of Biochemical Physics, Russian Academy of Sciences, Moscow, Russia

[2]Institute of the Russian Academy of Sciences, N.N. Semenov Institute of Chemical Physics, Russian Academy of Sciences, Moscow, Russia

CONTENTS

ABSTRACT

The binary blends of butyl rubber (BR) and low-density polyethylene (PE) containing 0, 10, 20, 30, 40, 60, 70, 80, and 100 mass% of PE were investigated. Thermophysical parameters, such as the structure of PE and BR polymer chains, segmental mobility of the chains in the amorphous regions, proton transverse magnetic relaxation, and kinetic of oxidation, were determined. The relationship between the structure, molecular dynamics, and thermo-oxidative stability of low-density PE and BR blends was established.

2.1 INTRODUCTION

Studies aimed at developing new polymeric composites consisting of plastic and rubber, which do not require vulcanization and reinforcement, attract a significant interest nowadays. These studies are mainly based on the fact that the introduction of polyethylene in elastomers (butyl rubber, ethylene propylene diene rubbers, and others) makes it possible to receive the systems with a sufficiently high cohesive strength.[1–3] The method for creating these materials—mechanical melt mixing—makes their resistance to thermal and thermo-oxidative destruction an important problem. The same property is needed for processing and use of the products made of polymer composites. The main challenge facing the researchers and manufacturers engaged in the field of developing polymeric materials is to increase their thermo-oxidative stability. The study was aimed at establishing a relationship between the structure, molecular dynamics, and thermo-oxidative stability of low-density polyethylene (LDPE) and butyl rubber (BR) blends.

2.2 EXPERIMENTAL DETAILS

Binary blends of BR and LDPE were investigated. We used butyl rubber 1675N (hereinafter BR) and polymer LDPE 273-76. The blends containing 0, 10, 20, 30, 40, 60, 70, 80, and 100 mass% of PE were prepared in a Brabender-type mixer at a temperature of 170°C. The rotor speed was 60 rpm. The mixing time was 15 min. Film samples were obtained by compressing on a laboratory press at a temperature of 170°C, followed by rapid cooling to room temperature.

Thermophysical parameters were determined by differential scanning calorimetry (DSC), using the microcalorimeter DCM-10. Indium was used as a standard (T_m = 156.5°C; specific enthalpy is 28.4419 J/g). The thermal melting effect of the samples was determined by the peak area between the DSC curve and the baseline. We calculated the enthalpy of melting, based on the obtained thermograms. To determine the degree of crystallinity of LDPE, the value of the specific melting heat of PE crystallites was assumed to be 288 J/g. The error in the determination of melting point did not exceed 1°C, the melting heat 10%.

The structure of PE and BR polymer chains was determined by IR spectroscopy. The error in the determination of structural parameters did not exceed 15%.

We used the paramagnetic probe method to determine the segmental mobility of the chains in the amorphous regions of blends. A stable nitroxyl radical 2,2,6,6-tetramethylpiperidin-1-oxyl was used as a paramagnetic probe, which was injected in the polymeric film of saturated vapor. The spectra of nitroxyl radicals introduced in the sample were obtained by Electron paramagnetic resonance EPR. The radical probe correlation time (τ_c) was calculated from the spectrum by the formula[4]

$$\tau = \Delta H + (\sqrt{(I_+ / I_-)} - 1)\ 6.65\times10^{-10},$$

where ΔH_+ is the width of the spectral components, located in the weak field, and $(\sqrt{(I_+/I_-)} - 1)$ is the intensity component in the weak and strong field, respectively. Error in the determination of τ was within 5%.

Proton transverse magnetic relaxation of PE–BK samples was performed on a Bruker Minispek PC-120 spectrometer. This spectrometer operates at a proton resonance frequency of 20 MHz. The length of the 90° pulse is 2.7 μs and dead time is 7 μs. Two different pulse sequences were used for the measurement of T_2 relaxation time and the amount of rigid (crystalline) and soft (amorphous) phase components. To evaluate the spin–spin relaxation time T_2 and the fraction of protons with different degrees of mobility, we used techniques for studying induced signal decay after a 90° pulse (FID, free induction decay) and CPMG (Carr–Purcell–Meiboom–Gill). The error in the determination of parameters did not exceed 5%.

Kinetic oxidation curves of the mixtures were obtained using a manometric device with a circulating pump and by freezing volatile oxidation products at a temperature of 180°C and an oxygen pressure of 300 mm Hg (40 kPa). The error in the determination of kinetic parameters did not exceed 10%.

2.3 RESULTS AND DISCUSSION

The purpose of the research was to study binary blends of BR and LDPE, containing 0, 10, 20, 30, 40, 60, 70, 80, and 100 mass% of PE. The DSC method was used to determine the structural parameters for the crystalline regions of the sample of blends. Thermal parameters were determined by the same method. The results of research are presented in Table 2.1 for all the mixtures. The melting endotherms of the sample of mixtures have a single melting peak of PE. As seen from the table data, the temperature at the maximum PE melting peaks is insignificantly shifted toward low temperatures with a decrease in the PE content of the mixture, which may be associated with the formation of defective or smaller crystalline structures in the LDPE phase. At the same time, the degree of crystallinity of polyethylene (χ) decreases only in the samples with a high rubber content (70 mass%). This is most likely to be associated with the changes in the phase structure of the mixture, phase inversion. Rubber forms a dispersion medium in the samples of this composition, where PE is distributed in the form of small particles of the dispersed phase.

TABLE 2.1 Thermal Parameters of BR–PE Blends.

Composition of the BR/PE sample	0/100	20/80	40/60	60/40	70/30	80/20
$T_{m\ max}$ (°C)	129.0	129.0	128.0	128.0	127.0	127.0
ΔH_m of the blends	180.0	143.0	111.0	72.0	48.0	30.0
(J/g)	180.0	179.0	185.0	180.0	160.0	150.0
$\Delta H_{m\ PE}$ (J/g)	61.0	61.0	63.0	61.0	54.0	51.0
χ_{PE} (%)						

The changes in the structure of amorphous regions were determined by the IR spectroscopy method. The content of straightened and coiled conformers in the amorphous regions of the PE component was defined by the changes in the intensity of the bands responsible for deformation vibrations of the chains in straightened and folded conformations. The concentration of the former in PE was determined by the intensity of the band at 720 cm^{-1} and the D_{720}/D_{730} ratio. The band at 720 cm^{-1} is responsible for the fluctuations of methylene sequences, $-(CH_2)_n$, $n > 5$, in a trans-zigzag conformation in the amorphous regions of PE. The content of folded conformers of the TGT and GG types was determined according to the intensities of the bands at 1080 and 1306 cm^{-1}, respectively. The obtained data on the changes in the structure of PE chains depending on the composition of the PE–BR blends are shown in Table 2.2. As seen from the table, the content of straightened conformers T-T in the PE component increases, whereas the content of the coiled GG conformers decreases with the increased addition of rubber in PE.

TABLE 2.2 The Data on Changes in the Structure of PE and BR Chains Depending on the Composition of PE–BR Blends.

PE (mass%)	$\frac{D_{1305}}{D_{2740}}$	$\frac{D_{1080}}{D_{2740}}$	$\frac{D_{720}}{D_{730}}$	$\frac{D_{1230}}{D_{2740}}$	$\frac{D_{853}}{D_{2740}}$
0	–	–	–	26.4	0.764
30	–	–	1.263	42.0	0.979
40	0.062	–	1.22	48.7	1.254
60	0.750	0.520	1.19	65.4	1.838
80	0.875	0.417	1.18	96.6	2.720
100	1.229	0.312	1.11	–	–

The nature of changes in the intensities of the above-mentioned bands in the PE component suggests the presence of structural rearrangements by the type of conformational transitions GG → TGT → TT; the mechanism of this transition was previously established for the PE subjected to orientation drawing. Tension generated in the transduction chains under the influence of deformations as a result of mixing components may lead to similar changes in the conformational composition of chains in the volume of a polymer. Another reason for the enrichment of PE transduction chains with straightened conformers is implementation of the interphase phenomenon, in other words, the emergence of tension as a consequence of intermolecular interactions at the phase separation border.

The character of the changes in the BR structure can be established from the changes in the intensity of the bands at 1230 and 853 cm^{-1}. They grow monotonically with the increased content of BR, which apparently points to an increase in the size of its particles and, hence, the length of the PE–BR phase separation border. This leads to a speculation that the enrichment of PE matrix chains by straightened conformers with a decrease in its content in the mixture occurs on the boundary with the surface of the rubber particles. Changes in the structure of components and their chains must lead to a change in the molecular dynamics of polymers.

Changes in the structure of components and their chains should result in the alterations of the molecular dynamics of polymers. To determine the segmental mobility of the chains in the amorphous regions of blends, we used the paramagnetic probe method. A stable nitroxyl radical was used as a paramagnetic probe, which was injected in the polymeric film of saturated vapor. The spectra of nitroxyl radicals introduced in the sample were obtained by EPR. The radical correlation time (τ_c) was calculated from the spectrum. Segmental mobility is characterized by the value reciprocal to τ_c. The curve showing the change in τ_c of the blend is shown in Figure 2.1. As can be seen, the curve of the relationship between τ_c and the BR content in the mixture has a complex shape. With the increase in the content of BR in the mixture up to 40%, increase in the correlation time of the radical probe, i.e., a drop of the segmental mobility, is observed in the sample with 40% BR. The correlation time of the probe decreases with the growth of the rubber content in the mixture, thus increasing the mobility of the chains. Analysis of the data on the solubility of radical in mixtures shows that the highest concentration of the nitroxyl radical at the highest correlation time is observed in the region of the phase inversion (Fig. 2.2). It is speculated that the rearrangement of the conformational structure of the polymer chains in the PE matrix provides an increase in the free volume and simultaneously reduces the segmental mobility of the chains. Rubber fills the elements in the free volume of the blend and forms its own phase with the growth of its content. This leads to the reduction of the solubility of the nitroxyl radical, but decreases the radical correlation time. Given that the intermolecular interactions are absent, the radical correlation time characterizing the chain mobility will be determined by the flexibility of PE chains and its changes under the influence of the rubber introduced in polyethylene. Obviously, the maximum concentration of the radical corresponds to the phase inversion.

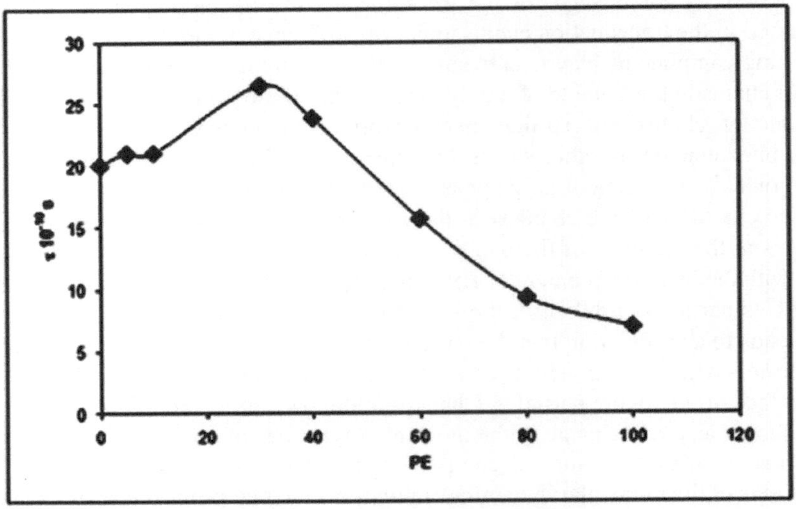

FIGURE 2.1 Dependence of the correlation time on the PE content.

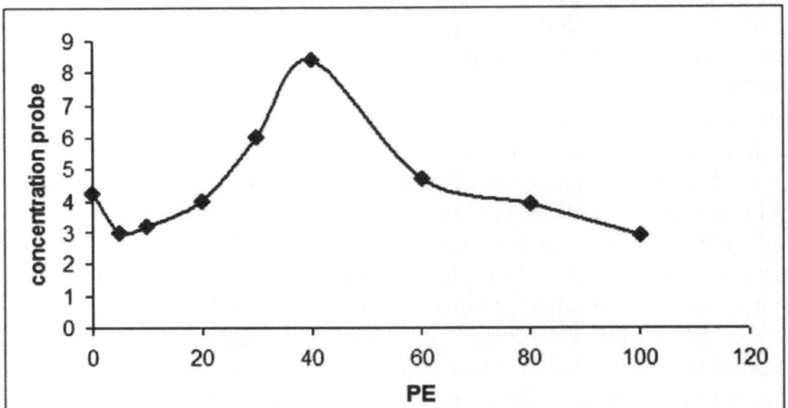

FIGURE 2.2 Dependence of the concentration probe on the PE content.

Molecular dynamics of PE–BR sample blends studied by the proton magnetic relaxation method on NMR was performed on a Bruker Minispek PC-120 spectrometer. This spectrometer operates at a proton resonance frequency of 20 MHz. The length of the 90° pulse is 2.7 μs and dead time is 7 μs. Two different pulse sequences were used to measure the relaxation time T_2 and amount of firm (crystalline) and soft (amorphous) phase components:

1. FID after a single pulse (90°) excitation;
2. CPMG—multiple-echo pulse sequence 90°–(τ–180°–τ)$_n$.

The FID pulse sequence[5,6] was used to register a quickly decaying part of the slow-moving protons (solid phase). The time of relaxation of these protons is a few microseconds and the duration of the collapse heavily depends on the inhomogeneity of the magnetic field B_0 in the sample. It is not possible to use this method for the precise determination of T_2, which lasted for more than 100 μs. The CPMG method removes the effect of the inhomogeneity of the magnetic field, but it can be used only for the registration of the slow part of collapse.

Our experiments were performed under the following conditions:

FID

- The temperature (*t*) 40°C
- The recycle delay between scans (RD) 1 s
- The beginning of the decay measurement (BD) 10 μs
- The end of the decay measurement (ED) 100 μs
- The number of experimental points (*n*) 50
- The number of scans (NS) 25

CPMG

- *t* = 40°C, RD = 1 s, *n* = 30, BD = 48 μs, ED = 1440 μs
- NS: 25–144
- Time between 90 and 180° pulses (τ) 12 μs

Our results of FID and CPMG experiments are presented in Table 2.3. The FID experiment for pure PE and PE–BR compositions (80 and 20 wt.% PE) demonstrated low mobile proton fraction for all samples (T_2 = 4–5 ms). The relative amount of these protons was 90–95%. When the part of BK in composition increased up to 80 mass%, the amount of mobile protons decreased to 47%. This fact is likely to indicate the increase in the segmental mobility of the chains in the mixtures as compared with pure PE and PE matrix. Since PE crystallites can serve as clamps for transition chains and, hence, inhibit the relaxation of protons, one of the reasons for the growing mobility of the chains in the blend may be reduction in the crystalline regions of polyethylene due to dilution of the polymer mass by an amorphous rubber.

TABLE 2.3 Magnetic Relaxation Characteristics of PE–BR Blends.

Composition of the BR/PE sample	χ of $_{PE,}$ $\chi$of the Blend %	SIS		CPMG	
		Proton spin–spin relaxation time T_2 (μs)	Relative content (%)	Proton spin–spin relaxation time T_2 (μs)	Relative content (%)
0/100	61	5.2	92.0	1357	49.0
	61	194	8.0	147	51.0
20/80	61	3.3	95.0	1010	60.0
	50	128	5.0	150	40.0
40/60	63	–	–	689	74.0
	38			151	26.0
60/40	61	–	–	636	61.0
	25			240	39.0
70/30	54	5.2	–	651	80.0
	16	214		239	20
80/20	51	–	47.0	649	85.0
	10		53.0	230	15.0
100/0	0	–	–	653	100

The crystallinity of PE for the samples with its different content (from 40 up to 100 mass%) did not change, but the volume of the crystalline phase in PE–BK compositions decreased more than twice. The molecular mobility of polymer chains also depends on the structural amorphous phase of composition and on a crystal–amorphous interface, which can be detected either as crystalline or as amorphous fraction depending on the method used. The deconvolution of proton transverse relaxation curves for different polymers (polyethylene, polypropylene, and their composition) reveals several components, which can be attributed to the crystalline, amorphous phases and crystal–amorphous interface.[7,8] It is possible to suggest that the less mobile proton fraction in our experiment (T_2 = 5 μs) is the fraction of the crystalline phase in the PE–BK composition. CPMG investigations of the PE–BR composition demonstrated two exponential components of the experimental relaxation curves—slow (S) and fast (F) decay components. The S-component could characterize the amorphous phase and the F-component (as a slow part of decay in the FID experiment) intermediate phase. When BR content in samples increased, the contribution of the S-component in experimental curves of decay increased too (Fig. 2.3). It is possible that the S-component and F-component reflect the rubber and polyethylene proton mobility, respectively. S-component relaxation times T_2 for BR samples (40, 60, 70, and 80 mass% BR) are close to T_2 of pure BR (T_2 = 650 μs) and it confirms our assumption. The growth BR fraction in the composition led to an increase in T_2 of the F-component and it suggested an increase in proton mobility in this intermediate phase. Obviously, this fact points to the formation of amorphous regions in the blends enriched with rubber, in which intermolecular interactions between PE and BR are manifested. The regions such as these represent interphase layers. The decline in the crystallinity degree of the polyethylene component also says about the formation of interphase layers. PE crystallites in these samples are surrounded by the amorphous regions containing rubber macrochains, which prevents formation of crystallites and reduces the degree of crystallinity of the polymer. Changes in the molecular mobility of chains have an impact on the thermo-oxidative resistance of PE–BR sample blends.

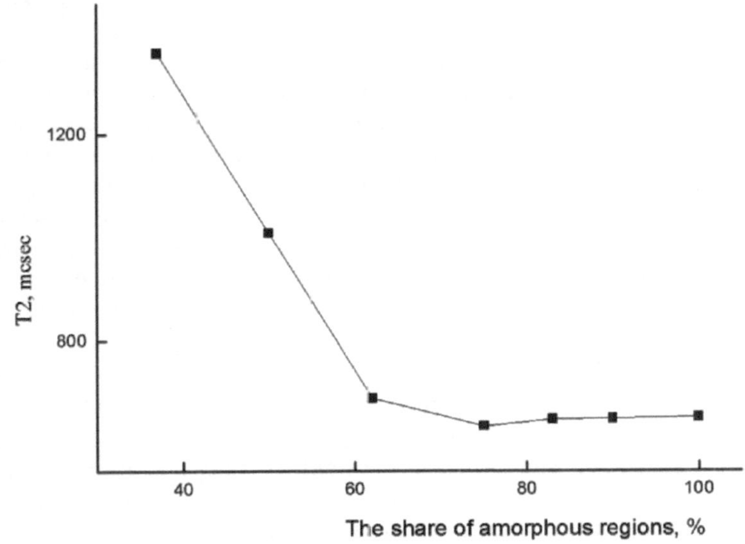

FIGURE 2.3 Proton relaxation time T_2 (microseconds) of the slow (S) component depending on the content of the amorphous phase in the PE–BR blend.

Figure 2.4 shows the kinetic curves of oxygen absorption by the tested samples. As seen from the figure, mixing polyolefin and rubber leads to the changes in the kinetics of oxygen absorption. Moreover, depending on the composition of the blend, the reaction rates not only vary but also correlate with the changes in the segmental mobility of the chains. For example, the nature of the curve showing the relationship between the radical probe correlation time τ_c and the composition of the blend corresponds to the changes in the rate of polymer oxidation. The introduction of rubber in PE leads to a slower oxidation process of the PE component in all the blends and of the BR component in the blends containing 30 and 40 mass% of PE. The oxidation reaction for PE starts with a delay—the induction period was about 50 min, while for a mixture containing 80% of PE, it was 20 min, and for 60%, 13 min. Other blends, as well as BR, are oxidized without an induction period. The dependence of the maximum rates of oxidation (Wo_2) of the samples on their composition has a complex nature (Table 2.4). Figure 2.5 (curve 1) shows the dependence of the maximum oxidation rate on the composition of the samples, which was obtained from the experiment, taking into consideration the oxidation of amorphous regions of oxygen absorption. The two regions can be distinguished on the curve (compare with the correlation time of the radical). The first region corresponds to the change in the composition from 5 to 40 mass% of PE. Moreover, a clear drop in the maximum rate of oxygen absorption is observed with increase in the content of PE. In the second region of the curve, Wo_2 monotonously increases with the increase of the PE content up to 40 mass% or more. We used three models to describe the kinetics of oxygen absorption by the PE–BR sample blends. The first model was based on the following approach: since the rate of BR oxidation is lower than the rate of PE oxidation, the first value was neglected. We assumed that the kinetic curves describing the absorption of oxygen by sample blends of different composition could be considered the curves of PE component oxidation. A kinetic curve for each blend was transformed into a curve, which took into account oxidation of PE, that is, was calculated per 100% of its content. The dependence of the values of oxidation rates obtained from the transformed curves on the content of PE is shown in Figure 2.5 (curve 2). Another approach took into consideration independent oxidation of PE and BR. At the same time, the oxidation rate of the sample was determined on the basis of the additive dependence as the sum of the rates of oxygen consumption by its components. Theoretical curves of oxygen absorption by pure polymers taking into account the reactivity of the amorphous phase of PE and BR and the composition of the mixture were obtained in the study. Based on the curves, gross values of the rates were defined, and the dependence of these rates on the composition of the samples was obtained (Fig. 2.5, curve 3). The comparison of the theoretical curves (2 and 3) and the experimental curve (1) shows significant differences in their shapes. The curve describing the localization of oxidation in the PE component has the form of the exponent. Moreover, a sharp increase in the rate of oxygen absorption is observed for the samples with a low content of the polyethylene component. All this fundamentally distinguishes curve 2 from the experimental curve. In accordance with curve 3, which takes into account the additive contribution of the components in the oxidation of the blend, the oxidation rate of the samples should increase monotonously with the increase in the PE content. This also contradicts the observed experimental dependence.

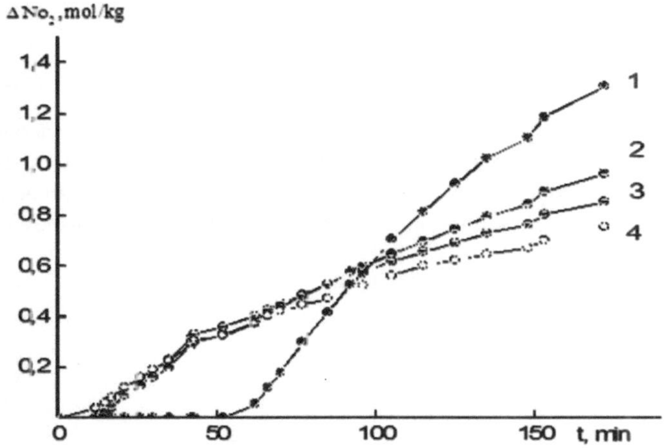

FIGURE 2.4 Kinetic oxidation curves for different films: PE (1), BR (2), blends of BR/PE (70/30) (3), and BR/PE (60/40) (4) at 180°C; the oxygen pressure is 300 mm Hg.

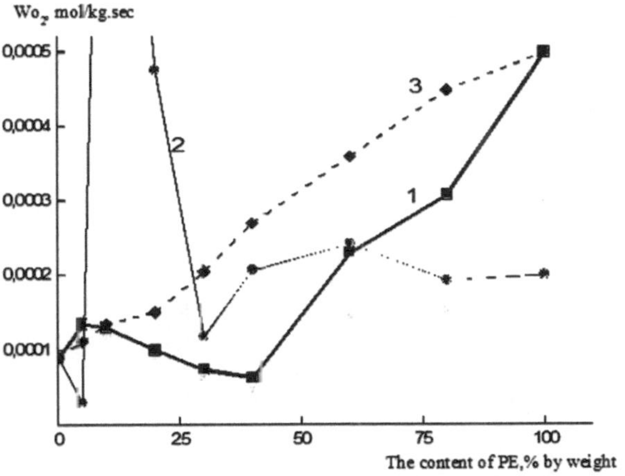

FIGURE 2.5 Relationships of the oxidation maximum rate and the composition of sample blends obtained from the experiment, taking into account the oxidation of the amorphous regions (1), and theoretical ones, obtained from PE oxidation conditions (2), and from participation in the oxidation of both components according to the additive scheme (3).

TABLE 2.4 Dependence of the Rate of Oxidation at 180°C Blends PE and BR on the Composition.

Composition of the BR/PE sample	0/100	20/80	40/60	60/40	70/30	80/20	90/10	95/5	100/0
The rate of oxidation (×*10⁴ mol/kg s)	2.0	1.5	1.4	0.5	0.6	0.9	1.2	1.1	0.9

Data presented in Figure 2.5 show that a model which takes into account cross-radical reactions between components is needed to describe the kinetics of oxidation of PE–BR mixtures. In this case, the oxidation of PE–BR mixtures can be described, taking into account the oxidation peculiarities of rubbers, by the following kinetic scheme:

Initiation of the kinetic oxidation chain:

$R_{pe}H \longrightarrow R_{pe}{*}$

$R_{br}H \longrightarrow R_{br}{*}$

Development of the kinetic chain:

For PE:

$$R_{pe}{*} + O_2 \xrightarrow{k_{11}} R_{pe}O_2{*} \quad (2.1)$$

$$R_{pe}O_2{*} + R_{pe}H \xrightarrow{k_{p11}} R_{pe}OOH + R_{pe}{*} \quad (2.2)$$

For BR:

$$R_{br}{*} + O_2 \xrightarrow{k_{22}} R_{br}O_2{*} \quad (2.3)$$

$$R_{br}O_2{*} + R_{br}H \xrightarrow{k_{p22}} R_{br}OOH + R_{br}{*} \quad (2.4)$$

$$R_{br}O_2{*} \longrightarrow R_{1br}CHO + R_{2br}O{*} \quad (2.5)$$

$$R_{2br}O{*} + RH \longrightarrow R_{2br}OH + R{*} \quad (2.6)$$

BR chain transfer reactions:

$$R_{2br}O^* + -CH=CH- \longrightarrow -CH-C^*H- \tag{2.7}$$

OR_{2br}

Secondary reactions of BR:

$$R_{2br}OH + O_2 \longrightarrow \text{aldehydes} \tag{2.8}$$

$$R_{1br}CHO + O_2 \longrightarrow \text{acids} \tag{2.9}$$

Cross-reactions of the chain development:

$$R_{pe}O_2^* + R_{br}H \xrightarrow{k_{p12}} R_{pe}OOH + R_{br}^* \tag{2.10}$$

$$R_{br}O_2^* + R_{pe}H \xrightarrow{k_{p21}} R_{br}OOH + R_{pe}^* \tag{2.11}$$

Chain termination (a quadratic break on peroxide radicals takes place at high oxygen pressure; alkyl and alkoxide radicals may participate in the break in the rubber):

$$2R_{pe}O_2^* \xrightarrow{k_{t11}} \text{products} \tag{2.12}$$

$$2R_{br}O_2^*(R^*_{BR}, RO^*_{br}) \xrightarrow{k_{t22}} \text{inactive products or cross-links} \tag{2.13}$$

Cross-termination reaction:

$$R_{pe}O_2^* + R_{br}O_2^*(R^*_{br}, RO^*_{br}) \xrightarrow{k_{t12}} \text{inactive products or cross-links} \tag{2.14}$$

Free valence, which emerged in the PE component of the blend, was released from the "cell" and localized in the interphase layer or at the interphase boundary, can participate in the BR component in the processes that are competitive to oxidation of hydrocarbon radicals, for example, in the destruction by reaction (2.5) or cross-linking of polymer chains by reaction (2.14). Moreover, this chain process of hydrocarbon oxidation may be blocked, which can be represented as a linear break:

$$R_{pe}OO^* (R_{pe}^*) + RbrH \xrightarrow{k'_{t12}} \text{inactive products} \tag{2.15}$$

or

$$R_{pe}OO^* + R_{br}H \longrightarrow [R_{br}^* + R_{pe}OOH] \xrightarrow{k'_{t12}} R_{br}\text{–O–}R_{pe} \tag{2.16}$$

Intermolecular interactions are necessary for cross-reactions. They can be realized through the contact of different polymer chains at the interphase boundary or interphase layer. One might speculate that the complex (bimodal) dependence of the oxidation rate on the blend composition, which we observed in the study, is due to the peculiar features of the structure of PE–BR sample blends. The slowdown of the oxidation rate of the sample mixtures is associated with the formation of interphase boundaries and layers, that is, transition of molecules between the polyethylene phase and elastomeric matrix, thus creating a structure to ensure the development of cross-reactions.

The authors speculate that the radicals arising from cross-reactions lead to the transfer of the free valence of macromolecules from rigid-chain PE to flexible-chain BR; for this reason, they can pass into an inactive form, or die in the reaction of rubber destruction or cross-linking. Obviously, the greatest contribution of cross-reactions leading to a decrease in the oxidation activity of the blends is made in the samples enriched with rubber. In blends with a high content of PE, the contribution of cross-reactions decreases, chain reactions of polyethylene oxidation in its phase dominate, and the total rate of the blend oxidation starts to increase up to the oxidation rate of a pure polymer. Thus, the oxidation ability of PE–BR blends depends on the contribution of cross-reactions, the mechanism for the reaction of the free valence transfer from one component to another, whose implementation is determined by intermolecular contacts of PE and BR macromolecules in the interphase layers, or at the interphase boundaries.

KEYWORDS

- **low-density polyethylene**
- **butyl rubber**
- **blends**
- **thermal oxidation**
- **molecular dynamics**

REFERENCES

1. Zakharchenko, P. I.; Yashunskaya, F. I.; Evstratov, V. F.; Orlovsky, P. N. In *Spravochnik Rezinshchika*; Chemistry: Moscow, RU, 1971; pp 342–395.
2. Schwartz, A. G.; Dinzburg, B. N. In *The Combination of Rubbers with Plastics*; Khimiya: Moscow, RU, 1972; p 224.
3. Xakimulin, Yu. N; Volfson, S. I.; Kimel′blat, B. I. *Caoutchouc Rubber* 2007, *28*, p 32.
4. Piotrovsky, K. B.; Tarasova, Z. N. In *Aging and Stabilization of Synthetic Rubbers and of Vulcanizates;* Khimiya: Moscow, RU, 1980.
5. Carr, H. Y.; Purcell, E. M. *Phys. Rev.* **1954**, *94*, 630.

6. Meiboom, S.; Gill, D. *Rev. Sci. Instrum.* **1958**, *29*, 688.
7. Hedesiu, C.; Dan, E.; Demco, D. E.; Kleppinger, R.; Buda, A. A.; Blümich, B.; Remerie, K.; Victor, M.; Litvinov, V. M. *Polymer* **2007;** 48. 3, 763.
8. Blom, H.P.; the, J.W.; Bremner, T.; Rudin, A. *Polymer* 1998. *39* (17), 4011.

CHAPTER 3

A STUDY ON POSSIBILITIES FOR CELLULOSE-BASED TEXTILE WASTE TREATMENT INTO POWDER-LIKE FILLERS FOR EMULSION RUBBERS

V. M. MISIN[1], S. S. NIKULIN[2], and I. N. PUGACHEVA[2]

[1]N. M. Emanuel Institute of Biochemical Physics, Russian Academy of Sciences, Moscow, Russia

[2]Voronezh State University of the Engineering Technologies, Voronezh, Russia

CONTENTS

ABSTRACT

The main objective of this chapter is to review the possibility of filling butadiene-styrene rubber of SKS-30 ARK grade with powder-like fillers made of cotton fiber. Another objective of this chapter is incorporation of the rubber at the stage of latex as well as the estimation of the effect of fillers on the process of coagulation and the properties of the composites. This chapter will investigate the properties of composite materials on the basis of rubber latexes and fibrous materials.

3.1 INTRODUCTION

Fiberfills have a wide diverse raw material base that is practically unlimited. Various fiberfill waste products are formed at textile enterprises, garment workshops, and so on. Therefore, an important and actual practical task is to search for most perspective directions of their usage.[1]

In some of the published works, it was shown that fiberfills can be applied in composite structures of different intended purposes. Special attention is paid to the use of fiberfills in polymer composites. They are used in the production of mechanical-rubber articles (MRA). Incorporation of fiberfills and additives into MRA are performed with rolling mills in the process of producing rubber compounds. This way of incorporation does not allow attaining their uniform distribution in the bulk of rubber compound that will further have a negative effect on the properties of the obtained vulcanizates. A uniform distribution of fiberfills in the bulk of polymer matrix can be obtained by the change of the way of their incorporation. For example, incorporation of fiberfills into the latex of butadiene-styrene rubber before its supply to coagulation allows us to attain their uniform distribution in the obtained rubber crumb, and this results in an increase of such quality factors of vulcanizates as their immunity to the thermal-oxidation effect, multiple deformations, and so on.[2,3]

Results of the investigations of the influence of small doses of fiberfills (up to 1 mass% in a rubber) on the process of rubber extraction from latex and the properties of the obtained composites are presented in Refs. 2 and 3. In Ref. 4, a technological difficulty was noted that was related to the incorporation of fiberfills into the process in the dosage of more than 1 mass% in the rubber.

In this situation, it would be interesting to transform fiberfills into the powder state. This will make possible to incorporate a greater amount of filler into the rubber just at the stage of its production, and thus to attain its uniform distribution in the rubber matrix.

Powder-like fillers have rather wide application in the production of tires and in mechanical rubber industry.[5] The overwhelming amounts of the used powder-like fillers are of inorganic nature and they are incorporated into rubber compounds with the use of rolling mills during their production. This way of incorporation, just as in the case of fibers, does not make it possible to attain a uniform distribution of the filler in the rubber compound that further affects the properties of produced articles. Therefore, the elaboration of the new ways of incorporation of fillers into polymer composites in order to obtain the articles having a set of new properties is necessary both from scientific and practical viewpoints.

The aim of this work is to study the possibility of filling butadiene-styrene rubber of SKS-30 ARK grade with powder-like fillers made of cotton fiber. Another aim is the choice of the way of incorporation of the rubber at the stage of latex as well as the estimation of the effect of fillers on the process of coagulation and the properties of the composites.

3.2 EXPERIMENTAL SECTION

At first, we elaborated the technique of obtaining powder-like cellulose fillers from fibrous materials from cellulose-containing textile waste products.

For this, cotton fiber was subjected to rough crumbling and treated with sulfuric acid according to the following technique. Fibers of 0.5–3.0 cm in size were treated with 1.5–2.0 parts of sulfuric acid under stirring (acid concentration 20–30 mass%). Next, the pasty mass (fibers + sulfuric acid solution) was filtrated. The obtained powder-like filler was dried for 1–2 h. After the final drying, the powder-like mass was subjected to the additional crumbling up to a more highly dispersed state. Thus, the acid powder-like cellulose (APC) filler was obtained. To obtain the neutral powder-like cellulose (NPC) filler, APC was neutralized with the aqueous solution of sodium hydroxide with the concentration of 1.0–2.0 mass%.

At the second stage of investigations, some ways of incorporation of powder-like cellulose fillers into butadiene-styrene rubber were estimated and proposed for the use.

Thereto APC, NPC, and microcrystalline cellulose (MCC) were applied with the dosage for every sample (1, 3, 5, 10 mass%) in the rubber. In one of the studied ways of incorporation, the powder-like fillers were incorporated into the latex of butadiene-styrene rubber just before its supply to coagulation. An aqueous solution of NaCl (24 mass%), $MgCl_2$ (12 mass%), or $AlCl_3$ (10 mass%) was applied as a coagulant, and an aqueous solution of sulfuric acid with a concentration of 1–2 mass% was used as an acidifying agent. Powder-like cellulose fillers were incorporated in the following ways:

- in the dry form just into latex immediately before its supply for coagulation;
- in the dry form into latex, involving coagulant;
- simultaneously with the aqueous solution of coagulant in latex;
- with serum at the completion phase of extraction of rubber from latex.

At the third stage, we studied the influence of powder-like cellulose fillers on the process of coagulation.

The coagulation process was performed in the following way. Latex SKS-30 ARK (20 ml; dry residue of ~18%) was loaded into a coagulator, which was made in the form of a capacitance provided with a stirring device, and then it was thermally stabilized for 15–20 min at a temperature of 60°C. After that, 24% aqueous solution of coagulant (NaCl, $MgCl_2$, $AlCl_3$) was supplied. The coagulation process was completed by the addition of 1–2% aqueous solution of sulfuric acid to the mixture up to pH ≈2.0–2.5. The formed rubber crump was separated from the aqueous phase by filtration, washed up with water, dried in a desiccator at a temperature of 80–85°C, and then weighed. The mass of the formed rubber crump was calculated based on the dry residue of the original latex. After sedimentation of filtrate (or as it is named in the industry of synthetic rubber—"serum"), a possible presence of high-dispersive rubber crump in the sample was determined visually. Powder-like cellulose fillers were incorporated into the latex of butadiene-styrene rubber using all of the above-mentioned ways.

At the fourth stage, the influence of powder-like cellulose fillers on the properties of obtained rubbers, rubber compounds, and vulcanizates was estimated. To make the estimations, first rubber compounds were prepared with the use of conventional ingredients with their further vulcanization at the standard facilities.[6,7] The produced vulcanizates were subjected to physical–mechanical tests and the simultaneous study of vulcanization kinetics and swelling ability kinetics.

Swelling kinetics ability of vulcanizates filled with different fibers was studied in solvents of different polarities according to the following technique. Samples of vulcanizates were cut in the form of squares of size 1 × 1 cm and weighed. The number of samples for each series of measurements was 5. The samples were put into solvents for 8 h. Every hour when they were extracted from the solvents, their sizes were measured and they were weighed. The size and mass of a sample were measured until 24 h. After that we processed the obtained data:

- in order to determine the swelling degree a (mass%), we subtracted the mass of the original sample from that of the swollen sample: the obtained solvent mass was divided by the mass of the original sample and the result was multiplied by 100%; from the five obtained results for each of the samples, the greatest (equilibrium) value a_{max} was chosen;
- swelling constant rate was determined as

$$k = (1/\tau)\ (\ln[\alpha_{max}/(\alpha_{max} - \alpha_{\tau})]),$$

where τ is time (h) and α_{τ} is the value of the current swelling degree at the time τ.

3.3 RESULTS AND DISCUSSION

The obtained APC involved the rest of sulfuric acid. However, this disadvantage is transformed into an advantage in the case of the use of this filler in the production of emulsion rubbers where acidification of the system takes place at the stage of rubber extraction from latex.

One can expect that the use of APC fillers in the technological process of butadiene-styrene rubber production should reduce a total consumption of sulfuric acid and stabilize the coagulation stage. It should be noted that the process performed in the real industrial scale the separation stage of the obtained powder-like filler from sulfuric acid solution and its drying can be eliminated since extraction of butadiene-styrene rubber from the latex is accompanied by acidification of the system by a solution of sulfuric acid. Therefore, the obtained pasty mixture composed of a sulfuric acid solution and a powder-like filler on the basis of cellulose is reasonable to dilute with water in order to decrease the sulfuric acid concentration up to 1–2 mass% and to perform incorporation of the obtained dispersed mass into the coagulated latex instead of the "pure" solution of sulfuric acid. In order to make more overall estimate of the influence of powder-like cellulose fillers on the coagulation process and the properties of the obtained composites, comparative tests with the samples of MCC were performed.

The fractional composition of APC, NPC, and MCC fillers is presented in Figure 3.1. On the basis of the fractional composition of the powder-like fillers, the weight–average size of their particles was determined: APC ≈ 0.57 mm; NPC ≈ 0.14 mm; MCC ≈ 0.15 mm. The calculated specific surface of the particles in these powder-like fillers was of 70, 286, and 267 cm^2/g, respectively, accounting for the cellulose density $\rho = 1.5$ g/cm^3.

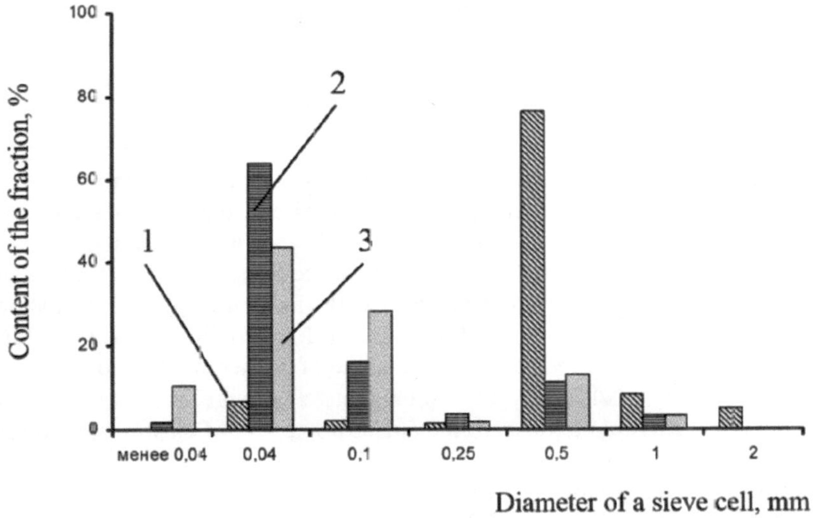

FIGURE 3.1 Fractional composition of the powder-like cellulose fillers: APC (1), NPC (2), MCC (3).

The images of the powder-like cellulose fillers obtained with the use of scanning electron microscopy JSM-6380 LV (magnification by 220–250 times) are presented in Figure 3.2.

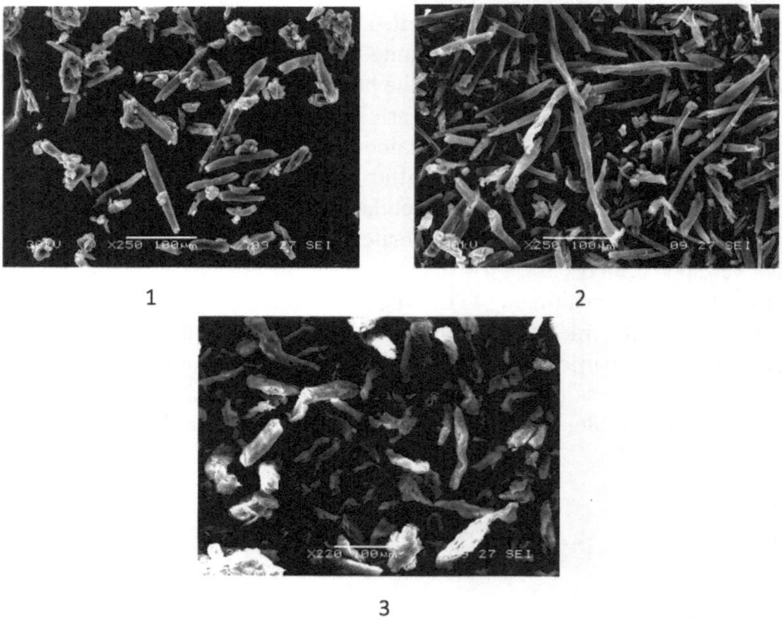

FIGURE 3.2 Electron microphotographs of APC (1), NPC (2), MCC (3).

These images were obtained in the secondary-electron emission mode. In order to prevent thermal destruction of the samples and their electrical charging under the impact of the electron beam, a gold layer with a thickness of 10 nm was deposited on the samples. The particles of cellulose powders were mainly presented by crystals with the shape factor l/d (the ratio of length l to the diameter d) varying within the interval of 1–9 for APC, 1–25 for NPC, and 1–8 for MCC. This was also confirmed indirectly by the value of packed density for the fillers: MCC = 0.79, NPC = 0.44, and APC = 0.68 g/cm^3. In the presence of such needle-like fillers, it is possible to observe the anisotropy effect of elastic-strengthening factors for vulcanizates.

In turn, the analysis of the particle elemental composition demonstrated the presence of sulfate groups in APC and their absence in NPC and MCC (Table 3.1).

TABLE 3.1 Elemental Composition of Cellulose Powder-Like Fillers.

Name of the element	Content of the elements (mass%)		
	NPC	APC	MCC
C	43.76	39.02	44.53
O	53.80	53.47	55.47
Na	0.62	0	0
S	0	5.49	0
Other	1.82	2.02	0

Comparing the possible ways of incorporation of powder-like cellulose fillers into butadiene-styrene rubber with the account of their properties, one can make a conclusion that incorporation of APC is appropriate to perform with a coagulant while MCC and NPC are reasonable to incorporate in the form of dry powders into latex just before its supply to coagulation.[8,9] However, since incorporation of the fillers in a dry form is connected with certain technological problems, in what follows, NPC and MCC are incorporated jointly with a coagulant.

Tables 3.2–3.4 present the results of the study of the influence of powder-like cellulose fillers on the yield of rubber crump obtained from latex in the presence of different coagulants: NaCl, $MgCl_2$, and $AlCl_3$. The analysis of the obtained results demonstrated that incorporation of all the above-mentioned powder-like cellulose fillers resulted in an increase in the yield of the formed rubber crump.

TABLE 3.2 Influence of Coagulant (NaCl) Consumption on the Yield of the Formed Rubber Crump for Different Amounts of Investigated Powder-Like Fillers

NaCl consumption (kg/ton of rubber)	Yield of the formed rubber crump (mass%)												
	Without filler	Amount of APC (mass% per rubber)				Amount of NPC (mass% per rubber)				Amount of MCC (mass% per rubber)			
		1	3	5	10	1	3	5	10	1	3	5	10
1	35.2	52.5	48.2	52.7	49.2	45.4	44.7	50.1	48.0	41.3	48.1	49.2	45.1
5	45.9	59.8	58.4	61.4	61.7	56.9	59.0	58.5	68.6	58.6	60.9	56.5	57.9
10	56.8	69.8	71.1	79.6	77.5	78.1	78.6	76.3	79.0	76.4	77.5	78.2	72.6
25	80.0	87.2	85.6	92.3	90.5	90.7	91.8	94.9	90.8	92.5	93.6	91.3	89.2
50	92.8	93.1	92.9	95.7	92.7	95.5	95.8	97.9	92.3	95.9	96.5	94.9	92.7
75	95.3	98.7	98.6	98.1	96.4	97.3	97.9	98.1	96.0	98.0	98.4	98.5	95.6
100	97.8	98.9	99.0	99.2	97.5	97.9	99.2	99.2	98.2	99.1	99.0	99.2	98.2

TABLE 3.3 Influence of Coagulant ($MgCl_2$) Consumption on the Yield of the Formed Rubber Crump for Different Amounts of Investigated Powder-Like Fillers.

$MgCl_2$ consumption (kg/ton of rubber)	Yield of the formed rubber crump (mass%)												
	Without filler	Amount of APC (mass% per rubber)				Amount of NPC (mass% per rubber)				Amount of MCC (mass% per rubber)			
		1	3	5	10	1	3	5	10	1	3	5	10
1	37.9	50.1	60.7	62.8	60.3	44.2	45.7	48.8	49.1	45.2	44.5	43.2	47.6
2	48.9	54.5	70.2	70.1	73.3	50.7	54.9	51.7	52.3	52.4	50.3	51.4	59.5
3	61.7	62.5	76.7	85.6	83.8	66.4	61.1	64.9	65.3	60.8	61.3	60.4	61.4
6	75.2	88.5	87.6	90.9	97.1	79.4	77.9	80.1	83.1	80.5	81.5	83.9	83.0
9	80.4	91.2	97.5	98.2	97.5	84.1	82.3	86.3	92.9	91.6	91.0	89.9	91.7
10	89.6	95.6	98.5	98.6	98.3	90.3	95.2	95.8	95.9	98.6	95.9	96.4	97.5
15	92.8	98.2	99.1	99.0	98.9	96.1	98.2	97.8	98.2	98.9	97.4	98.5	98.6
20	95.7	99.2	99.2	99.2	99.1	96.2	98.9	98.9	98.9	99.0	98.6	98.9	99.2

TABLE 3.4 Influence of Coagulant ($AlCl_3$) Consumption on the Yield of the Formed Rubber Crump for Different Amounts of the Investigated Powder-Like Fillers.

$AlCl_3$ consumption (kg/ton of rubber)	Yield of the formed rubber crump (mass%)												
	Without filler	Amount of APC (mass% per rubber)				Amount of NPC (mass% per rubber)				Amount of MCC (mass% per rubber)			
		1	3	5	10	1	3	5	10	1	3	5	10
0.3	43.0	57.9	68.9	66.0	74.3	56.7	54.6	56.1	55.3	57.1	56.8	59.7	55.0
0.7	52.3	79.9	82.7	82.8	83.4	74.0	70.9	72.0	75.1	70.0	69.3	69.3	65.1
1	71.2	92.3	93.7	92.7	91.1	83.0	82.2	80.8	81.9	81.4	80.1	81.2	81.7
2	90.4	95.0	97.1	96.3	95.3	94.1	92.1	93.3	90.9	92.8	91.8	91.2	91.5
3	93.5	98.6	98.2	97.5	96.2	97.5	98.1	97.2	97.3	97.7	94.0	95.7	93.7
4	96.5	99.1	98.6	99.2	99.1	98.2	99.2	98.2	99.2	98.2	98.3	97.3	95.9

This can be connected with a decrease of rubber losses in the form of highly dispersed crump with serum and rinsing water. Incorporation of powder-like cellulose fillers allowed us to reduce the consumption of the coagulant and acidifying reagent required for a complete separation of the rubber from latex. In case of the application of APC filler with a dosage of more than 7 mass% in the rubber, complete coagulation of latex can be attained without additional incorporation of the acidifying reagent—sulfuric acid solution—into the process. At the same time, powder-like cellulose fillers can absorb surface active materials on their surface as well as coagulant and the components of the emulsion system, thus facilitating a decrease of environmental pollution by sewage water, for example, APC (Table 3.5). Similar data were obtained both for NPC and MCC.

TABLE 3.5 Elemental Composition of APC Before and After Its Application in the Process of Rubber Separation from Latex.

Name of the element	Content (mass%)			
	APC composition before coagulation process	APC composition treated with the components of emulsion system in the presence of different electrolytes		
		NaCl	$MgCl_2$	$AlCl_3$
C	39.02	54.98	50.28	52.09
O	53.47	34.44	44.63	44.34
S	5.49	0.48	0.45	0.41
Cl	0	5.75	3.37	2.16
Na	0	4.06	0	0
Mg	0	0	0.97	0
Al	0	0	0	0.84
K	0	0.16	0.29	0.15
Additives	2.02	0.13	0.01	0.01

Proposals on the changes in any technological parameters of the rubber coagulation process should not have any negative effects on the properties of the rubber and its vulcanizates. Therefore, we investigated the process of vulcanization for the rubber samples separated with the application of different amounts of the studied fillers (3, 5, 10 mass% in the rubber). Results of these investigations are presented in Table 3.6. The properties of the rubber compounds and physical–mechanical factors for the filled vulcanizates are given in Table 3.7.

TABLE 3.6 Characteristics of the Vulcanization Process of the Rubber Compounds Involving Fillers.

Factor	Reference sample (without a filler)	Value of the factor for the rubbers with different fillers								
		Content of APC (mass% in a rubber)			Content of NPC (mass% in a rubber)			Content of MCC (mass% in a rubber)		
		3	5	10	3	5	10	3	5	10
M_L (dN m)	7.5	6.5	7.0	7.5	7.0	7.3	7.7	7.0	6.9	7.5
M_H (dN m)	32.8	31.5	32.9	33.0	34.0	34.3	37.5	33.0	34.8	36.7
t_S (min)	3.0	4.3	3.8	3.0	4.0	3.9	2.3	4.4	4.4	4.0
$t_{C(25)}$ (min)	9.9	8.3	8.9	8.7	10.0	10.0	8.7	8.8	8.1	8.3
	12.6	10.7	11.4	11.4	12.6	12.7	11.4	11.9	10.8	10.8
$t_{C(50)}$ (min)	22.0	22.0	22.5	21.8	22.8	23.1	21.9	22.8	22.1	21.4
$t_{C(90)}$ (min)										
R_υ (min^{-1})	5.3	5.6	5.3	5.3	5.3	5.2	5.1	5.4	5.6	5.7

Note: M_L, minimal torsion moment; M_H, conditional maximal torsion moment; t_S, time of vulcanization start; $t_{C(25)}$, time of attaining 25% vulcanization; $t_{C(50)}$, time of attaining 50% vulcanization; $t_{C(90)}$, time of attaining 90% vulcanization; R_υ, vulcanization rate.

TABLE 3.7 Properties of the Rubber Compounds and Vulcanizates Involving Fillers.

Factor	Reference sample (without a filler)	Value of the factor for the rubbers with different fillers											
		Content of APC (mass% in a rubber)				Content of NPC (mass% in a rubber)				Content of MCC (mass% in a rubber)			
		1	3	5	10	1	3	5	10	1	3	5	10
Mooney viscosity MB 1 + 4 (100°C) (a.u.)	57.0	52.0	53.0	54.0	57.0	54.0	54.0	52.0	58.6	54.0	54.0	54.0	55.0
Plasticity (a.u.)	0.40	0.40	0.40	0.40	0.40	0.35	0.35	0.32	0.35	0.28	0.29	0.28	0.29
Elastic recovery (mm)	1.10	0.80	0.70	0.70	1.10	1.50	1.40	1.60	1.29	1.88	1.97	1.86	1.82

TABLE 3.7 *(Continued)*

Factor	Reference sample (without a filler)	Value of the factor for the rubbers with different fillers											
		Content of APC (mass% in a rubber)				Content of NPC (mass% in a rubber)				Content of MCC (mass% in a rubber)			
		1	3	5	10	1	3	5	10	1	3	5	10
M_{300} (MPa)	8.1	8.0	8.1	8.0	7.8	8.0	7.9	8.5	9.2	8.8	8.6	8.8	8.6
f_p (MPa)	22.8	20.5	22.0	24.4	21.9	24.0	23.6	22.7	21.7	20.0	20.0	20.3	22.3
ε_p (%)	620	620	620	670	623	630	637	620	570	500	520	544	562
ε_{ocr} (%)	14	14	14	16	14	14	14	15	17	12	13	14	16
Ball drop resilience (%)	38	42	42	41	38	40	40	40	38	40	42	38	39
Shore hardness *A* (a.u.)	57	57	57	55	57	58	57	57	62	58	59	60	61

Note: M_{300}, elongation stress 300%; f_p, conditional disruption strength; ε_p, relative elongation under disruption; ε_{ocr}, relative residual deformation after disruption.

With an increase of content of all cellulose fillers from 3 to 10 mass% in rubber, the rise of the minimal (M_L) and conditional maximal (M_n) torsion moments was observed in the rubber compound under vulcanization. The presence of powder-like cellulose fillers increased the time of the vulcanization start for the rubber compounds but actually did not have an effect on the time of vulcanization ending (Table 3.6).

The presence of NPC and MCC in the amount 1–10 mass% in the rubber increased the hardness of vulcanizates due to the effect of the presence of fillers as a result of their large shape factor (1–25 for NPC), as well as their large specific surface (83 cm^3/100 g for MCC).[10] However, the presence of NPC and MCC did not influence the rubber ball drop resilience.

The reinforcing effect in the polymer matrix provided a linear increase of M_{300} for rubbers with an increase of content of NPC with a large shape factor (1–25). In turn, the increase of the content of the defect centers in the polymer matrix due to the accumulation of the coarse grains of particles (Fig. 3.1) reduced the strength and relative elongation of vulcanizates under disruption. An increase of MCC content in vulcanizates provided the rise of stress under 300% elongation (M_{300}), with a simultaneous decrease of the swelling rate of vulcanizates in solvents (Table 3.8). Moreover, an increase of MCC content multiplied the concentration of the transverse bonds and increased vulcanization rate along with a decrease of the strength and relative elongation at the disruption due to the accumulation of the defect centers in the polymer matrix, which can be explained by the absence of wetting for MCC particles with a rubber (Table 3.7).

TABLE 3.8 Influence of the Amount of Fillers and Nature of a Solution on the Ability to Vulcanizates Swelling.

Solvent	Swelling rate of vulcanizates in solvents (0–10% wt. on rubber)							
	Without fillers		3%		5%		10%	
	b	a	b	a	b	a	B	A
				APC				
Toluene (1)	−0.62	2.4	−0.54	2.3	−0.61	2.4	−0.58	2.4
Nefrac (2)	−0.38	1.9	−0.67	1.9	−0.89	1.9	−0.84	1.9
$1/Q_{max}^{tl} \times 10^3$	4.3		4.1		4.5		4.8	
				NPC				
Toluene (1)	−0.62	2.4	−0.82	2.4	−0.80	2.4	−0.58	2.4
Nefrac (2)	−0.38	1.9	−0.77	1.8	−0.64	1.9	−0.54	2.0
$1/Q_{max}^{tl} \times 10^3$	4.3		4.6		4.0		3.9	
				MCC				
Toluene (1)	−0.62	2.4	−0.34	2.4	−0.31	2.3	−0.30	2.3
Nefrac (2)	−0.38	1.9	−0.27	1.8	−0.27	1.7	−0.27	1.7
$1/Q_{max}^{tl} \times 10^3$	4.3		4.2		4.0		3.9	

Note: b, swelling rate (h^{-1}); a, $\lg Q_{max}$, equilibrium swelling rate (%); $1/Q_{max}^{tl} \times 10^3$—concentration of transverse bonds.

The swelling process of the obtained vulcanizates was investigated in Ref. 11 in different environments allowing us to simulate real service conditions of vulcanizates. Aromatic toluene and aliphatic nefrac (benzene) were taken as solvents as mostly widespread coupling media.

Analysis of the reference literature sources[12] showed that for all of the solvents, swelling kinetics of vulcanizates in the presence of the fillers could be described in semilogarithmic coordinates of the descending direct line (Fig. 3.3) of the form

$$\lg(Q_{max} - Q_{\tau}) = \lg Q_{max} - b\tau.$$

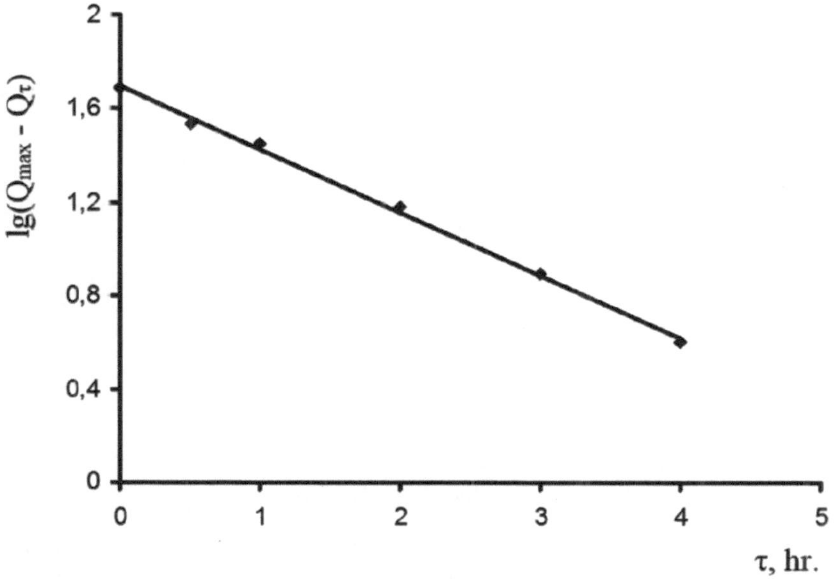

FIGURE 3.3 Swelling kinetics of vulcanizates in organic solvents.

By changing $\lg(Q_{max} - Q_{\tau}) = Y$, and $\lg Q_{max} = a$, the original equation can be written as

$$Y = a - b\tau,$$

where b is the swelling rate (h^{-1}); τ is the duration of swelling (h); Q_{max} is an equilibrium swelling degree (%); Q_{τ} is the current swelling rate (%).

Equilibrium swelling degree of vulcanizates, involving 3–10 mass%, in nefrac ($Q^{nf}_{mcc,max}$) was reduced up to 46–52% as compared with 84% for vulcanizate without MCC (Fig. 3.4, straight line 2). The swelling rate of vulcanizates in nefrac in the presence of 3–10 mass% of MCC was $b^{nf}_{mcc} = -0.27\ h^{-1}$; it did not depend on the value of MCC content but was less than the swelling rate of vulcanizate without MCC ($-0.38\ h^{-1}$) (Table 3.8).

In turn, the swelling rate of vulcanizates in toluene, b^{tl}_{mcc}, involving 3–10 mass% of MCC was 0.30–0.34 h^{-1}, and it was two times lower than that for vulcanizate without the filler ($-0.62\ h^{-1}$) (Table 3.8). An increase of MCC content within the investigated interval reduced as the equilibrium swelling degree $Q^{tl}_{mcc\ max}$ from 250 to 209% (Fig. 3.4, straight line 1), as swelling rate from -0.34 to $-0.30\ h^{-1}$.

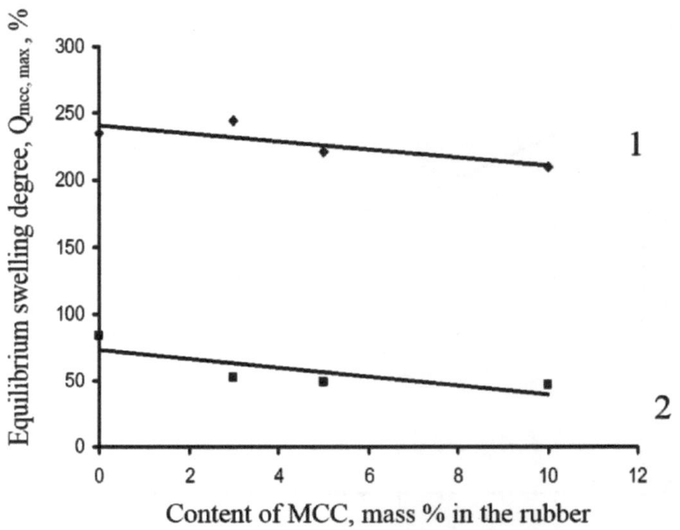

FIGURE 3.4 Influence of MCC content (mass%) and the nature of a solvent on the equilibrium swelling degree of vulcanizates ($Q_{mcc,max}$, %) in toluene (1) and nefrac (2).

Similar regularity is the characteristic of vulcanizates involving active fillers (e.g., technical carbon).[10]

The value of equilibrium swelling degree of vulcanizates involving 1–10 mass% of NPC ($Q^{nf}_{npc,max}$) in nefrac was 80–96% and actually did not differ (84%) from that in vulcanizates without the filler (Fig. 3.5, straight line 2).

The swelling rate of vulcanizates in nefrac (b^{nf}_{npc}) in the presence of 1–10 mass% of NPC increased up to the value of 0.48–0.77 h^{-1} as compared with the value of –0.38 h^{-1} for vulcanizate without the filler (Table 3.7). The dependence of the swelling rate b^{tl}_{npc} on the NPC content within the interval of 3–5 mass% was –0.80 to –0.82 h^{-1}. These values were higher than the swelling rate of vulcanizate without the filler, –0.62 h^{-1}.

An increase of $Q^{tl}_{npc,max}$ in toluene was observed with the rise of NPC content within the interval of 1–10 mass% (Fig. 3.5, straight line 1). Equilibrium swelling degree of vulcanizates in toluene, $Q^{tl}_{npc,max}$, with an increase of NPC content was grown up to 262% as compared with a vulcanizate without the filler (235%). The increase of $Q^{tl}_{npc,max}$ can be explained by the decrease of the density of transverse bonds in vulcanizate at the boundary between the phases of "polymer filler" and the greater value of the shape factor for NPC.

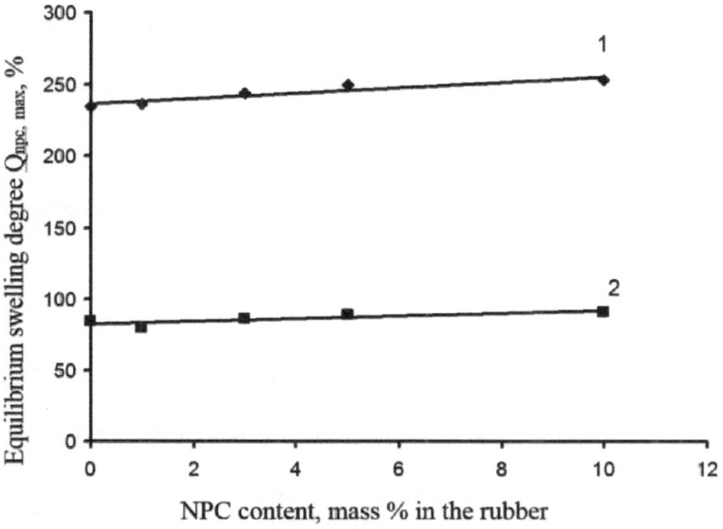

FIGURE 3.5 Influence of NPC content (mass%) and the nature of solvent on the equilibrium swelling degree of vulcanizate ($Q_{npc,max}$, %). 1—toluene; 2—nefrac.

For vulcanizates involving 1–10 mass% of APC in nefrac, the value of $Q^{nf}_{apc,max}$ was of 79–85% as compared with 84% for vulcanizates without the additives (Fig. 3.6, straight line 1), that is, it did not actually change.

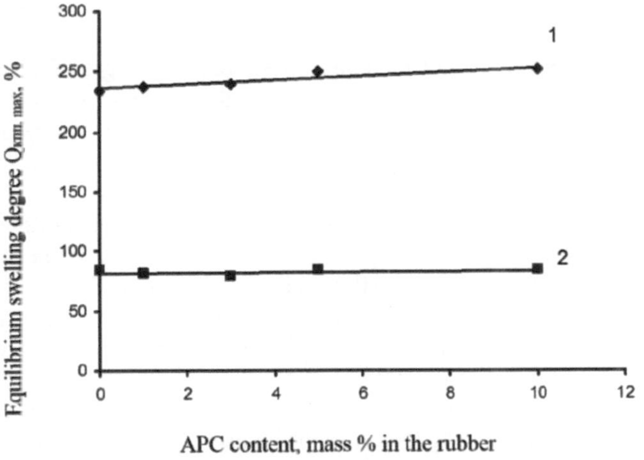

FIGURE 3.6 Influence of APC content (mass%) and the nature of solvent on the equilibrium swelling degree of vulcanizate ($Q_{apc,max}$, %). 1—toluene; 2—nefrac.

An increase of b^{nf}_{apc} values for vulcanizates involving APC up to −0.89 h^{-1} was observed as compared with the value of −0.38 h^{-1} for vulcanizates without the filler (Table 3.7) because of a bad wettability of the large APC particles with the rubber and occurrence of the tunnel effect. The equilibrium swelling degree of vulcanizates with 3–10 mass% of APC in toluene, $Q^{tl}_{apc,max}$ = 240–252%, was greater than that for vulcanizates without the additives (235%) (Fig. 3.6, straight line 1). This can be explained by a decrease of the density of transverse bonds in vulcanizate ($1/Q^{tl}_{apc,max}$) at the boundary of "polymer filler" due to a bad wettability of the particles of the acid powder-like filler.

3.4 CONCLUSION

1. Textile waste products can be used not only as fiberfills but also as powder-like fillers.
2. Differences in sizes, shape factor, specific surface, and fractional and chemical composition of the particles of powder-like cellulose fillers obtained in different ways were found in the work. Powders of the neutral and MCC are characterized by the higher specific surface (267–286 cm^2/g). The largest scattering interval of the shape factor was the characteristic of the neutral cellulose (l/d = 1–25). Greater content of sulfate groups was determined in the particles of the APC (5.49 mass% accounting for the bound sulfur).
3. Incorporation of powder-like cellulose fillers into SKS-30 ARK rubber does not have a negative effect on the physical–mechanical quality factors of vulcanizates. Thus, it is possible to get a considerable improvement of elastic-strength quality factors of butadiene-styrene rubber in the presence of these additives due to the choice of the required reactants for the interphase combination of cellulose with the rubber matrix.
4. The linear dependence of the equilibrium swelling degree for vulcanizates in the aliphatic and aromatic solvents on the content of powder-like cellulose fillers was found as a result of the work. With an increase of MCC content equilibrium swelling degree reduced, and for NPC and APC it was, on the contrary, enhanced.
5. Certain difference was found in swelling kinetics of vulcanizates in nefrac and toluene in the presence of MCC, APC, and NPC. In the presence of 3–10 mass% of MCC, the swelling rate of vulcanizates in solvents reduced by 1.25–2.0 times that is the characteristic for the rubbers with active fillers. Incorporation of 1–10 mass% of NPC increased the swelling rate of vulcanizate in nefrac by 1.25–2.0 times.
6. Diverse ways of incorporation of powder-like cellulose fillers into rubber emulsion before its coagulation with the use of different coagulants make it possible to improve the distribution of the additives in the rubber matrix, to reduce coagulant consumption, and in case of APC application—to de-

crease the consumption of the acidifying reactant up to its complete elimination from the coagulation process.

KEY WORDS

- **waste**
- **composite material**
- **rubber latex**
- **fibrous filling material**
- **physicomechanical properties**

REFERENCES

1. Nikulin, S. S.; Pugacheva, I. N.; Chernykh, O.N. In *Composite Materials on the Basis of Butadiene-Styrene Rubbers*; Academy of Natural Sciences, 2008; p 145.
2. Nikulin, S. S.; Akatova, I. N. Influence of Capron Fiber on Coagulation, Rubber Properties, Rubber Compounds and Vulcanizates. *Zhurnal Prikladnoi Khimii*, 2004, *77* (4), 696–698.
3. Nikulin, S. S.; Akatova, I. N. Influence of the Flax and Viscose Fiber on the Process of Separation of Butadiene-Styrene Rubber from the Latex. *Adv. Mod. Nat. Sci.* 2003, *6*, 10–13.
4. Akatova, I. N.; Nikulin, S. S. Influence of Fiberfill with High Dosages on the Process of Separation of Butadiene-Styrene Rubber from the Latex. *Product Usage Elastomers*, 2003, *6*, 13–16.
5. Mikhailin Yu. A. In *Constructional Polymer Composite Materials*; Scientific Foundations and Technologies, 2008; p 822.
6. Zakharov, N. D., Ed. In *Laboratory Practice in Rubber Technology*; Khimia, 1988; p 237.
7. Nikulin, S. GOST 15627-79. Synthetic Rubbers—Butadiene-Methylstyrene SKMS-30 ARK and Butadiene-Styrene SKS-30 ARK, 2007.
8. Pugacheva, I. N.; Nikulin, S. S. Application of Powder-Like Filler on the Basis of Cellulose in the Production of Emulsion Rubbers. *Mod. High-End Technol.* 2010, *5*, 52–56.
9. Pugacheva, I. N.; Nikulin, S. S. Application of Powder-Like Fillers in the Production Emulsion Rubbers. *Indust. Product. Use Elastomers*, 2010, *1*, 25–28.
10. Koshelev, F. F.; Korneev, A. F.; Bukanov, A. M. In *General Rubber Technology*; Khimia, 1978; p 528.
11. Zakharov, N. D.; Usachov, S. V.; Zakharkin, O. A.; Drovenikova, M. P.; Bolotov, V. S. In *Laboratory Practice in Rubber Technology. Basic Processes of Rubber Production and Methods of Their Control*; Khimia, 1977; p 168.
12. Tager, A. A. In *Physics and Chemistry of Polymers*; Scientific World, 2007; p 573.

CHAPTER 4

THEORETICAL ESTIMATION OF ACID FORCE OF MOLECULE *P*-DIMETHOXY-TRANS-STILBENE BY METHOD *AB INITIO*

V. A. BABKIN[1], D. S. ANDREEV[1], YU. A. PROCHUKHAN[3], K. Y. PROCHUKHAN[3], and G. E. ZAIKOV[2]

[1]Sebrykov Department, Volgograd State Architect-build University

[2]N.M. Emanuel Institute of Biochemical Physics, Russian Academy of Sciences, 4, Kosygin St., Moscow 119334, Russian Federation

[3]Bashkir State University, Kommunisticheskayaul., 19, Ufa 450076, Republic of Bashkortostan, Russia

CONTENTS

ABSTRACT

For the first time,quantum chemical calculation of a molecule of *p*-dimethoxy-*trans*-stilbene is executed by method *ab initio* with optimization of geometry on all parameters. The optimized geometrical and electronic structures of this compound areobtained. The acid power of *p*-dimethoxy-*trans*-stilbene is theoretically appreciated. As a result, it to relates to a class of very weak H-acids (pKa=+36, where pKa-universal index of acidity).

4.1 AIMS AND BACKGROUNDS

The aim of this work is to study the electronic structure of the molecule *p*-dimethoxy-*trans*-stilbene[1]and theoretical estimation of its acid power by the quantum-chemical method *ab initio*in base 6-311G**. The calculation was performed with the optimization of all parameters by the built-in standard gradient method in PC GAMES.[2] The calculation was executed in approach the insulated molecule in gas phase.The program MacMolPlt was used for the visual presentation of the model of the molecule.[3]

4.2 METHODICAL PART

Geometric and electronic structures, and general and electronic energies of the molecule *p*-dimethoxy-*trans*-stilbenewere obtained by the method *ab initio* in base 6-311G**and are shown in Figure 4.1 and in Table 4.1. The universal factor of acidity was calculated by the formulapKa=$49.04-134.6q_{max}^{H+4,5}$ (whereq_{max}^{H+} is a maximum positive charge on the atom of hydrogen;q_{max}^{H+}=+0.10 (for *p*-dimethoxy-*trans*-stilbeneq_{max}^{H+} alike Table 4.1)). This same formula is used in Ref. 6;pKa=36.

The quantum-chemical calculation of *p*-dimethoxy-*trans*-stilbene by the method *ab initio* in base 6-311G** was executed for the first time. The optimized geometric and electronic structures of this compound wereobtained. The acid power of *p*-dimethoxy-*trans*-stilbene was theoretically evaluated (pKa=36). This compound pertains to a class of very weak H-acids (pKa>14).

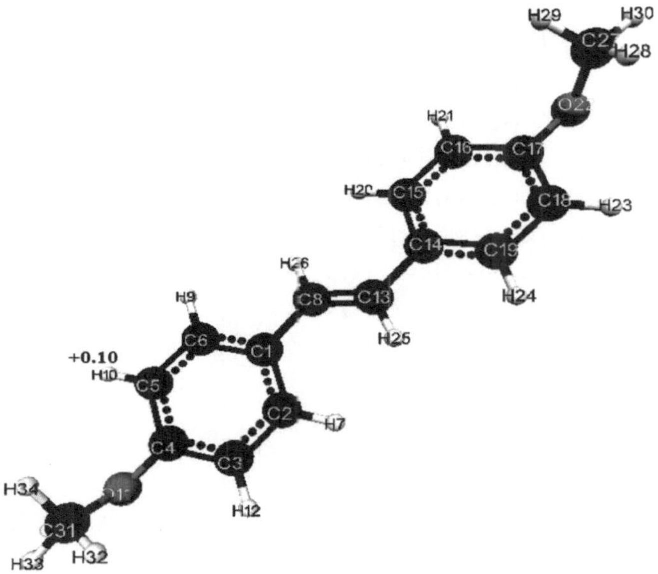

FIGURE 4.1 Geometric and electronic molecule structure of *p*-dimethoxy-trans-stilbene(E0= 2005.131 kDg/mol, Eel= −4981.689 kDg/mol).

TABLE 4.1 Optimized Bond Lengths, Valence Corners, and Charges on the Atoms of the Molecule *p*-dimethoxy-trans-stilbene.

Bond lengths	RA	Valence corners	Grad	Atom	Charges on atoms
C(2)–C(1)	1.40	C(5)–C(6)–C(1)	122	C(1)	−0.05
C(3)–C(2)	1.38	C(1)–C(2)–C(3)	121	C(2)	−0.07
C(4)–C(3)	1.39	C(2)–C(3)–C(4)	120	C(3)	−0.09
C(5)–C(4)	1.38	C(3)–C(4)–C(5)	120	C(4)	+0.21
C(6)–C(5)	1.38	O(11)–C(4)–C(5)	120	C(5)	−0.09
C(6)–C(1)	1.39	C(4)–C(5)–C(6)	120	C(6)	−0.07
H(7)–C(2)	1.07	C(2)–C(1)–C(6)	118	H(7)	+0.09

C(8)–C(1)	1.48	C(1)–C(2)–H(7)	120	C(8)	−0.08
H(9)–C(6)	1.08	C(2)–C(1)–C(8)	123	H(9)	+0.09
H(10)–C(5)	1.07	C(14)–C(13)–C(8)	127	H(10)	+0.10
O(11)–C(4)	1.36	C(5)–C(6)–H(9)	119	O(11)	−0.49
H(12)–C(3)	1.08	C(4)–C(5)–H(10)	119	H(12)	+0.10
C(13)–C(8)	1.33	C(3)–C(4)–O(11)	120	C(13)	−0.08
C(13)–C(14)	1.48	C(2)–C(3)–H(12)	121	C(14)	−0.06
C(14)–C(19)	1.39	C(1)–C(8)–C(13)	126	C(15)	−0.06
C(15)–C(14)	1.40	C(19)–C(14)–C(13)	119	C(16)	−0.09
C(16)–C(15)	1.38	C(15)–C(14)–C(13)	124	C(17)	+0.21
C(17)–C(16)	1.39	C(18)–C(19)–C(14)	122	C(18)	−0.09
C(18)–C(17)	1.38	C(19)–C(14)–C(15)	118	C(19)	−0.07
C(19)–C(18)	1.38	C(14)–C(15)–C(16)	121	H(20)	+0.09
H(20)–C(15)	1.07	C(15)–C(16)–C(17)	120	H(21)	+0.10
H(21)–C(16)	1.08	C(16)–C(17)–C(18)	120	O(22)	−0.49
O(22)–C(17)	1.36	O(22)–C(17)–C(18)	120	H(23)	+0.10
H(23)–C(18)	1.08	C(17)–C(18)–C(19)	120	H(24)	+0.09
H(24)–C(19)	1.08	C(14)–C(15)–H(20)	120	H(25)	+0.09
H(25)–C(13)	1.08	C(15)–C(16)–H(21)	121	H(26)	+0.09
H(26)–C(8)	1.08	C(16)–C(17)–O(22)	120	C(27)	0.00
C(27)–O(22)	1.41	C(17)–C(18)–H(23)	119	H(28)	+0.08
H(28)–C(27)	1.09	C(18)–C(19)–H(24)	119	H(29)	+0.08
H(29)–C(27)	1.09	C(8)–C(13)–H(25)	119	H(30)	+0.10
H(30)–C(27)	1.08	C(1)–C(8)–H(26)	114	C(31)	0.00
C(31)–O(11)	1.41	C(17)–O(22)–C(27)	116	H(32)	+0.08
H(32)–C(31)	1.09	O(22)–C(27)–H(28)	111	H(33)	+0.10
H(33)–C(31)	1.08	O(22)–C(27)–H(29)	111	H(34)	+0.08
H(34)–C(31)	1.09	O(22)–C(27)–H(30)	107		
		C(4)–O(11)–C(31)	116		
		O(11)–C(31)–H(32)	111		
		O(11)–C(31)–H(33)	107		
		O(11)–C(31)–H(34)	111		

KEYWORDS

- **quantum chemical calculation**
- **method *ab initio***
- ***p*-dimethoxy-*trans*-stilbene**
- **acid power**

REFERENCES

1. Kennedy, J. *In Cationic Polymerization of Olefins*;Mir Publisher: Moscow,RU, 1978;p 431.
2. Schmidt, M. W.;Baldridge, K. K.; Boatz, J. A., Elbert, S. T.; Gordon, M. S.; Jensen, J. H.; Koseki, S.; Matsunaga, N.; Nguyen, K. A.; Su,S.J.; et al. *J. Comput. Chem.***1993**, *14*, 1347–1363.
3. Bode, B. M.; Gordon, M. S. *J. Mol. Graphics Mod.***1998**, *16*, 133–138.
4. Babkin, V. A.; Fedunov, R. G.; Minsker, K. S.; et al. *OxidationCommun.***2002**, *25*(1), 21–47.
5. Babkin, V. A. et al. *Oxidation Commun.***1998,***21*(4), 454–460.
6. Babkin, V. A.; Zaikov, G. E. In*Nobel Laureates and Nanotechnology of the Applied Quantum Chemistry*;Nova Science Publisher: New York, USA, 2010; p 351.

CHAPTER 5

MODIFICATION OF POLYAMIDE 6 BY N,N'–BIS-MALEAMIDE ACID

N. R. PROKOPCHUK, E. T. KRUT'KO, and M. V. ZHURAVLEVA

Belarusian State Technological University, Sverdlova Str.13a , Minsk, Republic of Belarus. E-mail: v.polon k@belstu.by

CONTENTS

ABSTRACT

This chapter is based on receiving and researching of compositions on the basis of industrially made polyamide 6 modified by N,N'-bis-maleamide acid of meta-phenylenediamine. Existence in macromolecules of polyamides reactive carboxyl, amide and amino groups, capable to interact with multifunctional monomeric and oligomeric modifiers, gives the chance of receiving the materials possessing properties of sewed polymers. As a result, the use of bis-amino acid as a modifying additive in the system of aliphatic polyamide 6 provides an improvement in strength properties and thermal characteristics of a polymeric material.

5.1 INTRODUCTION

Due to the ever increasing demands of engineering industries of new techniques to polymeric materials, including created on the base of aliphatic polyamides, an urgent task of improving performance properties of these polymers, including high heat and heat resistance, resistance to thermal-oxidative and aggressive media, and adhesion characteristics to various substrates remains actual. Physical properties of aliphatic polyamides are mainly conditioned by strong intermolecular interactions at the expense of hydrogen bonds that are formed between the amide groups of neighboring macromolecules. In the preparation of polyamides, it is necessary to enter regulators (stabilizers) of molecular weight, in this case, acetic, adipic acids (as the cheapest and most available) and amines, salts of monocarboxylic acids and monoamines, or N-alkyl amides of monocarboxylic acids are used.[1] Polyamides as well as other polymers are polydispersed. In the polymer composition, there are large macromolecules and low molecular weight amides with a small number of elementary units in the molecules (oligomers). They contain terminal amino groups and carboxyl groups. Furthermore, the oligomers of amides may also exist in a cyclic form. The mechanism of their formation is still not clear. However, it is experimentally proved that in the absence of water, cyclic oligomers are not capable of polymerization, but they are easily converted into a polyamide in the presence of water. In this regard, the use of additives, which at processing of polyamides may release small amounts of moisture, can be very useful for improving the performance properties of materials, products, and coatings based on polyamides. One of the effective ways of targeted regulation of properties of commercially available polyamides is their chemical modification by polyfunctional reactive compounds.[2–5] Presence in the macromolecules of polyamides reactive carboxyl, amide and amino groups that are capable to react with monomeric and oligomeric polyfunctional modifiers enables the production of materials having the properties of cross-linked polymers. Thus, it was found that imido compounds, in particular N,N'–bis-maleimides, are effective modifiers of many polymers, also including polyamides.[6] Information about the use of N,N'–bis-maleinamido acids, as the modifiers in their synthesis of intermediates in the scientific literature, was not found.

5.2 MAIN PART

The aim of this work is to study and research the compositions based on polyamide 6 (PA 6), industrially produced by JSC "Grodno Azot" (Grodno, Belarus) (OST 6-06-09-93), intermediate product in the synthesis of modified –N,N'–m-phenylene bis-maleimide (FBMI) –N,N'–bis-maleamido metaphenylenediamine (FBMAK).

The modifying reagent was injected into the PA 6 at doses of 5–10 wt.%. FBMAK choice, as previous FBMI,[6] due to their high reactivity associated with the content of the reactant molecules of double bonds that can be disclosed under thermal or photographic processing to form a spatial cross-linked polymer structure.[6] Physical and mechanical properties of PA 6 are shown in Table 5.1.

TABLE 5.1 Physical and Mechanical Properties of PA 6.

Index name	Norm on the highest quality category
Color	From white color to light yellow color
Number of point inclusions per 100 g of product units	Not more than 8
Size of crumbs msm,	1.5–4.0
Moisture content (%)	0.03–2.00
Relative viscosity, dl/g	2.20–3.50
Mass fraction of extractables (%)	1.0–3.0
Melting point, °C	214–220

Receiving of FBMAK was performed by reaction of equimolar amounts of metaphenylenediamine with maleic anhydride at 20–25°C by gradual addition to a solution of diamine in a minimum amount of solvent—dimethylformamide stoichiometric amounts of maleic anhydride. To obtain the corresponding bis-maleimide, the second step (imidization) of bis-amino acid by heating it in a mixture of acetic anhydride and sodium acetate in the ratio 2.5:0.5 moles per mole of bis–amino acid at 70–90°C was performed. Upon completion of imidization (about 2.5 h of heating), FBMI was isolated and recrystallized from a mixture of ethyl and n-propyl alcohol, in the ratio 1:1. It should be noted that the yield of the intermediate product, FBMAK, in the first synthesis step is 85–90% while the final product after imidization (FBMI) is only 50–60%.

The melting point of the synthesized N,N'–m-phenylene-bis-maleimide was 203°C, which corresponded to the literature data.[7] FBMAK, when heated in the process of determining the melting point, becomes FBMI and the melting point of this compound is not clearly fixed.

Receiving of FBMAK and FBMI is performed according to Scheme 5.1. It is important to note that the perspective of use of polyamide 6 FBMAK as a modifier component instead of FBMI allows to exclude out the process of synthesizing FBMI, which is the step of high-cost chemical cyclodehydration (imidization) of FBMAK; moreover, the loss of the product during its conversion to the bis-imide of the final step of synthesis is eliminated, and during imidizing process, extracted water promotes destruction of cyclic structures in the system of the polyamide.

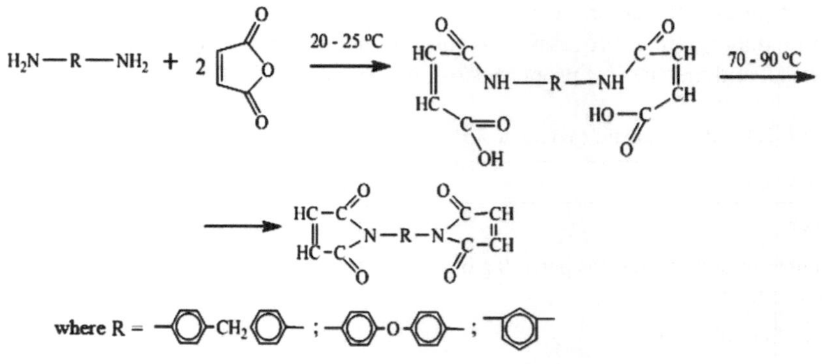

SCHEME 5.1

To study the structure of polyamide 6, the processes occurring in the polymer system, as well as assessing the completeness of spending FBMAK reactive groups by reacting with amine and amide functional groups of PA 6 during formation of three-dimensional structure when heated samples were performed IR spectroscopic study using FTIR spectrometer Nicolet 7101 (USA) in the range of 4000–300 cm^{-1} (resolution 1 cm^{-1}).[8] Furthermore, the possibility of formation of intermolecular cross-linking was confirmed by electron paramagnetic resonance (EPR).[9, 10] EPR spectra are recorded using the modified RE-1306 spectrometer with computer software. The heating process of polyamide compositions of modified FBMAK was carried out in the cavity of the spectrometer in the temperature range of 20–250°C. Before recording the spectra, the samples were cooled to room temperature. MnO containing ions of Mn was used as an external standard[9] and nitroxyl radicals were used as markers.

For example, evidence of more efficient structuring system in polyamide compositions containing as builders FBMAK compared with FBMI is that the time of correlation of stable nitroxyl radicals introduced into FBMAK modified PA 6 increased from 23·(source unmodified P6) to $50 \cdot 10^{-10}$ s with polyamide 6, modified by 5 wt.% FBMAK. For PA 6, modified with the same amount of MFBI, correlation time of nitroxide was significantly shorter period from 40^{-10} s.

Comparative analysis of the IR absorption spectra obtained in an airless environment of PA 6 compositions and FBMAK different composition (5–10 wt.% FBMAK builder) in the polymer melt, followed by heating of the samples at 150–210°C showed, for the modified polymer systems, absorbance decrease of absorption bands in the 1647 cm^{-1}, characteristic of the amide groups is observed.

It appears that at the process of chemical modification of PA 6 by bis-amino acid, as well as modifying it by its corresponding bis-imide reactions, an increase in the molecular weight of PA 6 take place. It occurs by the interaction of the carboxyl groups of the bis-amino acid modifier with amine-terminated polyamide groups (Scheme 5.2) and by the formation of interchain cross-links (Scheme 5.3) by opening the double bonds formed during the heat treatment at 150–200°C samples of polyamide compositions. This does not prevent homopolymerization of FBMI and that is due to the ability of the double bonds in FBMI to be activated by adjacent carbonyl groups of the imide cycle and disclosed by reacting with compounds containing labile hydrogen atom.

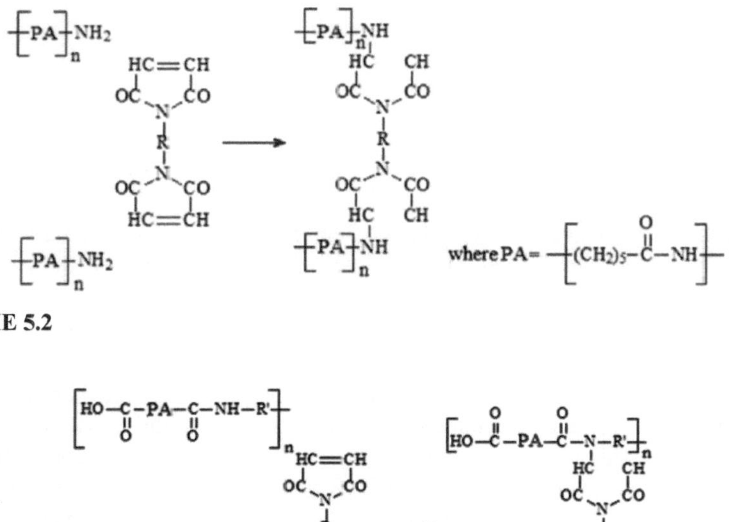

SCHEME 5.2

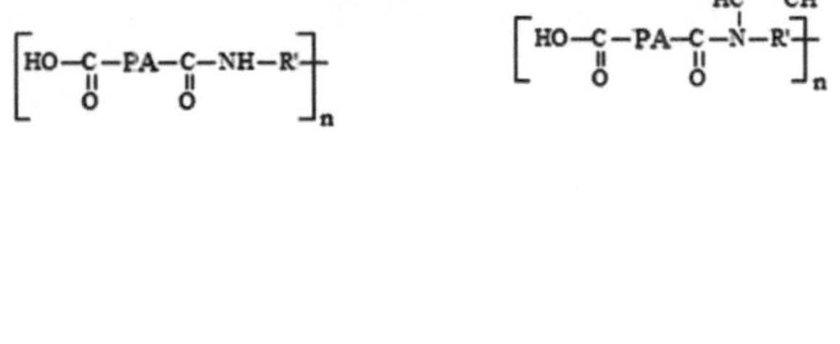

SCHEME 5.3

These processes all together result in a change of the supramolecular structure of the PA 6, causing the improvement of mechanical, thermal, and adhesive properties of the polymer.[6]

In practical use of FBMI, their thermochemical characteristics can have essential significance. In this connection, it is necessary and appropriate to examine the thermochemical characteristics of a number of different structures of BMI by differential scanning calorimetry.

DSC curves (Fig. 5.1) of all studied BMI, taken in the temperature range of 20–400°C at the rate of temperature rise of 10°C/min are clearly played two peaks endothermic, the melting process of BMI and exothermic corresponding to the process of disclosure of the double bonds.

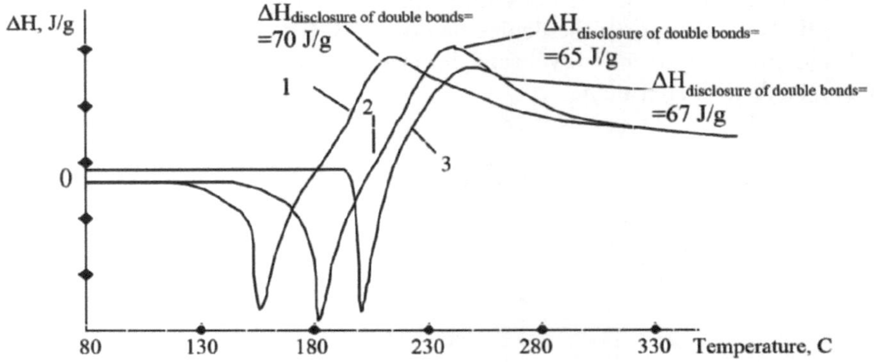

FIGURE 5.1 DSC curves of BMI. (1) N,N'-diphenylmethane-bis-maleimide; (2) N,N'-diphenyloxide-bis-maleimide; and (3) N,N'-m-phenylene-bis-maleimide (FBMI).

The position of the peaks in the respective process on the temperature scale depends on the chemical structure of the compounds. According to the value of enthalpy disclosure of double bonds, studied compounds in their thermochemical characteristics and probably related to reactivity in the reactions occurring with the opening of the double bonds are not significantly different. In this connection, the use of aliphatic polyamides FBMAK, FBMI is justified as modifying component. This component (MFBI), and accordingly MFBAK as its intermediate are produced in industrial conditions and are widely used for the production of a large range of heat-resistant composite materials. They are used to improve the adhesion of the cord to rubber mass in the manufacture of various kinds of tires for the automotive industry.

Effectiveness of modifying action of FBMAK in thermo-oxidative degradation processes of PA 6 compositions was evaluated on samples of three types: model-modified FBMAK powder compositions, films, and molded samples.

Model-modified powder compositions of PA 6 were prepared for the rapid analysis of possible stabilizing action FBMAK to the polymeric matrix. They were prepared as follows: PA 6 unstabilized granules were placed in a mixer, where nitrogen was supplied to provide oxygen-free environment. The temperature was raised to 240°C and a stirrer was switched on. With continuous stirring, the calculated amount of FBMAK was injected into the polymer melt. The mixture was being stirred for 5 min. Then it was poured into a mold and cooled. After curing, the sample was removed from the mold and crushed at a cryogenic temperature (after soaking for 10 min in liquid nitrogen at −180°C) to a fine powder. Investigation of the thermal stability was carried out by differential thermal and thermogravimetric analysis. The sample weight was 10 mg, a heating rate in the range of 20–500°C was 5°C/min. The activation energy of thermal oxidative degradation of the polymer E_d was calculated from the Broido method[11–13] according to dynamic thermogravimetry obtained by thermo-setting—module TA-400 of the Mettler Toledo company (Switzerland) in the temperature range of 330–400°C. Instrument calibration was performed on indium standard.

Laboratory samples of the polyamide films of 15–20 microns thick were cast from solutions of PA 6 with formic acid. To obtain solutions, a sample of the polymer (2 g) with a relative viscosity of 2.8 (determining in a solution of concentrated sulfuric acid) was dissolved in 28 ml of formic acid at room temperature. After that, the estimated quantity of FBMAK was injected into the solution and mixed thoroughly. Then degreased and dried on glass plates by pouring through a nozzle with an adjustable gap, the layer of film-forming composition was deposited. The glass plates were placed horizontally in a vacuum-drying oven, where at a residual pressure of 2 mm Hg. Art. and a temperature not above 30°C for 4.5 h, the solvent was removed. Formed on a glass substrate, polyamide film was removed and subjected to research.

Samples were also prepared by injection molding as follows: FBMAK modifier in an amount of 5–10 wt.% by weight of PA 6 was applied to the polymer granules without the heat stabilizer with a relative viscosity of 2.8 by dusting. Powdered pellets were peppered through the hopper to the injection molding machine, where they were melted. From the molten mass, bilateral blades with working part 50 × 10 × 3 mm were cast. Experimental samples, 1 day after their manufacturing, were placed in a heat chamber with a free circulation of air. They were maintained at 150°C for predetermined time (30–210 h). Determination of the breaking strength of the samples was performed by stretching according to GOST 112-80 using the tensile testing machine T 2020 DS 10 SH (Alpha Technologies UK, USA) at room temperature and velocity of the gripper 100 mm/min. Judging by the results of measurements of the tensile strength of the samples given in Table 5.2 generated in the step of heat treating of polyamide compositions meshwork of aliphatic polyamide 6 at the expense of the reactive groups of FBMAK (content in the composition of 5–10 wt.%) and functional groups of PA 6 predetermines the higher rates of heat resistance (temperature of thermo-oxidative degradation increases by 10–15°C) and deformation-strength properties (tensile strength increases by almost 10 MPa at initial modified samples, amounting 64.0–68.0 MPa) as compared to the unmodified polyamide 6 (59.0 MPa) and is comparable with PA 6 modified by the bis-maleimide (FBMI).

TABLE 5.2 Tensile Strength of PA 6 and Modified by MFBI MFBAK Samples of PA 6 after Thermal Oxidation in Air at 150°C for 30–210 h.

Material	Tensile strength, MPa								Td, °C	Ed, kJ/mol
	Initial	After thermal oxidation in air for, h								
		30	60	90	120	150	180	210		30
PA 6	59.0	50.0	49.0	41.0	35.5	30.5	23.0	18.6	275	120
PA 6 + 5 wt.% FBMAK	64.0	66.5	64.0	61.4	61.0	61.5	61.0	60.0	285	135
PA 6 + 10 wt.% FB-MAK	68.0	69.0	69.5	62.4	62.0	63.4	63.0	61.0	290	148
PA 6 + 5 wt.% FBMI6	63.4	64.6	63.0	62.6	62.0	62.0	61.4	60.7	287	141

Importantly, FBMAK as well as FBMI, provide a thermally stabilizing effect on polyamide 6 at various temperature and time of exposure of samples of polymer compositions. Calculation of the activation energy of thermal oxidative degradation of the samples (E_d) also indicates the formation of cross-links in a matrix structure of polyamide 6 at the expense of additives of MFBAM and MFBI, increasing depending on the modifier content of FBMAK from 135 to 148 kJ/mol.

As seen from Table 5.2, polyamide 6 modified with bis-maleamidoacid has not only improved strength characteristics as compared with the original PA 6, but is also more resistant to high temperature and time fields.

5.3 CONCLUSION

Thus, the use of bis-amino acid (FBMAK) as a builder in the system of aliphatic polyamide 6 improves strength properties and thermal characteristics of a polymeric material based on PA 6 virtually in the same range as the previously used for the same purpose FBMI.[6]

KEY WORDS

- **polyamide 6**
- **N,N′-diphenylmethane-bis-maleimide**
- **N,N′-diphenyloxide-bis-maleimide**
- **N,N′-m-phenylene-bis-maleimide**
- **modification**
- **energy activation**

REFERENCES

1. Kudryavtsev, G. I.; Noskov, M. P.; Volokhina, A.V. Polyamide Fibers. *Chemistry* **1976,** 259.
2. Hapugalle, G.; Prokopchuk, N. R.; Prakapovich, V. P.; Klimovtsova, I. A. New Thermostabilizers Polyamide 6 *NASB Vestsi. Ser. Him. Navuk.* **1999**, *1*, 114–119.
3. Vygodskii, Y. S. et al. The Anionic Polymerization of ε-Caprolactam and its Copolymerization with ω-Dodekalaktamom in the Presence of Aromatic Polyimides *Polym. Comp. Ser. A*. **2006**, *1* (6), 885–891.
4. Vygodskii, Y. S. et al. The Anionic Polymerization of ε-Karolaktama in the Presence of Aromatic Polyimides as Macromolecular Activators. *Polym. Comp. Ser. A*. **2003**, *45* (2), 188–195.
5. Vygodskii, Y. S. et al. The Anionic Polymerization of ε-Caprolactam in the Presence of Aromatic Diimides *Polym. Comp. Ser. A*. **2005**, *47* (7), 1077.
6. Biran, V. et al. *About Modifying the Action of N,N'-bis-imides of Unsaturated Dicarboxylic Acids, Aliphatic Polyamides*; Reports of the Academy of Sciences: Byelorussian SSR, 1983; Vol. 27 (8), pp 717–719.
7. Volozhin, A. I. et al. Synthesis of N, N'-Bis-Imides of Unsaturated Dicarboxylic Acids, Cycloaliphatic. *Vestsi BSSR. Ser. Him. Navuka*. **1974**, *1*, 98–100.
8. Nakanishi, K. *Infrared Spectra and Structure of Organic Compounds*, Wiley: New York, 1965; p 216.
9. Boldyrev, A. G. et al. *The Study by EPR of Free Radicals in Polyimides*; Report of the Academy of Sciences: USSR, 1965; T. *163* (5), pp 1143–1146.
10. Ingram, D. In *Electron Paramagnetic Resonance in Biology*; Agip, Y. I.; Kayushin, L. P., Eds.; Wiley: New York, 1972; p 296.
11. Broido, A.; Semple, A. Sensitive Graphical Method of Treating Thermogravimetric Analysis Date I. *Polym. Sci*. **1969**, Part A. 2, *7* (10), 1761–1773.
12. Prokopchuk, N. R. Study the Thermal Stability of Polymers by Derivatography *Vestsi BSSR. Ser. Him. Navuka*. **1984**, *4*, 119–121.
13. Polymer Construction. *Method for Determining the Durability of the Activation Energy for Thermal Oxidative Degradation of Polymeric Materials;* STB 13330-2002, Enter. 28.06.2002, Technical Committee on Technical Regulation and Standardization in Construction: the Ministry of Architecture and Construction of Belarus: Minsk, BY, 2002; p 8.

CHAPTER 6

PROTECTIVE PROPERTIES OF MODIFIED EPOXY COATINGS

N. R. PROKOPCHUK, M. V. ZHURAVLEVA, N. P. IVANOVA, T. A. ZHARSKAYA, and E. T. KRUT'KO

Belarusian State Technological University, Sverdlova Str.13a, Minsk, Republic of Belarus

CONTENTS

ABSTRACT

This chapter is devoted to the development and research of new film materials based on epoxy resins with improved properties. Evaluation of protective properties is done by electrochemical and physicomechanical methods of research that provide the most complete picture of the corrosion processes occurring under the paint film. These studies allowed adjusting the coating composition to the individual applications in order to achieve a high degree of protection of metal surfaces.

6.1 INTRODUCTION

Epoxy materials have superior performance properties, so they have been used to produce high-quality coatings. Coating materials based on epoxy oligomers are used to prepare coatings for various purposes, chemical-resistant, water-resistant, heat-resistant, and insulating coatings. They are characterized by high adhesion to metallic and nonmetallic surfaces, resistance to water, alkalis, acids, ionizing radiation, low porosity, small moisture absorbability, and high dielectric properties.[1] However, there are a number of outstanding issues to improve the barrier properties of paints based on epoxy resins, which limit their wider use in aeronautical engineering as well as in engineering and shipbuilding.[2] Chemical structure of epoxy resins provides ample opportunities to control their properties by introducing modifying additives to achieve maximum compliance with the requirements of the resulting material.

6.2 MAIN PART

The aim of this study was to develop and research new film materials based on epoxy diane resin with improved barrier properties.

The object of the study was commercially produced epoxy resin E-41 in solution (E-41s) (TU 6-10-607-78), which is a solution of the resin E-41 with a mass fraction 66 ± 2% in xylene (GOST 9410-78, GOST 9949-76) with acetone (GOST 2768-84) at a ratio of 4:3 by weight. Resin solution E-41 in a mixture of xylene and acetone (resin E-41s) is used for the manufacture of paints for various purposes. Resin E-41s refers to medium molecular weight (MW 900–2000) epoxy diane resin. Its density is 1.03–1.06 g/m^3. Product of copolycondensation of low molecular epoxy resin E-41 with difenidol propane is represented by the formula shown in Scheme 6.1.

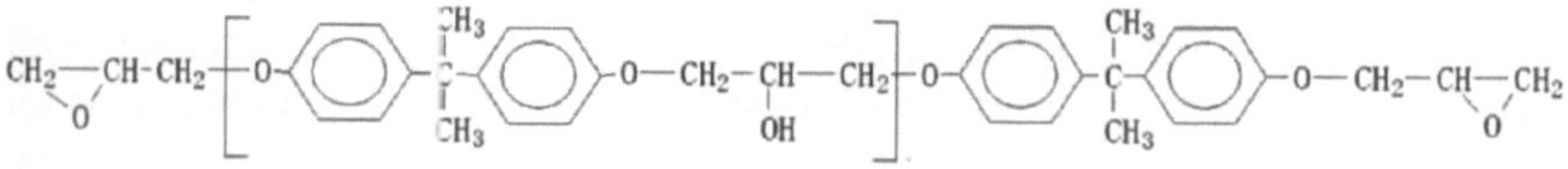

Scheme 1

Physicochemical characteristics of the resins E-41s are shown in Table 6.1.

TABLE 6.1 Physicochemical Characteristics of the Resin E-41s.

Index name	Norm on the highest quality category
Appearance of the resin solution	Homogeneous transparent liquid
Appearance of the film	Poured on glass is clean. Slight rash
Color on iodometric scale, mg I2/100 cm^3, not darker than	30
Viscosity by viscometer VZ-246 (OT-4) with a nozzle diameter of 4 mm at a temperature of 20.0 ± 0.5°C, c	80–130
Volume of solids, %	66 ± 2
Mass fraction of epoxy groups in terms of dry resin	6.8–8.3
Mass fraction of chloride ion (in terms of dry resin), %, not more than	0.0045
Mass fraction of saponifiable chlorine (in terms of dry resin), %, not more than	0.25

As a modifying component, p-aminophenol (pAPh) was used. It is known that aminophenols are corrosive inhibitors. The presence of the aromatic ring and the amide and hydroxyl functional groups in the molecule of p-aminophenol dictates the fundamental possibility of using this compound as a modifier of epoxy oligomer, allowing to increase the corrosion resistance of the coating.

The film-forming composition was prepared by introducing into E-41s 10% solution of p-aminophenol in dimethylformamide at a concentration range of 0.5–5.0 wt.%, followed by stirring until smooth. Hardener of brand E-45 (TU 6-10-1429-79 with changing No 2), low molecular weight polyamide resin solution in xylene, in the amount of 14% by weight of resin solids was used. Films on various substrates were cast from the above solutions. Curing of modified epoxy diane compositions were carried out at a temperature of 110°C for 140 min.

Adhesive strength of the formed coatings was determined by the standard method in accordance with ISO 2409 and GOST 15140-78 by the cross-cut incision with the kickback. The essence of the method in the application of cross-cut incisions on the finished coatings with the tool "Adhesion RN" and a visual assessment of the condition of the coatings after the impact action of the device "Kick Tester" exerted on the opposite side of the plate at the site of incision. Coating condition is compared with standard classification and measured in points.

Impact resistance of coating samples were evaluated using the device "Kick Tester" in accordance with ISO 6272 and GOST 4765-73.

The method for determining the impact strength of films (measured in centimeters) is based on the instantaneous deformation of the metal plate coated with the paint by the free fall of load on the sample and is realized through the device "Impact Tester," which is intended to control the impact strength of the polymer, powder, and paint coatings.

The hardness of paint coatings was determined with a pendulum device (ISO 1522). The essence of the method consists in determining the decay time (number of oscillations) of the pendulum in its contact with the painted coating. Hardness is determined by the ratio of the number of oscillations of the coated sample to the number of oscillations of the sample without coating.

Flexural strength of the coatings was determined by a device SHG1 (ISO 1519 GOST 6806-73) by bending of the coated sample around test cylinders since larger diameters at an angle of 180°. On one of the cylinder diameters, coating either cracks or breaks or peels. In this case, the paint has an elasticity of previous diameter test cylinder of the device in which it is not destroyed. Viewed count is in the radii of curvature in millimeters.

As can be seen from Table 6.2, at a curing temperature of 110°C of the obtained modified compositions with the modifier content of 0.5–2.0% hardness of the coating and adhesion is improving and impact strength is increasing.

TABLE 6.2 Adhesion and Strength Properties of Epoxy Diane Coatings on Steel Substrates (Warming up to 110°C, 140 min).

pAPh content, (%)	Hardness, rel. u.	Adhesion to steel, points	Tensile impact, cm
0.0	0.66	1	2.5
0.5	0.69	0	5.0
1.0	0.73	0	10.5
2.0	0.70	0	7.5
3.0	0.67	1	3.0
4.0	0.63	2	6.0
5.0	0.47	2	1.0

Protective properties of coatings are determined by the sum of physical and chemical properties, which can be reduced to four basic characteristics:

- Electrochemical and insulation properties of coatings;
- The ability of films to retard diffusion and carrying corrosive agents to a metal surface;
- The ability of coatings containing a film-forming, pigments or inhibitors, passivate, or electrochemically protect metal;
- Adhesion and mechanical properties of the coatings.

All these properties are interrelated and influence each other mutually. Deterioration of properties of the film as a diffusion barrier will immediately lead to a reduction in adhesion due to the development of the corrosion process under the film. Therefore, adhesion itself, how high it may be, cannot provide long-lasting protection from corrosion. Likewise, coatings with high diffusion limitations, but with poor adhesion cannot provide long-lasting protection.[3]

Paint application is one of the most common and reliable ways to protect metal surfaces from corrosion and give a decorative surface finish. It is known that the testing of protective properties of the coating in an operational environment takes a lot of time, which does not comply with any developers or manufacturers. Rapid tests provide information about the resistance of the coating under its compulsory destruction simulating natural mechanism of aging in a short trial time. Electrochemical methods are used as these accelerate the test methods.

Electrochemical methods are based on measurement of electrical parameters of the electrochemical phenomena occurring in the test solution. This measurement was performed using an electrochemical cell, which is a container with the test solution in which electrodes are placed. Electrochemical processes are accompanied by the appearance in the solution or changing of the potential difference between the electrodes or changing of the magnitude of current flowing through a solution.[4]

To assess the protective properties and select the modifier concentration in the polymer coating, the study of the time dependence of the stationary potential of the metal coating and removal of the anodic polarization curves was used. Potential measurement of the metal coating was being performed in 0.5% HCl at a temperature of 20 ± 2°C in the scale of silver chloride reference electrode for 24 h, then values were converted to a scale of the standard hydrogen electrode.

Removal of anodic polarization curves in a 0.5% HCl was being performed using a potentiostat 50 PI-1, and a driving voltage programmer 8 in potentiostatic mode. Tests were carried out in a three-electrode electrochemical cell. Before removing the anode polarization curve, value of the equilibrium potential of the metal cover was being measured for 5 min. Anodic polarization was carried out in potentiostatic mode at a potential step change in 20 mV with delayed current at each potential for 1 min.

By the slope of the Tafel plot of the polarization curve in the coordinates E–lgi (Fig. 6.1) value of the coefficient b was determined[5]:

$$b = \frac{2.303 \times R \times T}{\alpha \times n \times F}.$$

Corrosion current density in the metal coating was determined graphically by the intersection of stationary potential measured for 24 h (E24) in 0.5% HCI, and the straight portion of the anode polarization curve, which if necessary, was extrapolated. It was experimentally established that offset values of stationary potential of corrosive base covering systems in electronegative side over time may be due to relief of the anode ionization of the metal due to the coating moisture permeability and increase of its conductivity.

For samples with an epoxy polymer coating (EPC), stationary potential takes more electropositive value compared to carbon steel. With increasing concentration of modifying additive in the p-aminophenol coating, displacement capacity reaches 30 mV.

For samples with EPC with modifier content of 0.5–5.0% in the potential range −0.2 to 0.3 *b,* Tafel slopes of all areas of the anode curves are approximately the same (Table 6.3). This suggests that the mechanism of active dissolution of iron in the pores of the polymer coating does not change, and inhibition is due to the decrease of the effective surface of the dissolving metal.

Organic modifier p-aminophenol introduced into the coating has two functional groups of atoms containing lone pair electrons, facilitating its adsorption on iron, which applies to transition metals with a free *d* orbitals. The presence of p-aminophenol in the polymer coating decreases the effective surface of the dissolving metal and inhibits the anodic process.

Extrapolation of the linear portions of the polarization curves (Fig. 6.1) to the measured stationary capacity allows to determine the corrosion rate.

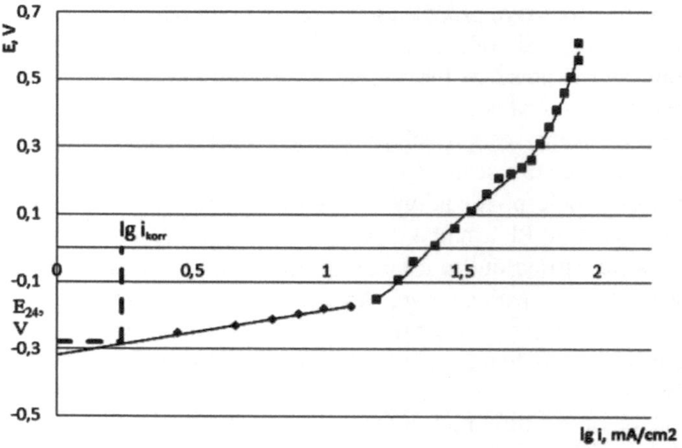

FIGURE 6.1 Anodic polarizat curve. Example: Steel 08 kp: EPC + 1% modifier. E_{24} = −0.275 V; $i_{кор}$ = 1.62 mA/cm², t = 19°C.

Table 6.3 shows the stationary potentials and the corrosion rate of the metal-covering systems.

TABLE 6.3 Stationary Potentials and the Corrosion Rate of the Steel 08 kp—Epoxy Coating in 0.5% HCl.

Tested sample	***E*, B**	**lg*i***	***i*, mA/cm²**	***b*, B**	**Percentage of modifier**
Carbon steel 08 kp	−0.288	0.29	1.95	0.1346	–
Carbon steel 08 kp: LPC + 0.5 M	−0.28	0.24	1.74	0.1130	0.5
Carbon steel 08 kp: LPC + 1.0 M	−0.275	0.21	1.62	0.0978	1
Carbon steel 08 kp: LPC + 2.0 M	−0.266	0.18	1.51	0.0911	2
Carbon steel 08 kp: LPC + 3.0 M	−0.26	0.15	1.41	0.0572	3
Carbon steel 08 kp: LPC + 4.0 M	−0.252	0.11	1.29	0.0434	4
Carbon steel 08 kp: LPC + 5.0 M	−0.238	−0.05	0.89	0.0328	5
Tested sample	*E*, B	lg*i*, mA/cm²	*i*, mA/cm²	*b*, B	Percentage of modifier

From these data, one can conclude that the polymer coatings inhibit corrosion of carbon steel 08 kp , which corrodes in 0.5% HCl at a speed of 1.95 mA/cm². Application of epoxy polymer coating containing modifier of n-aminophenol 0.5–5.0%, reduces the corrosion rate of steel in 1.1–2.2 times.

With increasing concentration of n-aminophenol modifier in the polymer coating corrosion resistance at the system increases, whereas the corrosion current density decreases (Fig. 6.2), and the polarization curves are shifted to lower currents. Basing on the obtained results, it can be concluded that the polymer coatings based on epoxy resins with modifier inhibit corrosion of carbon steel better than without it.

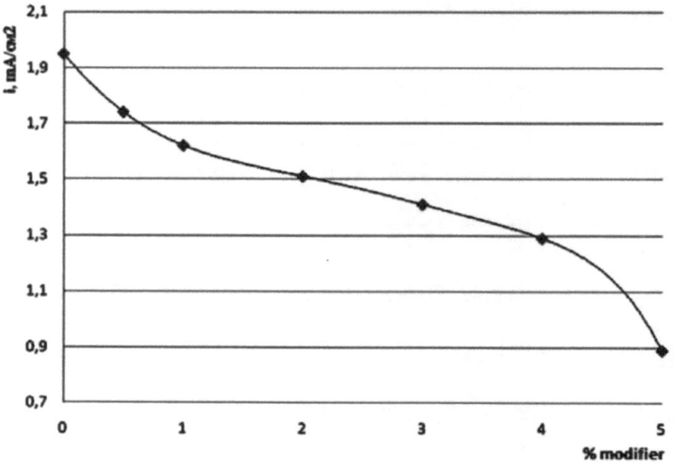

FIGURE 6.2 The dependence of the corrosion current on the percentage of modifier.

6.3 CONCLUSION

Introduction of n-aminophenol modifier at a concentration of 0.5–5.0 wt.% into a polymeric epoxy resin coating improves the corrosion resistance of the system to 0.5% HCl, while the current density of steel corrosion 08 kp decreases from 1.95 to 0.89 mA/cm^2 and the polarization curves are shifted to lower currents. Application of the epoxy polymer coating modifier in an amount of 5% reduces the corrosion rate of carbon steel in 08 kp 0.5% HCl 2.2 times. Addition of 0.5–2.0% modifier also improves physical and mechanical characteristics (improved strength, hardness, and adhesion of coatings).

Experimental studies have shown that additional evaluation of protective properties of coatings by electrochemical methods in conjunction with conventional for the paint industry methods of research yielded a better understanding of corrosion processes under cover, and allowed to assess their impact on the course of the concentrations of the modifier. These studies allowed to adjust the coating composition to the individual applications in order to achieve a high degree of protection of metal surfaces.

KEYWORDS

- **oligoepoxide**
- **aminophenol**
- **electrochemical method**
- **epoxy polymer**
- **coatings**
- **corrosion**
- **mechanical properties**

REFERENCES

1. Lee, H.; Neville, K. *Handbook of Epoxy Resins;* Energiya: Moscow, RU, 1973; p 268.
2. Kireyeva, V. G. *Coating Materials Based on Epoxy Resins;* NIITEKHIM: Moscow, RU, 1992; p 48 (Survey information/Ser "Paint industry").
3. Rosenfeld, I. L. *Corrosion Inhibitors;* Khimiya: Moscow, RU, 1977; p 352.
4. Semenova, I. V. In *Corrosion and Corrosion Protection*; Florianovich, G. M.; Khoroshilov, A. V., Eds.; Fizmatlit: Moscow, RU, 2006; p 328.
5. Damascus, B. B.; Peter, O. *Electrochemistry: A Textbook for Chem. Faculty. Universities;* Higher School: Moscow, RU, 1987; p 295.

CHAPTER 7

CHEMICAL MODIFICATION OF SYNDIOTACTIC 1,2-POLYBUTADIENE

M. I. ABDULLIN, A. B. GLAZYRIN, O. S. KUKOVINETS, A. A. BASYROV, and G. E. ZAIKOV

CONTENTS

ABSTRACT

Results of chemical modification of syndiotactic 1,2-polybutadiene under various chemical reagents are presented. Influence of the reagent nature, both on the reactivity of >*C*=*C*< double bonds in syndiotactic 1,2-polybutadiene macromolecules at its chemical modification and on the composition of the modified polymer products, is considered basing on the analysis of literature and the authors' own researches.

Obtaining polymer materials of novel or improved properties is considered a major direction in modern macromolecular chemistry.[1] Researches aimed at obtaining polymers via chemical modification methods have been quite prominent along with traditional synthesis methods of new polymer products by polymerization or polycondensation of monomers.

Polymers with unsaturated macrochains hold much promise for modification. The activity of carbon–carbon double bonds as related to many.

Syndiotactic 1,2-polybutadiene obtained reagents allow introducing substituents of different chemical nature in the polymer chain (including heteroatoms). Such modification helps to vary within wide range of physical and chemical properties of the polymer and render it new and useful features. By stereospecific butadiene polymerization in complex catalyst solutions[2–7] provides much interest for chemical modification.

In contrast to 1,4-polybutadiens and 1,2-polybutadiens of the atactic structure, the syndiotactic 1,2-polybutadiene exhibit thermoplastic properties combining elasticity of vulcanized rubber and ability to move to the viscous state at high temperatures and be processed like thermoplastic polymers.[8–11]

The presence of unsaturated >*C*=*C*< bonds in the syndiotactic 1,2-PB macromolecules creates prerequisites for including this polymer into various chemical reactions resulting in new polymer products. Unlike 1,4-polybutadiens, the chemical modification of syndiotactic 1,2-PB is insufficiently studied, though there are some data available.[12–15]

A peculiarity of the syndiotactic 1,2-PB produced nowadays is the presence of statistically distributed cis-and trans- units of 1,4-diene polymerization[16,17] in macromolecules along with the order of 1,2- units at polymerization of 1,3-butadiene. Their content amounts to 10–16%. Thus, by its chemical structure, syndiotactic 1,2-polybutadiene can be considered as a copolymer product containing an orderly arrangement of 1,2-units and statistically distributed 1,4-polymerization units of 1,3-butadiene:

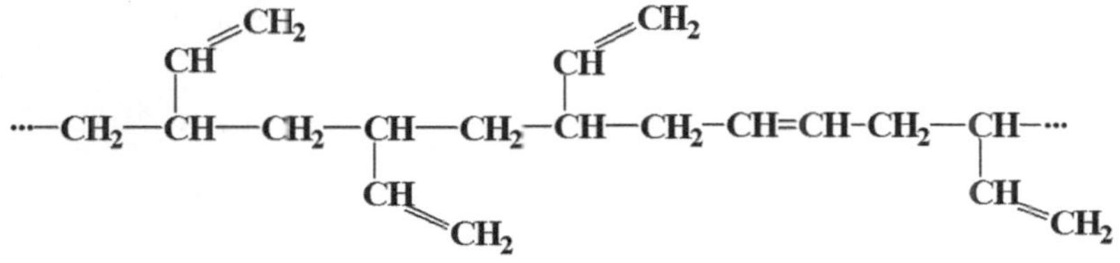

1

Taking into account, syndiotactic 1,2-PB microstructures and the presence of >*C*=*C*< various bonds in the polydiene macromolecules, the influence of some factors on the polymer chemical transformations has been of interest. The factors in question are determined both by the double bond nature in the polymer and the nature of the substituent in the macromolecules.

In the paper, the interaction of the syndiotactic 1,2-PB and the reagents of different chemical nature such as ozone, peroxy compounds, halogens, carbenes, aromatic amines, and maleic anhydride are considered.

A syndiotactic 1,2-PB with the molecular weight $M_n = (53–72) \times 10^3$; $M_w/M_n = 1.8–2.2$; 84–86% of 1,2-butadiene units (the rest being 1,4-polymerization units); syndiotacticity degree 53–86% and crystallinity degree 14–22% were used for modification.

7.1 OZONATION

At interaction of syndiotactic 1,2-PB and ozone, the influence of the inductive effect of the alkyl substituents at the carbon–carbon double bond on the reactivity of double bonds in 1,2- and 1,4-units of butadiene polymerization is vividly displayed. Ozone first attacks the most electron-saturated inner double bond of the polymer chain. The process is accompanied by the break in the >*C*=*C*< bonds of the main chain of macromolecules and a noticeable decrease in the intrinsic viscosity and molecular weight of the polymer (Fig. 7.1) at the initial stage of the reaction (functionalization degree $\alpha < 10\%$).[17–20] Due to the ozone high reactivity, partial splitting of the vinyl groups is accompanied by spending double bonds in the main polydiene chain.

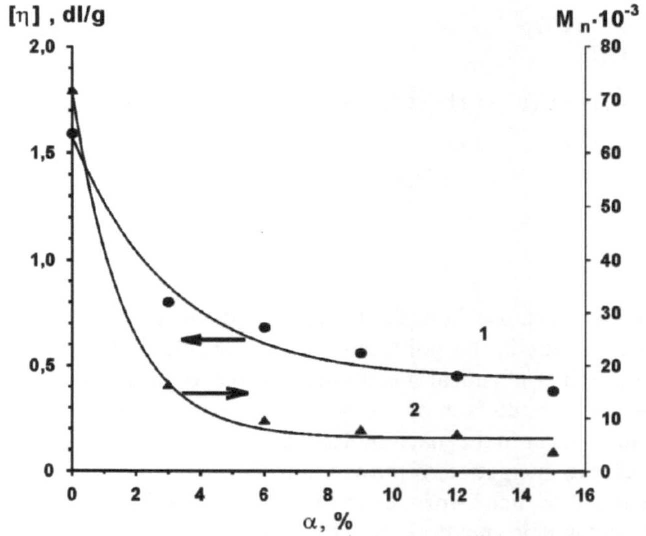

FIGURE 7.1 Dependence of the intrinsic viscosity [h] (1) and the average molecular weight M_n (2) of the formyl derivative of the syndiotactic 1,2-PB from the functionalization degree of the α-polymer (with chloroform as a solvent, 25°C).

However, it does not affect the average molecular weight of the polymer up to the functionalization degree of 15% (Fig. 7.1). Depending on the chemical nature of the reagent used for the decomposition of the syndiotactic 1,2-PB ozonolysis products (dimethyl sulfide or lithium aluminum hydride),[17–20] the polymer products containing aldehyde or hydroxyl groups are obtained (Scheme 7.1).

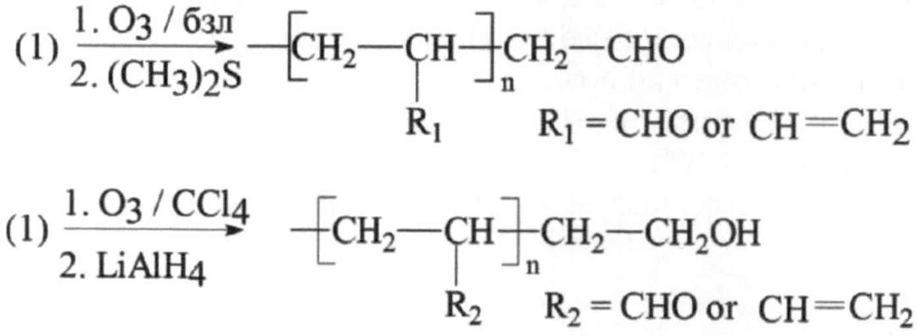

SCHEME 7.1

The structure of the modified polymers is set using IR and NMR spectroscopy.[17] The presence of C-atom characteristic signals connected with aldehyde (201.0–

201.5 ppm) or hydroxyl (56.0–65.5 ppm) groups in ^{13}C NMR spectra allows identifying the reaction products.

Thus, the syndiotactic 1,2-PB derivatives with oxygen-contained groupings in the macromolecules may be obtained via ozonation. Modified 1,2-PB with different molecular weight and functionalization degrees containing hydroxi or carbonyl groups are possible to obtain by regulating the ozonation degree and varying the reagent nature used for decomposing the products of the polymer ozonation.

7.2 EPOXIDATION

Influence of the double bond polymer nature in the reaction direction and the polymer modification degree is vividly revealed in the epoxidation reaction of the syndiotactic 1,2-PB, which is carried out under peracids (performic, peracetic, meta-chloroperbenzoic, and trifluoroperacetic ones),[21–27] *tert*-butyl hydroperoxide,[21–23] and other reagents[28–30] (Table 7.1).

TABLE 7.1 Influence of the Epoxidizing Agent on the Functionalization Degree α of Syndiotactic 1,2-PB and the Composition of the Modified Polymer*.

Epoxidizing agent	α, mol. (%)	Content in the modified polymer, mol.%			
		Epoxy groups		>C=C<bonds	
		1,2-units	1,4-units	1,2-units	1,4-units
**R^1COOH/H_2O_2	11.0–16.0	–	11.0–16.0	84.0	0–5.0
**R^2COOOH	32.1	16.1	16.0	67.9	–
**R^3COOOH	34.6	18.6	16.0	65.4	–
Na_2WO_4/H_2O_2	31.0	15.0	16.0	69.0	–
Na_2MoO_4/H_2O_2	23.7	7.7	16.0	76.3	–
$Mo(CO)_6$/*t*-BuOOH	18.0	18.0	–	66.0	16.0
NaClO	16.0	–	16.0	84.0	–
$NaHCO_3/H_2O_2$	16.0	–	16.0	84.0	–

* The polymer with the content of 1,2- and 1,4-units with 84 and 16%, respectively;
** where R^1–H–, Me–, Et–, $CH_3CH(OH)$–; R^2–*m*–ClC_6H_4; R^3–CF_3–.

Depending on the nature of the epoxidated agent and conditions of the reaction,[21–27] polymer products of different composition and functionalization degree can be obtained (Table 7.1).

As established earlier,[21–24,26] at interaction of syndiotactic 1,2-PB and aliphatic peracids obtained in situ under hydroperoxide on the corresponding acid, the >*C*=*C*< double bonds in 1,4-units of macromolecules are mainly subjected to epoxidation (Scheme 7.2). Higher activity of the double bonds of 1,4-polymerization units in the epoxidation reaction is revealed at interacting syndiotactic 1,2-PB and sodium hypochlorite as well as percarbonic acid salts obtained in situ through the appropriate carbonate and hydroperoxide.[31–35]

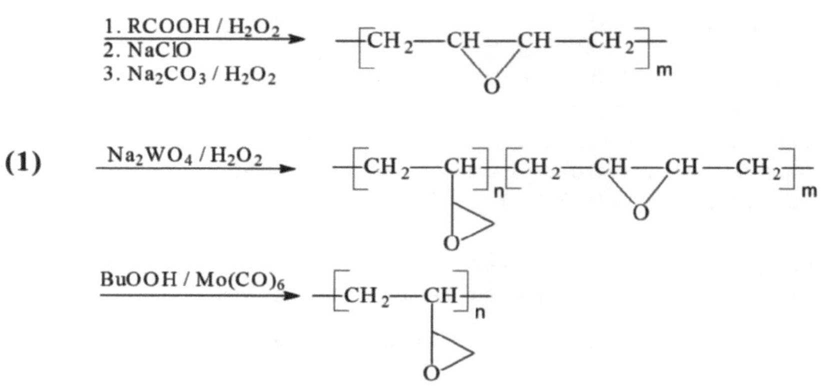

SCHEME 7.2

It should be noted that the syndiotactic 1,2-PB epoxidation by the sodium hypochlorite and percarbonic acid salts carried out in the alkaline enables to prevent the disclosure reactions of the epoxy groups and a gelation process of the reaction mass observed at polydiene epoxidation by aliphatic peracids.[31–32] The functionalization degree of 1,2-PB in the reactions with the stated epoxidizing agents ($\alpha \leq 16\%$, Table 7.1) is determined by the content of inner double bonds in the polymer.

To obtain the syndiotactic 1,2-PB modifiers of a higher degree of functionalization (up to 35%) containing oxirane groups, both in the main and side chain of macromolecules (Scheme 7.2), it is necessary to use only active epoxidizing agents meta-chloroperbenzoic (MCPBA), trifluoroperacetic acids (TFPA),[23] and metal complexes of molybdenum and tungsten, obtained by reacting the corresponding salts with hydroperoxide[25] (Table 7.1). From the epoxidizing agents given, a trifluoroperacetic acid is the most active (Fig. 7.2).[23,26]

However, at the syndiotactic 1,2-PB epoxidation by the trifluoroperacetic acid, a number of special conditions is required to prevent the gelation of the reaction mass, namely usage of the base (Na_2HPO_4, Na_2CO_3) and low temperature (less than 5°C).[23]

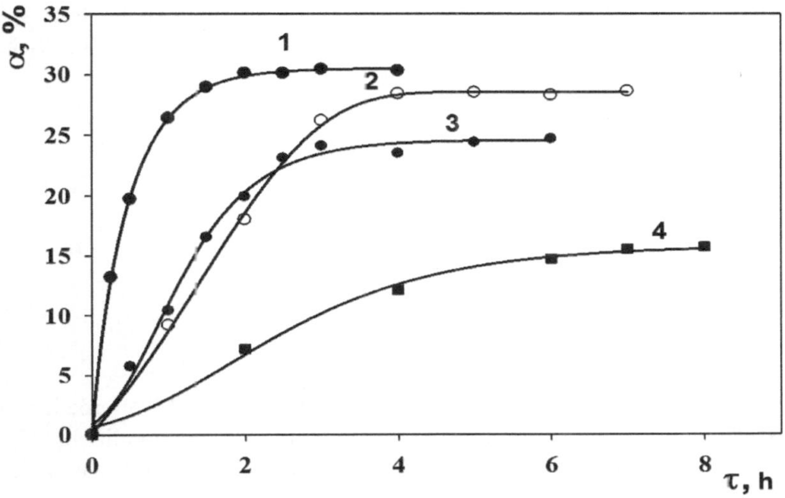

FIGURE 7.2 Influence of the peracid nature on the kinetics of oxirane groups accumulation at the syndiotactic 1,2-PB epoxidation: 1—TFPA ($[Na_2HPO_4]/[TFPA] = 2$; 0°C); 2—MCPBA (50°C); 3—Na_2MoO_4/H_2O_2 (55°C); 4—$HCOOH/H_2O_2$ (50°C).

At reacting the syndiotactic 1,2-PB and the catalyst complex [*t*-BuOOH–$Mo(CO)_6$], a steric control at approaching the reagents to the double polymer bond is carried out.[21–24] This results in participation of less active but more available vinyl groups of macromolecules in the reaction (Table 7.1, Scheme 7.2).

Thus, modified polymer products with different functionalization degrees (up to 35%) may be obtained on the syndiotactic 1,2-PB basis according to the epoxidizing agent nature. The products in question contain oxirane groups in the main chain (with aliphatic peracids, percarbonic acid salts, and NaClO as epoxidizing agents), in the side units of macromolecules [*t*-BuOOH–$Mo(CO)_6$] or in 1,2- and 1,4-units (TFPA, MCPBA, and metal complexes of molybdenum and tungsten).

7.3 HYDROCHLORINATION

In adding hydrogen halides and halogens to the >*C*=*C*< double bond of 1,2-PB, the functionalization degree of the polymer is mostly determined by the reactivity of the electrophilic agent. Relatively low degree of polydiene hydrochlorination (10–15%) at interaction of HCl and syndiotactic 1,2-PB[16,36,37] is caused by insufficient reactivity of hydrogen chloride in the electrophilic addition reaction by the double bond (Table 7.2). Due to this, more electron-saturated >*C*=*C*< bonds in 1,4-units of butadiene polymerization are subjected to modification.

The process is intensified at polydiene hydrochlorination under $AlCl_3$ due to a harder electrophile $H^+[AlCl_4]^-$ formation at its interaction with HCl.[37] In this case, double bonds of both 1,2- and 1,4-polidiene units take part in the reaction (Scheme 7.3):

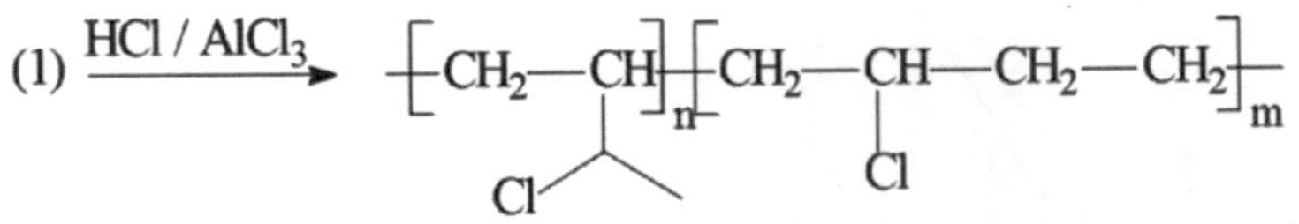

SCHEME 7.3

Usage of the catalyst $AlCl_3$ and a polar solvent medium (dichloroethane) allows speeding up the hydrochlorination process (Table 7.2) and obtaining polymer products with chlorine contained up to 28 mass% and the functionalization degree α up to 71%.[37]

By the ^{13}C NMR spectroscopy method,[37] it is established that the >*C*=*C*< double bond in the main chain of macromolecules is more active at 1,2-PB catalytic hydrochlorination. Its interaction with HCl results in the formation of the structure (a) (Scheme 7.4). At hydrochlorination of double bonds in the side chain, the chlorine atom addition is controlled by formation of the most stable carbocation at the intermediate stage. This results in the structure (b) with the chlorine atom at carbon β-atom of the vinyl group (Scheme 7.4):

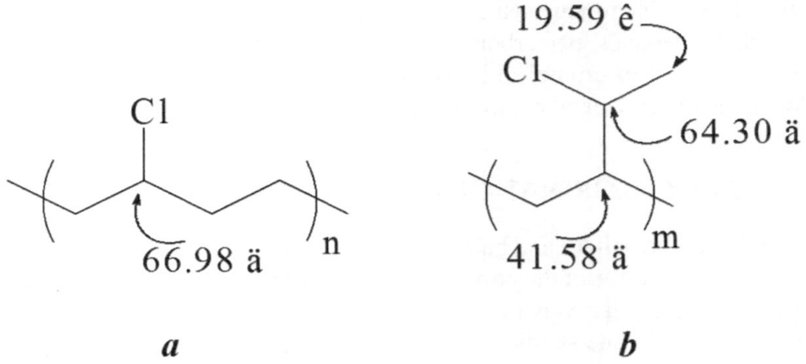

SCHEME 7.4

7.4 HALOGENATION

Effective electrophilic agents like chlorine and bromine easily join double carbon–carbon bonds,[16,38–40] both in the main chain of syndiotactic 1,2-PB and in the side chains of macromolecules (Scheme 7.5):

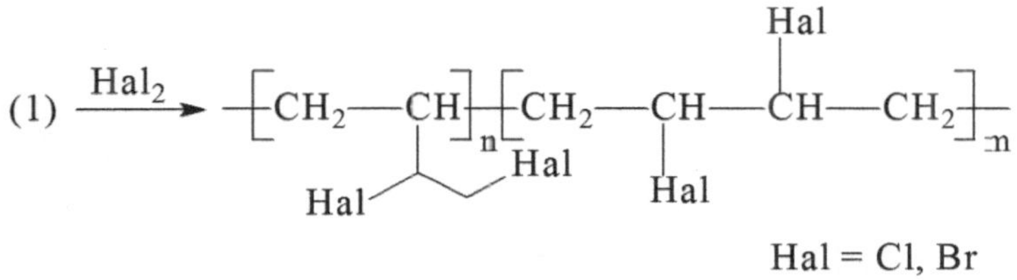

SCHEME 7.5

The reaction proceeds quantitatively: syndiotactic 1,2-PB chlorine derivatives with chlorine w(Cl) up to 56 mass% (α ~98%) (Fig. 7.3) and syndiotactic 1,2-PB bromo-derivatives with bromine up to 70 mass% (α ~94%) are obtained.

According to the ^{13}C NMR spectroscopy, polymer molecules with dihalogen structural units (Scheme 7.6) and their statistic distribution in the macro chain serve as the main products of syndiotactic 1,2-PB halogenation.[16,40]

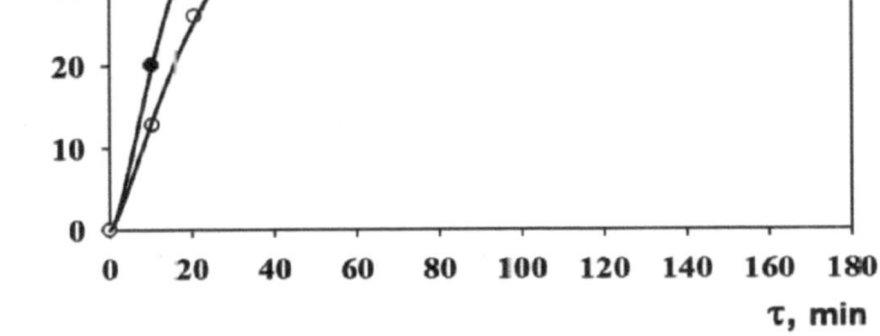

FIGURE 7.3 Kinetics of syndiotactic 1,2-PB chlorination. Chlorine consumption: **1**—1 mol/h per mol of syndiotactic 1,2-PB; **2**—2 mol/h per mol of syndiotactic 1,2-PB; 20°C, with $CHCl_3$ as a solvent.

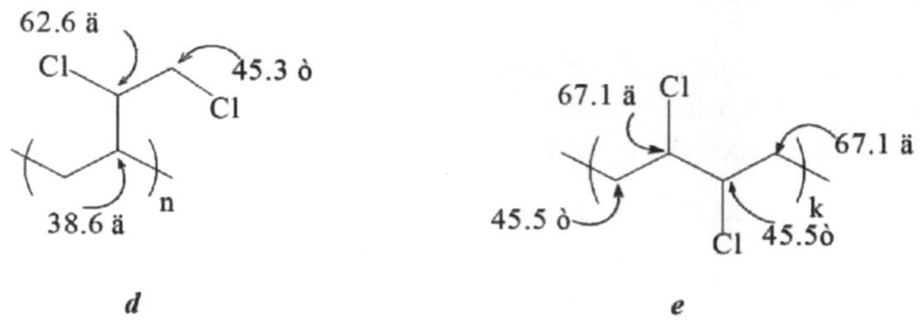

SCHEME 7.6

7.5 DICHLOROCYCLOPROPANATION

There is an alternative method for introducing chlorine atoms into the 1,2-PB macromolecule structure, namely a dichlorocyclopropanation reaction. It is based on generating an active electrophile agent in the reaction mass at dichlorocarbene modification, which is able to interact with double carbon–carbon polymer bonds.[41–43] The syndiotactic 1,2-PB dichlorocyclopropanation is quite effective at dichlorocarbene generating by Macoshi by the chloroform reacting with an aqueous solution of an alkali metal hydroxide. The reaction is carried out at the presence of a phase transfer catalyst and dichlorocarbene addition in situ to the double polydiene links[44–46] according to Scheme 7.7:

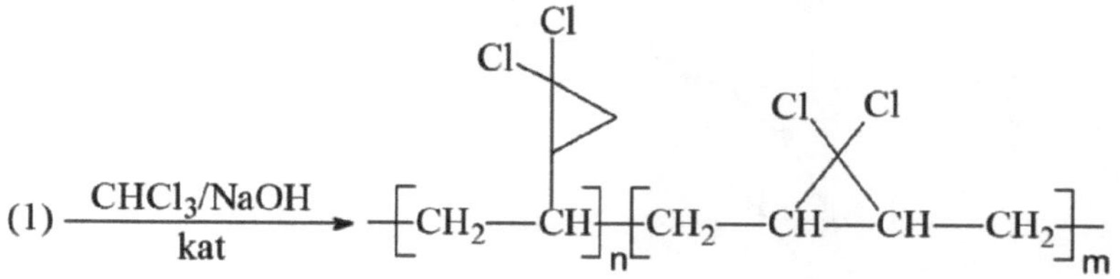

SCHEME 7.7

The ^{13}C NMR spectroscopy results testify the double bonds dichlorocyclopropanation, both in the main chain and side chains of polydiene macromolecules (Scheme 7.8).

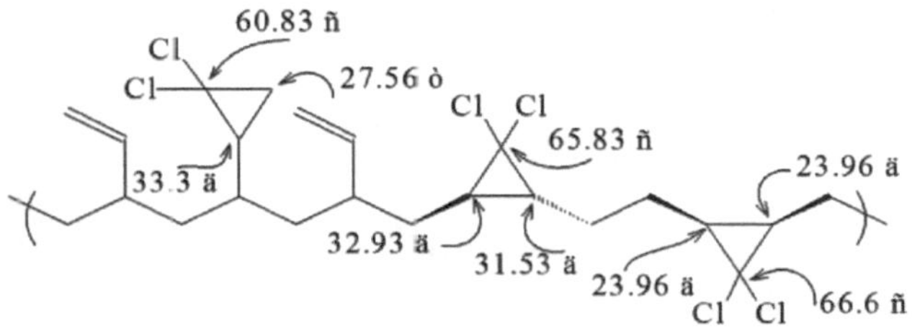

SCHEME 7.8

Cis-and trans-double bonds in the 1,4-addition units[45] are more active in the dichlorocarbene reaction.

The polymer products obtained contain chlorine up to 50 mass%, which corresponds to the syndiotactic 1,2-PB functionalization degree ~97%, that is, in the reaction by the Makoshi method, full dichlorocarbenation of unsaturated >*C*=*C*< polydiene bonds is achieved.[23,45]

7.6 INTERACTION WITH METHYL DIAZOACETATE

Modified polymers with methoxycarbonyl substituted cyclopropane groups[47–51] are obtained by interaction of syndiotactic 1,2-PB and carb methoxycarbonyl generated at a catalytic methyl diazoacetate decomposition in the organic solvent medium (Scheme 7.9):

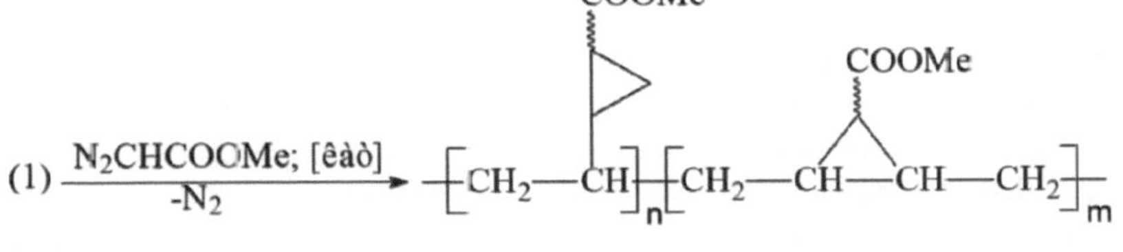

SCHEME 7.9

The catalytic decomposition of alkyl diazoacetates comprises the formation of the intermediate complex of alkyl diazoacetate and the catalyst.[51,52] The generated (alcoxicarbonyl)carbene at further nitrogen splitting is stabilized by the catalyst with the carbine complex formation,[51,52] interaction of which with the alkene results in the cyclopropanation products (Scheme 7.10):

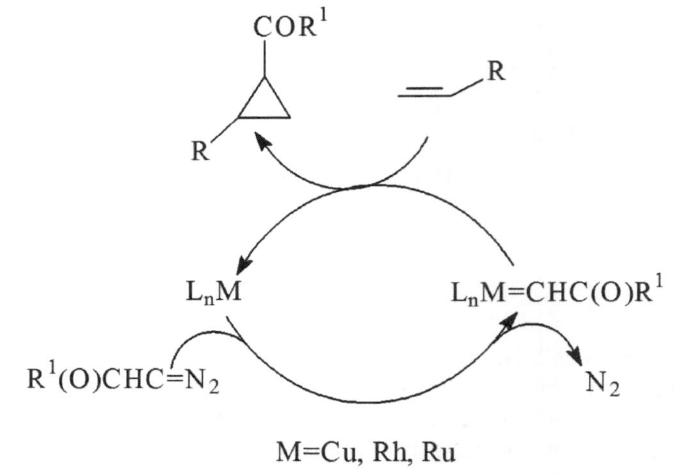

SCHEME 7.10

The output of the cyclopropanation products is determined by the reactivity of the >*C*=*C*< double bond in the alkene as well as the stability and reactivity of the carbine complex $L_nM{=}CH(O)R^1$, which fully depend on the catalyst used.[52]

The catalysts applied in the syndiotactic 1,2-PB cyclopropanation range as follows: $Rh_2(OAc)_4$ ($\alpha = 38\%$) > [Cu OTf]·0,5 C_6H_6 ($\alpha = 28\%$) > $Cu(OTf)_2$ ($\alpha = 22\%$).[51]

By the ^{13}C NMR spectroscopy methods, it is established that in the presence of copper (I), (II) compounds, double bonds, both of the main chain and in the side units of syndiotactic 1,2-PB macromolecules are subjected to cyclopropanation, whereas at rhodium acetate $Rh_2(OAc)_4$, mostly the >*C*=*C*< bonds in the 1,4-addition units undergo it.[51]

Thus, catalytic cyclopropanation of syndiotactic 1,2-PB under methyl diazoacetate allows obtaining polymer products with the functionalization degree up to 38% and their macromolecules containing cyclopropane groups with an ester substituent. The determining factor influencing the cyclopropanation direction and the syndiotactic 1,2-PB functionalization degree is the catalyst nature. By using catalysts of different chemical nature, it is possible to purposefully obtain the syndiotactic 1,2-PB derivatives containing cyclopropane groups in the main chain (with rhodium acetate as a catalyst) or in 1,2- and 1,4-polydiene units (copper compounds), respectively.

Along with the electronic factors determined by different electron saturation of the >*C*=*C*< bonds in 1,2- and 1,4-units of polydiene addition and the catalyst nature used in modification, the steric factors may also influence the reaction and the syndiotactic 1,2-PB functionalization degree. The examples of the steric control may serve the polydiene reactions with aromatic amines and maleic anhydride apart from the above-mentioned epoxidation reactions of syndiotactic 1,2-PB by *tert*-butyl hydroperoxide.

7.7 INTERACTION WITH AROMATIC AMINES

Steric difficulties prevent the interaction of double bonds of the main chain of syndiotactic 1,2-PB macromolecules and aromatic amines (aniline, N, N-dimethylaniline, and acetanilide). In the reaction with amines[17,20,23] catalyzed by $Na[AlCl_4]$, the vinyl groups of the polymer enter the reaction and form the corresponding syndiotactic 1,2-PB arylamino derivatives (Scheme 7.11):

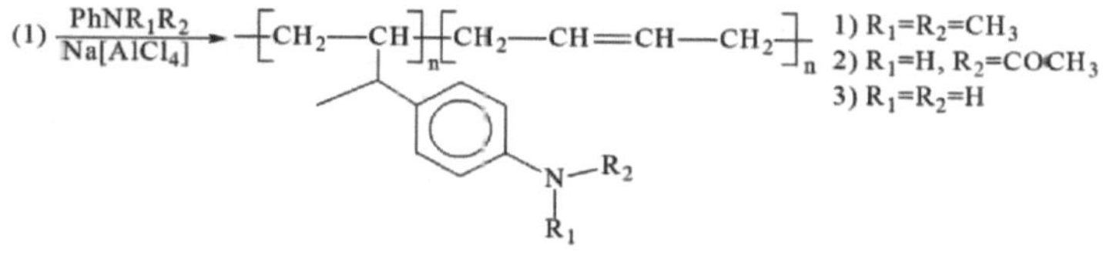

SCHEME 7.11

From the NMR spectra analysis, it is seen that the polymer functionalization is held through the β-atom of carbon vinyl groups.[17]

Introduction of arylamino groups in the syndiotactic 1,2-PB macromolecules leads to increasing the molecular weight M_w (Fig. 7.4) and the size of macromolecular coils characterized by the mean-square radius of gyration $(\overline{R^2})^{1/2}$.[17,20]

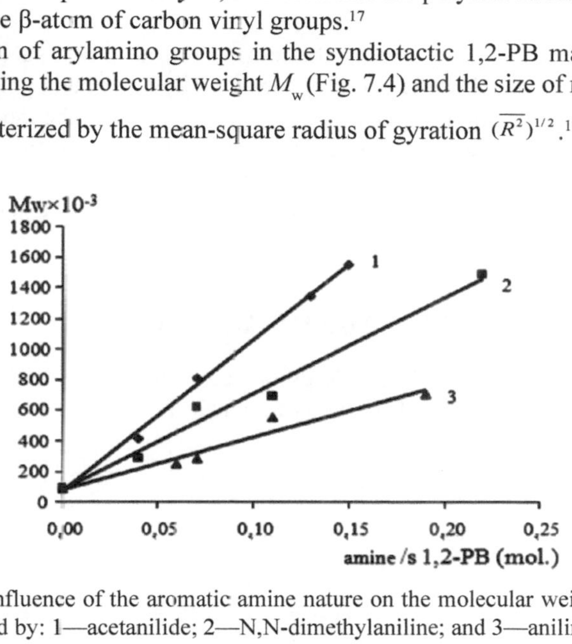

FIGURE 7.4 Influence of the aromatic amine nature on the molecular weight (M_w) of the polymer modified by: 1—acetanilide; 2—N,N-dimethylaniline; and 3—aniline.

The results obtained indicate the intramolecular interaction of monomer units modified by aromatic amines with vinyl groups of polydiene macromolecules at syndiotactic 1,2-PB modification. This leads to the formation of macromolecules of the branched and linear structure (Scheme 7.12):

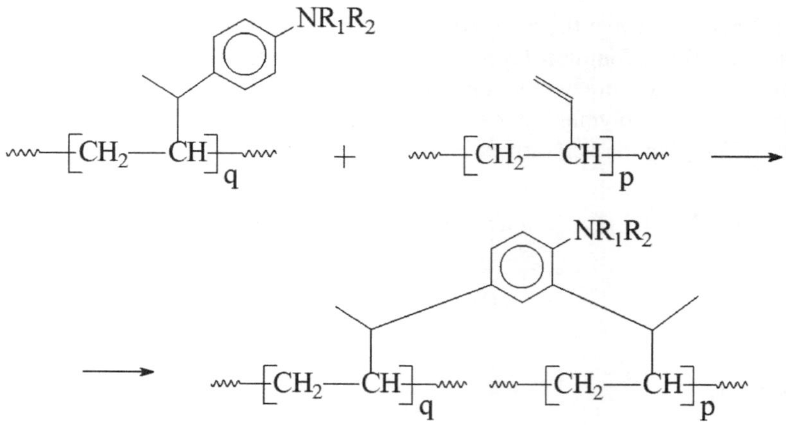

SCHEME 7.12

Steric difficulties determined by the introduction of bulky substituents in the polydiene units ("a neighbor effect"[53,54]) does not allow to obtain polymer products with high functionalization degree as the arylamino groups in the modified polymer does not exceed 8 mol.%. At the same time, secondary intermolecular reactions are induced in the synthesis process involving arylamino groups of the modified macromolecules and result in the formation of linear or branched polymer products with high molecular weight.

7.8 INTERACTION WITH MALEIC ANHYDRIDE

The polymer products with anhydride groups are synthesized by thermal addition (190°C) of the maleic anhydride to the syndiotactic 1,2-PB[23,24] (Scheme 7.13):

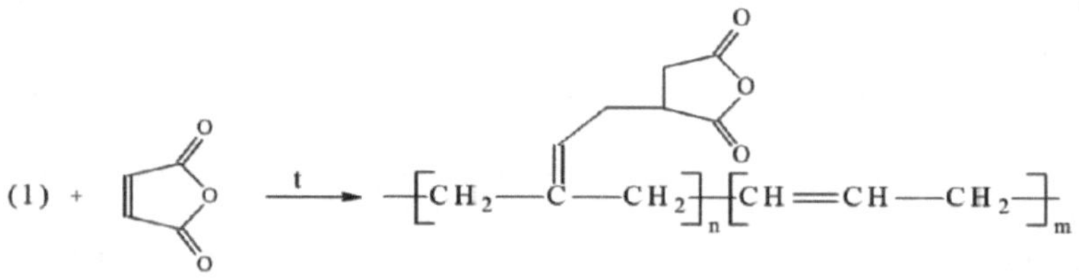

SCHEME 7.13

The ^{13}C NMR spectroscopy results show that the maleic anhydride addition is carried out as an ene-reaction[55] by the vinyl bonds of the polymer without the cycle disclosure and the double bond is moved to the β-carbon atom of the vinyl bond.[23,24] The maleic anhydride addition to the >*C*=*C*< double bonds of 1,4-units of polydiene macromolecules does not take place. As in synthesis of the arylamino derivatives of syndiotactic 1,2-PB, it is connected with steric difficulties preventing the interaction of bulk molecules of the maleic anhydride with inner double bonds of the polymer chain.[23]

At syndiotactic 1,2-PB modification by the maleic anhydride, the so-called "neighbor effect" is observed, that is, the introduction of bulk substituents into the polymer chain prevents the functionalization of the neighboring polymer units due to steric difficulties. For this reason, the content of anhydride groups in the modified polymer molecules do not exceed ~15 mol. %.

Thus, >*C*=*C*< double bonds in 1,2- and 1,4-units of syndiotactic 1,2-PB macromolecules considerably differ in the reactivity due to the polydiene structure. The inductive effect of the alkyl substituents resulting in the increase of the electron density of the inner double bonds of macromolecules determines their high activity in the considered reactions with different electrophilic agents.

At interaction of syndiotactic 1,2-PB with strong electrophiles (ozone, halogens, and dichlorocarbene), both inner double bonds and side vinyl groups of polydiene macromolecules are involved in the reaction. It results in polymer products formation with quite a high functionalization degree. In the case when the used reagent does not display enough activity (interaction of syndiotactic 1,2-PB and hydrogen chloride and aliphatic peracids), the process is controlled by electronic factors: more active double bonds in 1,4-units of the polymer chain are subjected to modification, whereas the formed polymer products are characterized by a relatively low functionalization degree.

Polymer modification reactions are mostly carried out through vinyl groups at appearance of steric difficulties. They are connected with formation of a bulky intermediate complex or usage of reagents of big-sized molecules (reactions with aromatic amines, maleic anhydride, *t*-BuOOH/$Mo(CO)_6$). Such reactions are controlled by steric factors. The predominant course of the reaction by the side vinyl groups of polymer macromolecules is determined by their more accessibility to the reagent attack. However, in such reactions, high functionalization degree of syndiotactic 1,2-PB cannot be achieved due to steric difficulties arousing through the introduction of bulky substituents into the polymer chain. They limit the reagent approaching to the reactive polydiene bonds.

Thus, a targeted chemical modification of the polydiene accompanied by obtaining polymer products of different content and novel properties can be carried out using differences in the reactivity of >*C*=*C*< double bonds of syndiotactic 1,2-PB. Various polymer products with a set of complex properties is possible to obtain on the syndiotactic 1,2-PB basis varying the nature of the modifying agent, a functionalization degree of the polymer, and synthesis conditions.

KEYWORDS

- **syndiotactic 1,2-polybutadiene**
- **chemical modification**
- **functionalization degree**
- **polymer products**

REFERENCES

1. Kochnev, A. M.; Galibeev, S. S.; *Khim. Khim. Tekhnol.* **2003**, *46* (4), 3–10.
2. Byrihina, N. N.; Aksenov, V. I.; Kuznetsov, E. I. Patent R.F. 2177008, 2001.
3. Ermakova, I.; Drozdov, B. D.; Gavrilova, L. V.; Shmeleva, N. V. Patent R.F. 2072362, 1998.
4. Luo, S. Patent US. 6284702, 2002.
5. Wong, T. H.; Cline, J. . Patent US. 5986026, 2000.
6. Ni, S.; Zhou, Z.; Tang, X. *Chinese J. Polym. Sci.* **1983**, *2*, 101–107.
7. Monteil, V.; Bastero, A.; Mecking, S. *Macromolecules* **2005**, *38,* 5393–5399.
8. Obata, Y.; Tosaki, Ch.; Ikeyama, M. *Polym. J.* **1975**.. *7* (2), 207–216.
9. Glazyrin, A. B.; Sheludchenko, A. V.; Zaboristov, V. N.; Abdullin, M. I. *Plasticheskiye Massy* **2005**, *8*, 13–15.
10. Abdullin, M. I.; Glazyrin, A. B.; Sheludchenko, A. V.; Samoilov, A. M.; Zaboristov, V. N. *Zhurnal Prikl.Khimii.* **2007**, *80* (11), 1913–1917.
11. Xigao *J. Polym. Sci.* **1990**, *28* (9), 285–288.
12. Kimura, S.; Shiraishi, N.; Yanagisawa, S. *Polym.-Plast. Technol. Eng.* **1975**, *5* (1), 83–105.
13. Lawson, G. Patent US. 4960834, 2003.
14. Gary, L. Patent US. 5278263, 2001.
15. Dontsov, A. A.; Lozovik, G. Y. *Chlorinated Polymers;* Khimiya: Moscow, RU,. 1979; p 232.
16. Asfandiyarov, R. N. Synthesis and Properties of Halogenated 1,2-Polybutadienes; Ph.D. Chem. Science: Ufa, Bashkir State University, 2008; p 135.
17. Kayumova, *M. A. Synthesis and properties of the oxygen-and aryl-containing derivatives of 1,2-syndiotactic polybutadiene*; Ph.D. Chem. Science.: Ufa, Bashkir State University, 2007; p115.
18. Abdullin, M. I.; Kukovinets, O. S.; Kayumova, M. A.; Sigaeva, N. N.; Ionova, I. A.; Musluhov, R. R.; Zaboristov, V. N. *Vysokomol. Soyed.* **2004**, *46* (10), 1774–1778.

19. Gainullina, T. V.; Kayumova, M. A.; Kukovinets, O. S.; Sigaeva, N. N.; Muslukhov, R. R.; Zaboristov, V. N.; Abdullin, M. I. *Polym. Sci. Ser. B.* **2005**, *47* (9), 248–252.
20. Abdullin, M. I.; Kukovinets, O. S.; Kayumova, M. A.; Sigaeva, N. N.; Musluhov, R. R. *Bashkirskiy khimicheskiy zhurnal. Plastics* **2006**, *13* (1), 29–30.
21. Gainullina, T. V.; Kayumova, M. A.; Kukovinets, O. S.; Sigaeva, N. N.; Muslukhov, R. R.; Zaboristov, V. N.; Abdullin, M. I. *Vysokomol. Soyed.* **2005**, *47* (9), 1739–1744.
22. Abdullin, M. I.; Gaynullina, T. V.; Kukovinets, O. S.; Khalimov, A. R.; Sigaeva, N. N.; Musluhov, R. R.; Kayumova, M. A. *Zhurnal Prikl. Khimii.* **2006,** *79* (8), 1320–1325.
23. Abdullin, M. I.; Glazyrin, A. B.; Kukovinets, O. S.; Basyrov, A. A. *Khimi. Khimi. Tekhnol.* **2012**. *55* (5), 71–79.
24. Kukovinets, O. S.; Glazyrin, A. B.; Basyrov, A. A.; Dokichev, V. A.; Abdullin, M. I. *Izvestiya Ufimskogo Nauchnogo Tsentra Rossiyskoy Akademii Nauk* **2013**, *1*, 29–37.
25. Valekzhanin, I. V.; Abdullin, M. I.; Glazyrin, A. B.; Kukovinets, O. S.; Basyrov, A. A. *Aktualnyye Problemy Gumanitarnykh i Yestestvennykh Nauk* **2012**, *3*, 13–14.
26. Abdullin, M. I.; Basyrov, A. A.; Kukovinets, O. S.; Glazyrin, A. B.; Khamidullina, G. I. *Polym. Sci., Ser. B* **2013**, *55* (5–6), 349–354.
27. Abdullin, M. I.; Glazyrin, A. B.; Kukovinets, O. S.; Basyrov, A. A. *Mezhdunarodnyy Nauchno-issledovatelskiy Zhurnal* **2012**, *4*, 36–39.
28. Abdullin, M. I.; Glazyrin, A. B.; Kukovinets, O. S.; Valekzhanin, I. V.; Klysova, G. U.; Basyrov, A. A. Patent R.F. 2465285, 2012.
29. Abdullin, M. I.; Glazyrin, A. B.; Kukovinets, O. S.; Valekzhanin, I. V.; Kalimullina, R. A.; Basyrov, A. A. Patent R.F. 2456301, 2012.
30. Kurmakova, I. N. *Vysokomol. Soyed.* **1985**, *21* (12), 906–910.
31. Hayashi, O.; Kurihara, H.; Matsumoto, Y. Patent US. 4528340, 1985.
32. Blackborow, J. R. Patent US. 5034471, 1991.
33. Jacobi, M. M.; Viga, A.; Schuster, R. H. *Raw Mater. Appl.* **2002**, *3*, 82–89.
34. Emmons, W. D.; Pagano, A. S. *J. Am. Chem. Soc.* **1955**, *77* (1), 89–92.
35. Xigao, J.; Allan, S. H. *J. Polym. Sci.: Polym. Chem. Ed.* **1991**, *29*, 1183–1189.
36. Abdullin, M. I.; Glazyrin, A. B.; Asfandiyarov, R. N. *Vysokomol. Soyed.* **2009**, *51* (8), 1567–1572
37. Glazyrin, A. B.; Abdullin, M. I.; Muslukhov, R. R.; Kraikin, V. A. *Polym. Sci., Ser. A.* **2011**, *53* (2), 110–115.
38. Abdullin, M. I.; Glazyrin, A. B.; Akhmetova, V. R.; Zaboristov, V. N. *Polym. Sci., Ser. B.* **2006**, *48* (4), 104–107.
39. Abdullin, M. I.; Glazyrin, A. B.; Asfandiyarov, R. N. *Zhurnal Prikl. Khimii.* **2007**, *10*, 1699–1702.
40. Abdullin, M. I.; Glazyrin, A. B.; Asfandiyarov, R. N.; Akhmetova, V. R. *Plasticheskiye Massy* **2006**, *11*, 20–22.
41. Lishanskiy, I. S.; Shchitokhtsev, V. A.; Vinogradova, N. D. *Vysokomol. Soyed.* **1966**, *8*, 186–171.
42. Komorski, R. A.; Horhe, S. E.; Carman, C. J. *J. Polym. Sci.: Polym. Chem. Ed.* **1983**, *21*, 89–96.
43. Nonetzny, A.; Biethan, U. *Angew. Makromol. Chem.* **1978**, *74*, 61–79.
44. Glazyrin, A. B.; Abdullin, M. I.; Kukovinets, O. S. *Vestnik Bashkirskogo Universiteta* **2009**, *14* (3), 1133–1140.
45. Glazyrin, A. B.; Abdullin, M. I.; Muslukhov, R. R. *Polym. Sci., Ser. B.* **2012**, *54*, 234–239.
46. Glazyrin, A. B.; Abdullin, M. I.: Khabirova, D. F.; Muslukhov, R. R. Patent RF. 2456303, 2012.
47. Glazyrin, A. B.; Abdullin, M. I.; Sultanova, R. M.; Dokichev, V. A.; Muslukhov, R. R.; Yangirov, T. A.; Khabirova, D. F. Patent RF. 2443674, 2012.

48. Glazyrin, A. B.; Abdullin, M. I.; Sultanova, R. M.; Dokichev, V. A.; Muslukhov, R. R.; Yangirov, T. A.; Khabirova, D. F. Patent RF. 2447055, 2012.
49. Gareyev, V. F.; Yangirov, T. A.; Kraykin, V. A.; Kuznetsov, S. I.; Sultanova, R. M.; Biglova, R. Z.; Dokichev, V. A. *Vestnik Bashkirskogo Universiteta* **2009**, *14* (1), 36–39.
50. Gareyev, V. F.; Yangirov, T. A.; Volodina, V. P.; Sultanova, R. M.; Biglova, R. Z.; Dokichev, V. A. *Zhurnal Prikl. Khimii.* **2009**, *83* (7), 1209–1212.
51. Glazyrin, A. B.; Abdullin, M. I.; Dokichev, V. A.; Sultanova, R. M.; Muslukhov, R. R.; Yangirov, T. A. *Polym. Sci., Ser. B.* **2013**, *55*, 604–609.
52. Shapiro, Ye. A.; Dyatkin, A. B.; Nefedov, O. M. *Diazoether*; Nauka: Moscow, 1992; p 78.
53. Fedtke, M. *Chemical Reactions of Polymers*; Chemistry: Moscow, RU, 1990; p 152.
54. Kuleznev, V. N.; Shershnev, V. A. *The Chemistry and Physics of Polymers*; Kolos: Moscow, RU, 2007; p 367.
55. Vatsuro, K. V.; Mishchenko, G. L. *Named Reactions in Organic Chemistry*; Khimiya: Moscow, RU, 1976; p 528.

CHAPTER 8

NORMALIZATION OF THE MAINTENANCE OF METALS IN OBJECTS OF AN ENVIRONMENT

J.A. TUNAKOVA, S.V. NOVIKOVA, R.A. SHAGIDULLINA, and M.I. ARTSIS[1]

Kazan National Research Technological University, 65 Karl Marx str., Kazan 420015, Tatarstan, Russia. E-mail: juliaprof@mail.ru

[1]N.M. Emanuel Institute of Biochemical Physics, Russian Academy of Sciences, 4 Kosygin str., Moscow 119334, Russia. E-mail: Chembio@sky.chph.ras.ru

CONTENTS

ABSTRACT

Results of modeling of the maintenance of the basic polluting substances (metals) in hair of the person are examined depending on their maintenance in objects of an environment. On the basis of the constructed models, normalization of the maintenance of metals in objects of an environment is carried out in view of receipt of metals in an organism, both air and by a waterway.

8.1 INTRODUCTION

The system of sanitary and hygienic normalization working in territory of the Russian Federation takes into account the combined action (simultaneous action of several substances at the same way of receipt) at calculation of summing effect at their simultaneous receipt in atmospheric air. But in any way, it does not take into account the effects of complex receipt of harmful substances in an organism in various ways and with various environments with air, water, food, etc.

Metals concern to priority polluting substances, supervision for which are obligatory in all environments due to their high risk to health of the person in rather low concentration, and also ability to bioaccumulaytion. Danger of constant receipt of metals is caused by impossibility their independent removing from an organism. Metals act in an organism with water, the foodstuff, inhaled air.[1–3]

8.2 SETTLEMENT—EXPERIMENTAL PART

We develop methodology of normalization which basis were levels of accumulation of metals in an organism of the person - basic protected object in territory of city, in relation to the established[4] regional specifications of the maintenance in bioenvironments of the population of city.

We had led the analysis of amounts of receipt of metals by air and water—food way on the materials stated in Ref. 5. It has been established that for metals such as iron, copper, and lead is characteristic both air and water–food ways of receipt to an organism with domination of the last. For other metals—the water—food way of receipt or an air way of receipt considerably prevails is not estimated.

Basis of construction of models were results of long-term experimental researches of the maintenance of metals in hair of children in the territory of Kazan, potable water (the characteristic a waterway of receipt), snow, and soil covers (the characteristic of an air way of receipt, in view of existential restrictions of regular supervision over the maintenance of metals in atmospheric air). Nonlinear methods of plural regress were applied to modeling on the basis of criterion of the least squares.

At the first stage, the maintenance of lead in an organism of the person (hair) was investigated at air (ground) and water (potable water) ways of receipt to an organism.

A linear regress model of dependence of the maintenance of lead in hair was used from the maintenance of lead in ground and potable water:

$$Pb\_v = 254{,}880 + 3{,}669{,}371*Pb\_vod + 0.03691*Pb\_p \quad (1)$$

Factor of plural correlation $R = 0.1406$; factor of determination $R_2 = 0.01976$ (about 2% of an explained dispersion). The resulted factors of regress: Pb_vod = 0.022928; Pb_p = 0.153831.

The model specifies direct dependence of the maintenance of lead in hair from his maintenance in potable water and in ground. The degree of influence of variability of the maintenance of lead in ground is seven times than in water. Small values of factors of plural correlation and determination speak about small efficiency of model. In this connection, there is a task of construction nonlinear plural regress models.

Then a series of experiments on modeling nonlinear regress models of dependence Lead in hair (Lead in potable water - Lead in ground) has been lead. A polynom model was examined:

Polynom models have shown results close among themselves (factors of plural correlation about 0.47 and factors of determination about 0.22), and have allowed to raise factor of determination more than 10 times in comparison with linear model. Only the models were effective, including the maximal degrees of independent variables (except for linear members). Among the considered models of the most adequate, it is necessary to count square-law polynom model of plural regress with factor of determination 0.235.

Efficiency constructed not polynom models is comparable to efficiency polynom: the maximal factor of determination (0.236597) parabolic models not on surpasses the best result for square-law polynom models (0.234906) many.

Summary results are resulted in Table 8.1.

TABLE 8.1 Results of Modeling for Lead.

The name of model	Model	Factor of determinations
Square law	$Z = 5549 - 7968*X + 0.61*Y + 264{,}014*X^2 - 0.009*Y^2$	0.235
Cubic	$Z = 3981 - 4154*X + 0.32*Y - 6{,}020{,}931*X^3 - 0.0001*Y^3$	0.229
A polynom fourth	$Z = 3195 - 28{,}83{,}836{,}474*X + 0.229273*Y + 203{,}319{,}032*X^4 - 0.000001*Y^4$	0.222

TABLE 8.1 *(Continued)*

The name of model	Model	Factor of determinations
A polynom of the fifth degree	$Z = 27,267,639 - 2,258,425*X + 0.181962*Y + 8,144,045,245*X^5 - 2.39 - 0.8*Y^5$	0.215
Parabolic	$Z = 19,478 + 14,809*X - 1.11*Y - 3627*\sqrt{X} + 11.99*\sqrt{Y}$	0.236
Logarithmic	$Z = -590,466 + 7,190,986*X - 0.539*Y - 107,638*ln(X) + 15.188*ln(Y)$	0.234115
Hyperbolic	$Z = -8198 + 3361*X - 0.251*Y + 0.748*\frac{1}{X} - 178.273*\frac{1}{Y}$	0.224212
Cubic-hyperbolic model	$Z = -6,043,589 - 1,344,461,912*X + 0.117143*Y + 3,790,787,976,566*X^3 - 0.000066*Y^3 + 0.282468*\frac{1}{X} - 67,768,974*\frac{1}{Y}$	0.231261
Square-law-logarithmic model (without linear members)	$Z = -28,496,762 + 12,549,000,68,251*X^2 - 0.004610*Y^2 - 56,696,124*ln(X) + 8,157,432*ln(Y)$	0.235711
Square-law-parabolic model (without linear members)	$Z = 104,104,117 + 171,775,113,216*X^2 - 0.006323*Y^2 - 1,269,670,921*\sqrt{X} + 4,275,894*\sqrt{Y}$	0.236
Logarithmic-parabolic model	$Z = -1329 - 208*ln(X) + 29.25*ln(Y) + 3417*\sqrt{X} - 11.17*\sqrt{Y}$	0.231

The models containing three and more various elementary function in structure, mattered F-statistics of less significance value that speaks about their notorious inefficiency. Thus, by results of the lead experiments carried out by us, it is established that all nonlinear (polynom, not polynom, and mixed) models have shown approximately equal efficiency; however, the most adequate should be counted the as parabolic nonlinear model, allowed to reach the maximal value of factor of determination 0.236597:

$$Pb\_v = 194,78 + 14809,24 \cdot Pb\_vod - 1,11 \cdot Pb\_p - \\ - 3627,04 \cdot \sqrt{Pb\_vod} + 11,99 \cdot \sqrt{Pb\_p} \quad (2)$$

On the basis of the received most adequate model, specifications of the maintenance of lead in objects of an environment are established. The regional specification of the maintenance of lead in hair makes 10 gramme.[4]. The normative maintenance of lead established by us in ground makes 27.5 mg/kg.[6]

Thus, the equation for definition of the maximal marginal level of the maintenance of lead in potable water becomes:

$$10 = 194{,}78 + 14809{,}24 \cdot Pb\_vod - 1{,}11 \cdot 27{,}5 - 3627{,}04 \cdot \sqrt{Pb\_vod} + 11{,}99 \cdot \sqrt{27{,}5} \tag{3}$$

The decision of the given equation is received numerically. To the decision, it was applied two-step algorithm of definition of intervals by a graphic method and specifications of roots by a method of Bolzano. In result, two valid positive roots are received:

$$Pb\_vod_1 = 0.0110 \text{ mg / liter}, \quad Pb\_vod_2 = 0.0200 \text{ mg / liter}$$

Thus, according to model (2), the maintenance of lead in drinking water should not exceed value of 0.02 mg/liter provided that lead also acts in an organism and airway. Further, the maintenance of iron in an organism of the person (hair) was investigated at air (a snow cover) and water (potable water) ways of receipt to an organism.

For check of a hypothesis about absence of linear connection between the maintenance of iron in hair and its maintenance in potable water and a snow (about zero values of factors of regress) Fisher's F-statistics was used. Value of F-statistics = 0.65641 at a significance value $p < 0.52535$, that is, a hypothesis about absence of linear connection deviates.

We applied linear regress model of dependence of the maintenance of iron in hair from the maintenance of iron in a snow cover and potable water (on a method of the least squares):

$$Fe\_v = 189{,}401 + 1.275.608 * Fe\_vod + 437.879 * Fe\_s/l \tag{3-a}$$

Factor of plural correlation $R = 0.195$; factor of determination R_2=0.04 (4% of an explained dispersion). The resulted factors of regress: Fe_vod = 0.04; Fe_s/l = 0.2.

Thus, direct influence of the maintenance of iron in atmospheric air (determined under the analysis of tests of a snow cover) on accumulation of iron in hair of children is revealed. The degree of influence of variability of the maintenance of iron in a snow cover is five times stronger than in potable water. Small values of factors of plural correlation and determination speak about small efficiency of model. In this connection, there is a task of construction nonlinear plural regress models. A series of experiments on modeling nonlinear regress models of dependence has been carried out: Iron in hair (Iron in potable water - Iron in a snow cover). Summary results are shown in Table 8.2.

TABLE 8.2 Results of Modeling for Iron.

The name of model	Model	Factor of determinations
Square-law	$Z=12,48+110,2*X+15,3*Y-688,58*X^2-6,64*Y^2$	0.071
Cubic	$Z=15,37+37,31*X+11,65*Y-2001,18*X^3-2,84*Y^3$	0.073
A polynom fourth	$Z=-60,8+3152,7*X-5,3*Y-31948,7*X^2+838375,7*X^4+18,4*Y^2-5,6*Y^4$	0.157
A polynom of the fifth degree	$Z=-26+1295*X+0*Y-152968*X^3+6812171*X^5+14*Y^3-5*Y^5$	0.162
Parabolic	$Z=-12,531-310*X-8,685*Y+175,990*\sqrt{X}+21,004*\sqrt{Y}$	0.064
Logarithmic	$Z=72,832-167,392*X-0,997*Y+13,4*ln(X)++3,1*ln(Y)$	0.061
Indicative	$Z=1298+1384*X+18,73*Y-1279*e^X-5,76*e^Y$	0.073
Hyperbolic	$Z=36,96-92,36*X+2,41*Y-0,57\ {}^1\!/_X-0,41\ {}^1\!/_Y$	0.059
Polynom-arabolic model of the third degree	$Z=-359,94-8176*X+35,93*Y+91118*X^3-6,18*Y^3+3512*\sqrt{X}-28,95*\sqrt{Y}$	0.148
Polynom-parabolic model of the fifth degree	$Z=-242-4772*X+25*Y+2624639*X^5-2*Y^5+2262*\sqrt{X}-20*\sqrt{Y}$	0.156

TABLE 8.2 *(Continued)*

The name of model	Model	Factor of determinations
Polynom-Logarithmic model	$Z=528-2352*X +16*Y+2294509*X^5 -1*Y^5+132*ln(X) -3*ln(Y)$	0.154
Polynom-Hyperbolic model	$Z=166,2-1395,6*X +12,7*Y+276475,7*X^4 -2,3*Y^4-4,0*1/X +0,3*1/Y$	0.147

Designations: *Z*—the Maintenance of iron in hair (Fe_v); *X*—the Maintenance of iron in potable water (Fe_vod); *Y*—the Maintenance of iron in a snow cover (Fe_s/l).

Polynom models have allowed the increasing factor of determination up to four times in comparison with linear regress model. The most effective should recognize model "a polynom of the fifth degree" with factor of plural correlation 0.402374 and factor of determination 0.161905. Also, it is possible to note that everything, except for square-law, polynom models appeared effective (mattered F-statistics above a significance value) at presence of zero factors in structure of a polynom (for cubic model—square-law factors; for a polynom of the fourth degree—cubic; for a polynom of the fifth degree—cubic and the fourth degree).

The mixed models containing three and more various elementary function in structure, mattered F-statistics of less significance value that speaks about their notorious inefficiency. All nonlinear mixed models have similar characteristics (factor of plural correlation at a level 0.38–0.4 and factor of determination at level 0.14–0.16). The most effective model of regress is the polynom–parabolic model of the fifth degree with factor of determination 0.156. However, the maximal plural correlation and as consequence, determination, provides polynom model of the fifth degree, which should be used for normalization of the maintenance of iron in objects of an environment.

Based on results of the lead experiments of the most effective appeared nonlinear mixed polynom model of regress of a kind:

$$Fe\_v = -26 + 1295 * Fe\_vod - 15968 * Fe\_vod^3 + \\ + 6,812,171 * Fe\_vod^5 + 14 * Fe\_s/l^3 - 5 * Fe\_s/l^5 \quad (4)$$

With factor of determination 0.16. The model can be recommended to use with a view of normalization of the maintenance of iron in an environment.

The regional specification of the maintenance of iron in hair makes 55 mkg/gramme[4]. The normative maintenance of iron received by us earlier in potable water makes 0.56 mg/liter.[6].

Thus, the equation for definition of the maximal marginal level of the maintenance of iron in a snow cover becomes:

$$55 = -26 + 1295 * 0.56 + 152,968 * 0.56^3 + 6,812,171 * 0.56^5 + 14 * Fe\_s/l^3 - 5 * Fe\_s/l^5 \quad (5)$$

The equation was solved numerically in two stages on the basis of transformation of the eq 5 to a task of minimization of square-law function: definition of intervals of a presence of roots by a graphic method, and specification of roots by a method of Bolzano. In result, the unique root of the eq 5 is received:

$$Fe\_s/l = 9357 \text{ mg/liter.}$$

Hence, according to model (4), the maintenance of iron in a snow cover should not exceed value of 9357 mg/liter provided that iron also acts in an organism with potable water. High value of the received parameter can be interpreted as weak influence of the maintenance of iron in a snow cover on its accumulation in hair. The same position is illustrated with absolute values of the resulted factors of regress for polynom–parabolic model: Fe_vod = 394,778; Fe_s/l = 001,735.

It is obvious, that influence of receipt of iron by a waterway on its accumulation in hair more than in 200 times is higher than influence of the maintenance of iron in a snow cover.

Further variability, the maintenance of copper in an organism, the person (hair) was investigated at air (a snow cover) and waterways (potable water) of receipt to an organism. For check of a hypothesis about absence of linear connection between the maintenance of copper in hair and its maintenance in potable water and a snow (about zero values of factors of regress) Fisher's F-statistics was used. Value of F-statistics = 19,197 at a significance value $p < 0.15821$, that is, a hypothesis about absence of linear connection deviates. It has been received linear regress model of dependence of the maintenance of copper in hair from the maintenance of copper in a snow cover and potable water (on a method of the least squares).

$$Cu\_v = 1985 - 440,827 * Cu\_vod + 1,32 * Cu\_s/l \quad (6)$$

Factor of plural correlation $R = 0.277555$; factor of determination R_2=0.077037 (about 8% of an explained dispersion); the resulted factors of regress:

$$Cu\_vod = -0.274142; \; Cu\_s/l = 0.0178092.$$

Thus, we establish direct influence of receipt of copper by airway on accumulation of copper in hair of children. A degree of influence of variability of the maintenance of copper in potable water on the maintenance of copper in hair negative. Also, small values of factors of plural correlation and determination speak about small efficiency of model. In this connection, there is a task of constructing non-linear plural regress models. A series of experiments on modeling nonlinear regress models of dependence Copper in hair (Copper in potable water - Copper in a snow cover) has been lead. Summary results are resulted in Table 8.2 (Z: the maintenance of copper in hair (Cu_v); X: the maintenance of copper in potable water (Cu_vod); Y: the Maintenance of copper in a snow cover (Cu_s/l)). Summary results are resulted in Table 8.3.

TABLE 8.3 Results of Modeling for Copper.

The name of model	Model	Factor of determinations
Square-law	$Z=8+12060*X-26*Y-4946586*X^2-77*Y^2$	0.116
Cubic	$Z=12+3488*X-18*Y-910344854*X^3+144*Y^3$	0.123
A polynom fourth	$Z=-45,673+96010,14*X+75,66*Y-4,22E+07*X^2+1,97E+12*X^4-69,02*Y^2+2186,72*Y^4$	0.177
A polynom of the fifth degree	$Z=-38,4+80317,82*X+ +65,05*Y-3,21E+07*X^2+4,68E+14*X^5-465,13*Y^2+3985,8*Y^5$	0.177
Parabolic	$Z=-32,3-29878,5*X-70*Y++2236,2*\sqrt{X}+29,8*\sqrt{Y}$	0.121
Logarithmic	$Z=203,3-17162,8*X-43,1*Y+55,6*log(X)+4,2*log(Y)$	0.128
Indicative	$Z=870,8-851*10^X- -4,639*10^Y$	0.077
Hyperbolic	$Z=43,63-10281*X-26,54*Y-0,02\ 1/X-0,03047\ 1/Y$	0.130
Polynom – Logarithmic model of the fourth degree	$Z=382-32395*X-141*Y+1,015+11*X^4+8724*Y^4+105*log(X)+11,72*log(Y)$	0.205

TABLE 8.3 *(Continued)*

The name of model	Model	Factor of determinations
Polynom – Logarithmic model of the fifth degree	$Z=389,380286 -32889*X-126*Y +4,01E+13*X^5 +39131*Y^5+108*log(X)+10,9*log(Y)$	0.212
Polynom – Hyperbolic model	$Z=49,29-11170*X -57,59*Y-4,7E+12*X^5 -0,024*Y^5-0,024*\frac{1}{X} -0,024*\frac{1}{Y}$	0.174
Polynom -parabolic model	$Z=-99,80 -72106*X -261,60*Y+6,49E+13*X^5 +49416,88*Y^5+5434,97*\sqrt{X} +106,46*\sqrt{Y}$	0.226

Polynom models have allowed to increase factor of determination by the order in comparison with linear regress model. The most effective should recognize model "a polynom of the fourth degree " with factor of plural correlation 0.421484 and factor of determination 0.177648. Also, it is possible to note that everything except for square-law, polynom models appeared effective (mattered F-statistics above a significance value) at presence of zero factors in structure of a polynom (for cubic model - square-law factors; for a polynom of the fourth degree - cubic; for a polynom of the fifth degree - cubic and the fourth degree).

However, the received values, nevertheless, are insufficient for the recommendation polynom models to practical application. Unfortunately, polynom regress models have not allowed factor of determination to increase. Therefore, was accepted decision to build mixed nonlinear regress models. In Table 8.3 models, which factor of determination has exceeded factor of determination polynom models of the fourth degree (0.177) are resulted only.

The mixed models containing three and more various elementary function in structure, mattered F-statistics of less significance value that speaks about their notorious inefficiency. All 4 nonlinear mixed models have similar characteristics (factor of plural correlation at a level 0.41–0.47 and factor of determination at a level 0.17–0.23). The most effective model of regress is the polynom–parabolic model. This model can be recommended for practical application.

By results of the lead experiments of the most effective, there was a nonlinear mixed polynom–parabolic model of regress of a kind:

$$Cu\_v = -99{,}81 - 72106{,}43 * Cu\_vod - 261{,}60 * Cu\_s/l + \\ + (6{,}5E+13) * Cu\_vod^5 + 49416{,}9 * Cu\_s/l^5 + \\ + 5434{,}97 * \sqrt{Cu\_vod} + 106{,}5 * \sqrt{Cu\_s/l} \quad (6\text{-a})$$

With factors of determination 0.226555. The model can be recommended to use with a view of normalization of the maintenance of copper in an environment.

The regional specification of the maintenance of copper in hair has made 25 mkg/gramme.[4] The normative maintenance (contents) of copper in the snow cover, determined by us, has made 0.154 mg/liter.[6]

Thus, the equation for definition of the normative maintenance of copper in a snow cover becomes:

$$25 = -99{,}81 - 72106{,}43 * Cu\_vod - 261{,}60 * 0{,}154 + \\ + (6{,}5E+13) * Cu\_vod^5 + 49416{,}9 * 0{,}154^5 + \\ + 5434{,}97 * \sqrt{Cu\_vod} + 106{,}5 * \sqrt{0{,}154} \quad (7)$$

The equation was solved numerically in two stages on the basis of transformation of the eq 7 to a task of minimization of square-law function: definition of intervals of a presence of roots by a graphic method, and specification of roots by a method of Bolzano. In result, the unique root of the eq 7 is received: Cu_vod = 0.314 mg/liter.

Hence, according to model (7), the maintenance of copper in potable water should not exceed value of 0.314 mg/liter provided that copper also acts in an organism and air way.

8.3 CONCLUSION

Thus, we establish quantitative characteristics of distribution of metals in the connected environments and scientifically proved specifications of the maintenance of metals in objects of the environment, taking into account joint receipt of metals in an organism of the person air and are received by a waterway.

KEYWORDS

- **multivariate regress nonlinear models**
- **metals**
- **normalization**
- **organism**
- **person**
- **complex**
- **environment**
- **objects**

REFERENCES

1. Levich, A. P.; Bulgakov, N. G.; Maxims, C. N. *Theoretical and Methodical Bases of Technology of the Regional Control of the Natural Environment According to Ecological Monitoring*; РЭФИА: Moscow, RU, 2004; p 271.
2. Maximov, V. N.; Bulgakov, N. G.; and Levich, A. P. *Environmental Indices: Systems Analysis Approach;* EOL SS Publishers: London, UK, 1999; pp 363–381.
3. P 2.1.10.1920-04 Management according to risk for health of the population at influence of the chemical substances polluting an environment.
4. Maltsev, S. V.; Valiev, V. S.; Zingareeva, G. G.; Valiev, B. C.; Ganeeva, L. A. 1 International Symposium “Modern problems of geochemical ecology of illnesses” materials. The first international symposium (Cheboksary, 17-20 сент. 2001) Cheboksary, 2001. With. 71.
5. Rocky, A. V. Chemical Element in Physiology and Ecologies of Person, the World, 2004, 215 with.
6. Tunakova, J. A.; Fajzullina, R. A.; Shmakov, J. U. *Magazine of the Kazan Technological University* **2012**, *16* (12) 71–73.

CHAPTER 9

INFLUENCE OF HYDROLYSIS ON DIFFUSION OFELECTROLYTES

A. C. F. RIBEIRO[1]*, L. M. P. VERHSSIMO[1], A. J. M. VALENTE[1], A. M. T. D. P. V. CABRAL[2], and V. M. M. LOBO[1]

[1]Department of Chemistry, University of Coimbra, 3004–535 Coimbra, Portugal;
Tel: +351-239-854460; fax: +351-239-827703;
*e-mail (corresponding author): anacfrib@ci.uc.pt;
luisve@gmail.com; ava ente@ci.uc.pt; vlobo@ci.uc.pt

[2]Faculty of Pharmacy, University of Coimbra, 3000-295 Coimbra, Portugal;
E-mail: acabral@ff.uc.pt

CONTENTS

ABSTRACT

The experimental mutual diffusion coefficients of potassium chloride, magnesium nitrate, cobalt chloride, and beryllium sulfate in aqueous solutions have been compared with those estimated by Onsager and Fuoss.From the observed deviations, we are proposing a model that allows to estimate thepercentage of H_3O^+ (aq) resulting inhydrolysis of some ions (i.e., beryllium, cobalt, potassium, and magnesium ions), contributing in this way to a better knowledge of the structure of these systems.

The diffusion of beryllium sulfate is clearlymost affected by the beryllium ion hydrolysis.

9.1 THEORETICAL ASPECTS

9.1.1 CONCEPTS OF DIFFUSION

Diffusion data of electrolytes in aqueous solutions are of great interest not only for fundamental purposes (providing a detailed comprehensive information—both kinetic and thermodynamic), but also for many technical fields such as corrosion studies.[1–3]The gradient of chemical potential in the solution is the force producing the irreversible process which we call diffusion.[1–3]However, in most solutions, that force may be attributed to the gradient of the concentration at constant temperature. Thus, the diffusion coefficient, *D,* in a binary system, may be defined in terms of the concentration gradient by a phenomenological relationship known as Fick's first law,

$$J = -D \,\mathrm{grad} c \tag{1.1}$$

or, considering only one dimension for practical reasons,

$$J = -D \frac{\partial c}{\partial x} \tag{1.2}$$

where *J* represents the flow of matter across a suitable chosen reference plane per unit area and per unittime, in a one-dimensional system, and *c* is the concentration of solute in moles per volume unit at the point considered; eq 1.1 may be used to measure *D*. The diffusion coefficientmay also be measured by considering Fick's second law, in one-dimensional system,

$$\frac{\partial c}{\partial t} = \frac{\partial}{\partial x}\left(D \frac{\partial c}{\partial x} \right) \tag{1.3}$$

In general, the available methods are grouped into two groups: steady and unsteady-state methods, according to eqs1.2 and 1.3. Diffusion is a three-dimensional phenomenon, but many of the experimental methods used to analyze diffusion restrict it to a one-dimensional process.[1–4]

The resolution of eq1.2 is much easier if we consider D as a constant. This approximation is applicable only when there are small differences of concentration, which is the case in our open-ended conductometric technique,[3,5] and in the Taylor technique.[3,6,7] In these conditions, it is legitimate to consider that our measurements of differential diffusion coefficients obtained by the above techniques are parameters with a well-defined thermodynamic meaning.[3–7]

9.1.2 EFFECT OF THE HYDROLYSIS ON DIFFUSION OF ELECTROLYTES IN AQUEOUS SOLUTIONS

A theory of mutual diffusion of electrolytes in aqueous solutions capable of accurately predicting diffusion coefficients has not yet been successfully developed, due to the complex nature of these systems. However, Onsager and Fuoss (eq1.4)[8]hasallowed the estimation of diffusion coefficients with a good approximation for dilute solutions and symmetrical electrolytes of the type 1:1. This equation may be expressed by

$$D=\left(1+c\frac{\partial \ln y_{\pm}}{\partial c}\right)\left(D^{0}+\Sigma\,\Delta_{n}\right) \tag{1.4}$$

where D is the mutual diffusion coefficient of the electrolyte, the first term in parenthesis is the activity factor, $y_{\pm}$ is the mean molar activity coefficient, c is the concentration in moldm^{-3}, D^0 is the Nernst limiting value of the diffusion coefficient (eq1.4), and Δ_n are the electrophoretic terms given by

$$\Delta_n = k_B T A_n \frac{\left(z_1^n t_2^0 + z_2^n t_1^0\right)^2}{\left|z_1 z_2\right| a^n} \tag{1.5}$$

where k_B is the Boltzmann's constant; T is the absolute temperature; A_n are functions of the dielectric constant, of the solvent viscosity, of the temperature, and of the dimensionless concentration-dependent quantity (k_a), k being the reciprocal of average radius of the ionic atmosphere; t_1^0 and t_2^0 are the limiting transport numbers of the cation and anion, respectively.

Since the expression for the electrophoretic effect has been derived on the basis of the expansion of the exponential Boltzmann function, because that function is consistent with the Poisson equation, we only would have to take the electrophoretic term of the first and the second order ($n = 1$ and $n = 2$) into account. Thus, the experimental data D_{exp} can be compared with the calculated D_{OF} on the basis of eq1.6

$$D = \left(1 + c\frac{\partial lny_{\pm}}{\partial c}\right)\left(D^0 + \Delta_1 + \Delta_2\right) \tag{1.6}$$

A good agreement between the observedvalues and those calculatedby eq1.6for potassium chloride[9] and magnesium nitrate[10] was observed and reported in the literature,but for cobalt chloride[11] and beryllium sulfate,[12] they are definitely higher than theory predicts. This is not surprising if we take into account the change with concentration of parameters, such as viscosity, hydration, and hydrolysis, factors not taken into account in Onsager–Fuoss model.[8] In fact, those differences may possibly be due to hydrolysis of the cations (eq 1.7),[13–15] which would be more pronouncedin cobalt chloride and beryllium sulfate than potassium chloride and magnesium nitrate.

$$M(H_2O)_x^{n+}(aq) \overset{\leftarrow}{\rightarrow} M(H_2O)_{x-1}(OH)^{(n-1)+}(aq) + H^+(aq) \tag{1.7}$$

We areproposing a model in order to allow the estimation of the amount of ion $H^+(aq)$ produced in each of the systems studied usingexperimental diffusion coefficients determined by our conductometric method and the respective predictive by that theory.

That is, having in mind the acidic character of some cations,[14,15] the percentagesof ion $H^+(aq)$ produced (or the amount of acid that would be necessary to add a given electrolyte solution in the absence of hydrolysis in order to simulate a more real system) are determined by the following system (eqs1.8 and 1.9),

$$D_{acid}z + D_{electrolyte}y = D_{exp,} \tag{1.8}$$

$$z + y = 1 \tag{1.9}$$

where $z\times 100$ and $y \times 100$ are the percentages of acid and electrolyte, respectively, and D_{exp} is the valueof the diffusion coefficients of electrolytes in aqueous solutions, and D_{acid} and $D_{electrolyte}$ are the Onsager and Fuoss values of the diffusion coefficients for the acids and electrolytes, respectively, in aqueous solutions.

9.2 EXPERIMENTAL ASPECTS

CONDUCTOMETRIC TECHNIQUE

An open-ended capillary cell (Fig. 9.1),[5] which has been used to obtain mutual diffusion coefficients for a wide variety of electrolytes,[5,9–13,16–19] has been described in great detail[5]. Basically, it consists of two vertical capillaries, each closed at one end by a platinum electrode, and positioned one above the other with the open ends separated by a distance of about 14 mm. The upper and lower tubes, initially filled with solutions of concentrations 0.75 and 1.25 *c*, respectively, are surrounded with a solution of concentration *c*. This ambient solution is contained in a glass tank (200 × 140 × 60 mm) immersed in a thermostat bath at 25°C. Perspex sheets divide the tank internally and a glass stirrer creates a slow lateral flow of ambient solution across the open ends of the capillaries. The experimental conditions are such that the concentration at each of the open ends is equal to the ambient solution value *c*, that is, the physical length of the capillary tube coincides with the diffusion path. This means that the required boundary conditions described in the literature[5] to solve Fick's second law of diffusion are applicable. Therefore, the so-called Δl effect[5] is reduced to negligible proportions. In our manually operated apparatus, diffusion is followed by measuring the ratio $w=R_t/R_b$ of resistances R_t and R_b of the upper and lower tubes by an alternating current transformer bridge. In our automatic apparatus, *w* is measured by a Solartron digital voltmeter (DVM) 7061 with 6 1/2 digits. A power source (Bradley Electronic Model 232) supplies a 30 V sinusoidal signal at 4 kHz (stable to within 0.1 mV) to a potential divider that applies a 250 mV signal to the platinum electrodes in the top and bottom capillaries. By measuring the voltages V' and V'' from top and bottom electrodes to a central electrode at ground potential in a fraction of a second, the DVM calculates *w*.

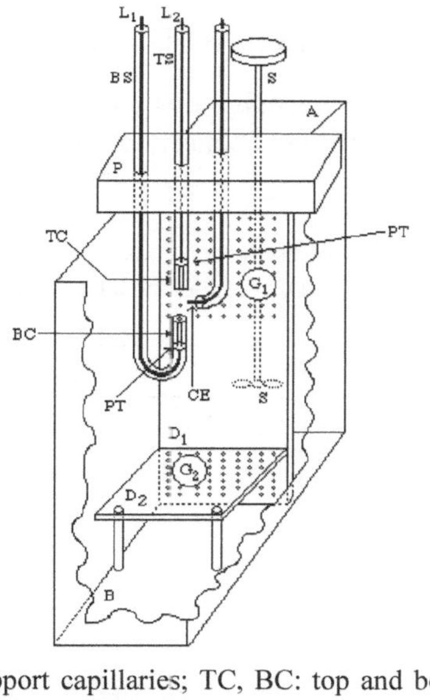

FIGURE 9.1 TS, BS: support capillaries; TC, BC: top and bottom diffusion capillaries; CE: central electrode; PT: platinum electrodes; D1, D2: perspex sheets; S: glass stirrer; P: perspex block; G1, G2: perforations in perspex sheets; A, B: sections of the tank; L1, L2: small diameter coaxial leads.[5]

In order to measure the differential diffusion coefficient D at a given concentration c, the bulk solution of concentration c is prepared by mixing 1 L of "top" solution with 1 L of "bottom" solution, accurately measured. The glass tank and the two capillaries are filled with c solution, immersed in the thermostat, and allowed to come to thermal equilibrium. The resistance ratio $w = w_\infty$ measured under these conditions (with solutions in both capillaries at concentration c accurately gives the quantity $\tau_\infty = 10^4/(1 + w_\infty)$.

The capillaries are filled with the "top" and "bottom" solutions, which are then allowed to diffuse into the "bulk" solution. Resistance ratio readings are taken at various recorded times, beginning 1000 min after the start of the experiment, to determine the quantity $\tau = 10^4/(1+w)$ asτ approaches τ_∞. The diffusion coefficient is evaluated using a linear least-squares procedure to fit the data and, finally, an iterative process is applied using 20 terms of the expansion series of Fick's second law for the present boundary conditions. The theory developed for the cell has been described previously.[5]

Interdiffusion coefficients of potassium chloride,[9] magnesium nitrate,[10] cobalt chloride,[11] and beryllium sulfate[12] in water at 298.15 K and at concentrations from 0.001 to 0.01 mol dm^{-3} have been measured using a conductometric cell and an automatic apparatus to follow diffusion. The cell uses an open-ended capillary method and a conductometric technique is used to follow the diffusion process by measuring the resistance of a solution inside the capillaries at recorded times.

9.3 EXPERIMENTAL RESULTS AND DISCUSSION

Table 9.1 shows the estimatedpercentage of hydrogen ions, z, resulting from the hydrolysis of different ions at different concentrationsat 298.15 K by using the eqs 1.8–1.9.

TABLE 9.1 Estimated Percentage of Hydrogen Ions, z, Resulting From the Hydrolysis of Some Cations in Aqueous Solutions at Different Concentrations,c, and at 298.15 KUsing Eqs 1.8 and 1.9.

c/	z%	z%	z%	z%
moldm^{-3}	**(KCl)**	**($Mg(NO_3)_2$)**	**($CoCl_2$)**	**($BeSO_4$)**
1×10^{-3}	2	–	5	–
3×10^{-3}	3	1	6	69
5×10^{-3}	2	2	8	48
8×10^{-3}	3	1	6	26
1×10^{-2}	3	1	6	34

From theanalysis cf this table, we see that the diffusion of cobalt ion and mainly berylliumion arethe most clearly affected bytheir hydrolysis. At the lowest Be^{2+}ion concentration (3×10^{-3} mol dm^{-3}), the effect of the hydrogen ions on the whole diffusion process has an important and main role($z = 69\%$, Table 9.1). However, we can, toa good approximation, describe thesesystemsas binaryif we considercertain-facts. For example, in the case of solutions ofberylliumsulfate, several studies[12]in-dicated thatthe predominant species present arethe solvent,Be^{2+}, SO_4^{2-},and H^{+}ions, as well as indicatedin the following schemeresulting fromhydrolysis of theBe^{2+}ion.

$$[Be(H_2O)_4]^{2+} \leftrightarrow [Be(H_2O)_3(OH)]^{-} + H^{+}$$
$$[Be(H_2O)_3(OH)]^{-} + [Be(H_2O)_4]^{2+} \leftrightarrow [(H_2O)_3\, Be\text{–}OH\text{–}Be\, (H_2O)_3]^{3+}$$

However, based on pH measurements (i.e., $2.9 \leq pH \leq 3.9$) for various $BeSO_4$so-lutions at concentrations ranging from 3×10^{-3}to1×10^{-2} mol dm^{-3}, and from the low equilibrium constantsrelative to the hydrolysis of theBe^{2+}ion,[15] we can then treat this system as being a binary system. Moreover, given the uniformity of the concentra-tion gradient imposed by our method, it is considered that the value of the diffusion coefficient, which is given to the scientific community, is a quantitative measure of all existing species that diffuse.

Concerning the other electrolytes (KCl and $Mg(NO_3)_2$),[9,10] the reasonable agree-ment obtained between the calculated and the experimental values of diffusion coef-ficient, should result in negligible percentages for the hydronium ion (Table 9.1).

9.4 CONCLUSION

We have estimated thepercentage of H_3O^+ (aq) resulting from the hydrolysis of some ions (i.e., beryllium, cobalt, potassium, and magnesium ions), using a new model and the experimental and calculated mutual diffusion coefficients for potas-sium chloride, magnesium nitrate, cobalt chloride, and beryllium sulfate in aqueous solutions, contributing in this way for a better knowledge of the structure of these systems.

Among the cited systems, we concluded that the diffusion of beryllium sulfate is clearly the most affected by the beryllium ion hydrolysis.

ACKNOWLEDGMENT

Financial support of the Coimbra Chemistry Centre from the FCT through project Pest-OE/QUI/UI0313/2014 is gratefully acknowledged.

KEYWORDS

- **diffusion coefficients**
- **aqueous solutions**
- **electrolytes**
- **hydration**

REFERENCES

1. Robinson, R.A.; Stokes, R.H. *Electrolyte Solutions*; 2nd ed., Butterworths: London, 1959.
2. Harned, H.S.; Owen, B.B. *The Physical Chemistry of Electrolytic Solutions*; 3rd ed., Reinhold Pub. Corp.: New York, 1964.
3. Tyrrell, H.J.V.; Harris, K.R. *Diffusion in Liquids: A Theoretical and Experimental Study;* Butterworths: London, 1984.
4. Lobo, V.M.M. *Handbook of Electrolyte Solutions*; Elsevier: Amsterdam, 1990.
5. Agar, J.N.; Lobo, V.M.M. Measurement of Diffusion Coefficients of Electrolytes by a Modified Open-Ended Capillary Method.*J. Chem. Soc.Faraday Trans.***1975;***71,* 1659–1666.
6. Callendar, R.; Leaist, D.G. Diffusion Coefficients for Binary, Ternary, and Polydisperse Solutions from Peak-Width Analysis of Taylor Dispersion Profiles.*J. Sol. Chem.* **2006;***35,* 353–379.
7. Barthel, J.; Feuerlein, F.; Neuder, R.; Wachter, R. Calibration of Conductance Cells at Various Temperatures. *J. Sol. Chem.***1980**;*9,* 209–212 .
8. Onsager, L.; Fuoss, R.M. Irreversible Processes in Electrolytes. Diffusion. Conductance and Viscous Flow in Arbitrary Mixtures of Strong Electrolytes. *J. Phys. Chem.***1932**;*36,* 2689–2778.
9. Lobo, V.M.M.; Ribeiro, A.C.F.; Verissimo, L.M.P. Diffusion Coefficients in Aqueous Solutions of Potassium Chloride at High and Low Concentrations. *J.Mol.Liq.* **2001**;*94,* 61–66.
10. Lobo, V.M.M.; Ribeiro, A.C.F.; Veríssimo, L.M.P. Diffusion Coefficients in Aqueous Solutions of Magnesium Nitrate. Ber. Bunsen Gesells. *Phys. Chem. Chem. Phys.***1994**;*98,* 205–208.
11. Ribeiro, A.C.F.; Lobo,V.M.M.; Natividade, J.J.S. Diffusion Coefficients in Aqueous Solutions of Cobalt Chloride at 298.15 K. *J. Chem. Eng. Data* **2002;***47,* 539–541.
12. Lobo, V.M.M.; Ribeiro, A.C.F.; Veríssimo, L.M.P. Diffusion Coefficients in Aqueous Solutions of Beryllium Sulphate at 298 K. *J. Chem. Eng. Data* **1994;***39,* 726–728.
13. Valente, A.J.M.; Ribeiro, A.C.F.; Lobo, V.M.M.; Jiménez A. Diffusion Coefficients of Lead(II) Nitrate in Nitric Acid Aqueous Solutions at 298.15 K. *J. Mol. Liq.* **2004**;*111,* 33–38.
14. Baes, C.F.; Mesmer, R.E. *The Hydrolysis of Cations*; John Wiley & Sons: NewYork, 1976.
15. Burgess, J. *Metal Ions in Solution*; John Wiley & Sons, Chichester, Sussex: England, 1978.
16. Ribeiro, A.C.F.; Lobo, V.M.M.; Oliveira, L.R.C.; Burrows, H.D.; Azevedo, E.F.G.; Fangaia, S.I.G.; Nicolau, P.M.G.; Fernando, A.D.R.A. Diffusion Coefficients of Chromium Chloride in Aqueous Solutions at 298.15 and 303.15 K. *J. Chem. Eng. Data* **2005;***50,* 1014–1017.
17. Ribeiro, A.C.F.; Esteso, M.A.; Lobo,V.M.M.; Valente, A.J.M.; Sobral, A.J.F.N.; Burrows, H.D. Diffusion Coefficients of Aluminium Chloride in Aqueous Solutions at 298.15 K, 303.15 K and 310.15 K. *Electrochim. Acta.* **2007,***52,* 6450–6455.
18. Veríssimo, A.L.M.P.; Ribeiro, A.C.F.; Lobo, V.M.M.; Esteso, M.A. Effect of Hydrolysis on Diffusion of Ferric Sulphate in Aqueous Solutions. *J. Chem. Thermodyn.***2012**;*55,* 56–59.

19. Ribeiro, A. A.C.F.; Gomes, J.C.S.; Veríssimo, L.M.P.; Romero L.H. Blanco, C.; Esteso, M.A. Diffusion of Cadmium Chloride in Aqueous Solutions at Physiological Temperature 310.15 K. *J. Chem.Thermodyn.* **2013**;*57*, 404–407.

CHAPTER 10

OZONE DECOMPOSITION

T. BATAKLIEV, V. GEORGIEV, M. ANACHKOV, S. RAKOVSKY, and G.E. ZAIKOV*

Institute of Catalysis, Bulgarian Academy of Sciences, Bonchev St. #11, Sofia 1113, Bulgaria

*N. M. Emanuel Institute of Biochemical Physics, Russian Academy of Sciences, 4, Kosigina St., Moscow119334, Russian Federation

CONTENTS

ABSTRACT

Catalytic ozone decomposition is of great significance because ozone is toxic substance commonly found or generated in human environments (aircraft cabins, offices with photocopiers, laser printers, sterilizers). Considerable work has been done on ozone decomposition reported in the literature. This review provides a comprehensive summary of this literature, concentrating on analysis of the physicochemical properties, synthesis, and catalytic decomposition of the ozone. This is supplemented by review on kinetics and catalyst characterization, which ties together the previously reported results. It has been found that noble metals and oxides of transition metals are the most active substances for ozone decomposition. The high price of precious metals stimulated the use of metal oxide catalyst and particularly the catalysts based on manganese oxide. The kinetics of ozone decomposition has been determined to be offirst order. A mechanism of the reaction of catalytic ozone decomposition was discussed based on detailed spectroscopic investigations of the catalytic surface, showing the existence of peroxide and superoxide surface intermediates.

10.1 INTRODUCTION

In recent years, the scientific research in all leading countries of the world is aimed primarily at solving the deep environmental problems on the planet, including air pollution and global warming. One of the factors affecting these processes negatively is the presence of ozone in ground atmospheric layers. This is a result of the wide use of ozone in lot of important industrial processes such as cleaning of potable water and soil, disinfection of plant and animal products, textile bleaching, complete oxidation of waste gases from the production of various organic chemicals, sterilization of medical supplies, and so on.[1]

The history of ozone chemistry as a research field began immediately after its discovery from Schönbein in 1840.[2]The atmospheric ozone is focused mainly on the so-called "ozone layer" at a height of 15–30 km above the earth surface wherein the ozone concentration ranges from 1 to 10 ppm.[3] The ozone synthesis at that altitude runs photochemically through the influence of solar radiation on molecular oxygen. The atmospheric ozone is invaluable to all living organisms because it absorbs the harmful ultraviolet radiation from the sun. The study of the kinetics and mechanism of ozone reactions in modern science is closely related to solving the ozone holes problem that reflect the trend of recent decades to the depletion of the atmospheric ozone layer.

Ozone has oxidation, antibacteriological, and antiviral properties that make it widely used in the treatment of natural, industrial, and polluted waters; swimming pools; contaminated gases; for medical use; and so on. Catalytic decomposition of residual ozone is imperative because from environmental point of view, the release of ozone in the lower atmosphere has negative consequences.

The presence of ozone in the surrounding human environment (airplane cabins, copiers, laser printers, sterilizers, etc.) also raises the issue of its catalytic decomposition as ozone is highly toxic above concentrations of 0.1 mg/m[34,5] and can damage human health. The use of catalysts based on transition metal oxides are emerging as very effective from an environmental perspective and at the same time practical and inexpensive method applied in the decomposition of residual ozone.

10.2 SOME PHYSICOCHEMICAL PROPERTIES OF OZONE

Ozone is an allotropic modification of oxygen that can exist in all three physical conditions. In normal conditions, ozone is colorless gas with a pungent odor while at very low concentrations (up to 0.04 ppm) it can feel like a pleasant freshness. A characteristic property of ozone odor is the fact that it is easily addictive and at the same time hazardous for men who work with ozone in view of its high toxicity at concentrations above the limit (>0.1 mg/m^3).

When the ozone concentration is more than 15–20%, it has blue color. At atmospheric pressure and temperature of 161.3 K, the ozone becomes fluid in deep blue color. It cures at 80.6 K by acquiring dark purple color.[6]Ozone is explosive in all three physical conditions. The work with ozone concentrations of 0–15% can be considered safe.[4] The danger of ozone explosion is a function of its thermodynamic instability ($\Delta G°298=-163$ kJ/mol) and ozone decomposition to diatomic oxygen is a thermodynamically favorable process with heat of reaction $\Delta H°298=-138$ kJ/mol.[7] The ozone molecule consists of three oxygen atoms located at the vertices of obtuse-angled triangle with a central angle of 116°8'±5' and length of O–O-bond: $\rho_{o-o} = 0.1278\pm0.0003$ nm (Fig. 10.1).

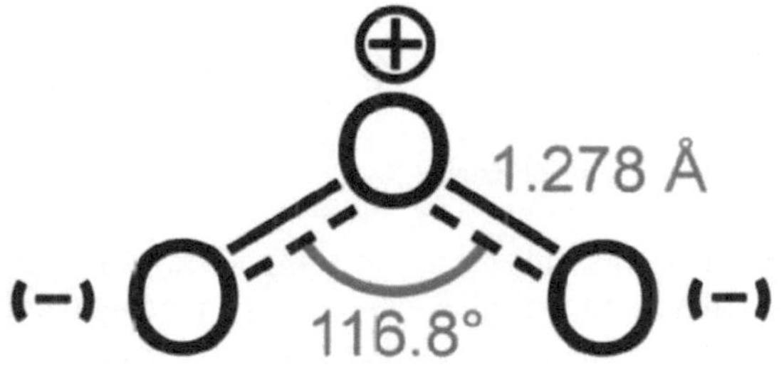

FIGURE 10.1 Structure of ozone molecule.

In gas phase, the ozone molecule is a singlet biradical while in liquid phase it reacts generally as a dipolar ion. The homogeneous reaction of pure gaseous ozone decomposition is characterized by certain velocity at sufficiently high temperatures. The kinetics study of the ozone thermal decomposition is also complicated by the fact that above a certain critical temperature, the steady kinetic decomposition is transformed into explosion, subsequently passing to detonation.[1] According to Thorpe,[8] the detonation of ozone is observed above 105°C. The gaseous ozone is characterized by different time of half-life depending on the temperature (Table 10.1):

TABLE 10.1 Half-Life of Ozone.

Temp. (°C)	Half-life
−50	3 months
−35	18 days
−25	8 days
20	3 days
120	1.5h
250	1.5s

The ozone structure is resonance stabilized, which is one of the reasons for its resistance against decomposition at low temperatures (Fig.10.2).

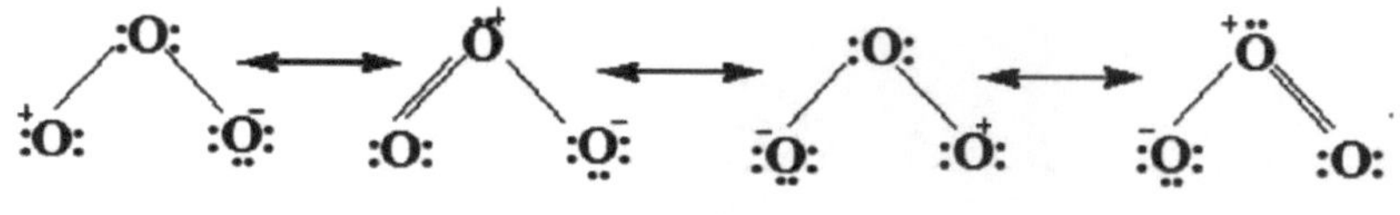

FIGURE 10.2 Resonance structures of ozone.

In most reactions of ozone with inorganic compounds, it reacts with participation of one oxygen atom and the other two are separated as O_2. Typically, the elements are oxidized to their highest oxidation states. For example, manganese is oxidized to $[MnO]^{4-}$, halogen oxides to metals (ClO_2, Br_2O_5), ammoniato NH_4NO_3, and nitrogen oxides pass into N_2O_5.[9–11]

Ozone has spectral characteristics from the IR-region to the vacuum UV-region, which are present in a significant number of works, such as the majority of them are carried out with gaseous ozone.[12–18] It should be noted that most researchers, who use the spectrophotometric method for analysis of ozone, work in the main area of the spectrum 200–310 nm, where is the wide band with maximum at ~255 nm for gaseous ozone.[13,15–17] That maximum is characterized with high value of the coefficient of extinction (Table 10.2):

TABLE 10.2 Coefficient of Extinction (ε, $l.mol^{-1}.cm^{-1}$) of Gaseous Ozone in UV-Region.[19]

λ, nm	[35]	[33]	[36]	[47]	λ, nm	[35]	[33]	[36]
253.6*	2981	3024	2952	3316	296.7**	150.5	153.4	156.8
270	–	–	–	2302	302.2	74.4	–	74.4
289.4	383	337.5	387.5	–	334.2	1.50	–	1.46

*The values of rare 1830 $l.mol^{-1}.cm^{-1}$at 254.0 nm[4]and3020 $l.mol^{-1}.cm^{-1}$ at 253.6 nm.[14]
** ε = 160 $l.mol^{-1}.cm^{-1}$ at 295 nm.[18]

10.3 OZONE SYNTHESIS AND ANALYSIS

Ozone for industrial aims is synthesized from pure oxygen by thermal, photochemical, chemical, and electrochemical methods, in all forms of electrical discharge and under the action of a particles stream. [1]The synthesis of ozone is carried out by the following reactions shown in the scheme:

$$O_2 + (e^-, h\nu, T) \rightarrow 2O\ (O_2^*) \tag{1}$$

$$O_2 + O + M \leftrightarrow O_3 + M \tag{2}$$

$$O_2^* + O_2 \rightarrow O_3 + O \tag{3}$$

where M is every third particle.

At low temperatures, the gas consists essentially of molecular oxygen and at higher,by atomic oxygen. There is no area of temperatures at normal pressure wherein the partial pressure of ozone is significant. The maximum steady-state pressure, which is observed at temperature of 350 K, is only 9.10^{-7} bar. The values of the equilibrium constant of reaction (2) at different temperatures are presented in Table 10.3.[20]

TABLE 10.3 Equilibrium Constant (K_e) of Reaction (2) Depending on the Temperature.

T, K	1500	2000	3000	4000	5000	6000
K_e, M	1662.10^{-11}	4413.10^{-7}	1264.10^{-2}	2.104	48.37	382.9

At high temperatures, when the concentration of atomic oxygen is high, the equilibrium of reaction (2) is moved to the left and the ozone concentration is low. At low temperatures, the equilibrium is shifted to the right, but through the low concentration of atomic oxygen, the ozone content is negligible. For synthesis of significant concentrations of ozone, it is necessary to combine two following conditions: (1) low temperature and (2) the formation of superequilibrium concentrations of atomic oxygen.It is possible to synthesizeupper equilibrium atomic concentrations of oxygen at low temperatures by using nonthermal processes, such as dissociation of oxygen with particles flow, electrons, hν, electrochemical, and chemical influences.Ozone will always be formed at low temperatures, when there is a process of oxygen dissociation. Higher concentrations of ozone may be obtained by thermal methods providing storage ("quenching") of superequilibrium concentrations of atomic oxygen at low temperatures. Photochemical synthesis of ozone occurs upon irradiation of gaseous or liquid oxygen by UV radiation with wavelength of $\lambda<210$ nm.[21] It is assumed[22] that the formation of ozone in the presence of radiation with wavelength in the range of $175<\lambda<210$ may be associated with the formation of excited oxygen molecules:

$$O_2 + h\nu \rightarrow (O_2^*) \tag{4}$$

$$O_2^* + O_2 \rightarrow O_3 + O \tag{5}$$

$$O_2 + O + M \leftrightarrow O_3 + M \tag{6}$$

The photochemical generation of ozone has an important role in atmospheric processes, but it is hard to apply for industrial applicationbecause of the high energy costs (32 kWh/kg ozone) for the preparation of high-energy shortwave radiation. Nowadays, the industrial synthesis of ozone happens mainly by any of the electric discharge methods by passing oxygen containing gas through high-voltage (8–10 kV) electrodes. The aim is creation of conditions in which oxygen is dissociated into atoms. This is possible in all forms of electrical discharges: smoldering, silent, crown, arc, barrier, and so on. The main reason for the dissociation is the hit of molecular oxygen with the acceleration in electric field electrons,[1,6] when part of the kinetic energy is transformed in energy of dissociation of the bond O–O and excitation of oxygen molecules. In general, molecular oxygen dissociates in two oxygen atoms, then on final stage, two molecules of oxygen form one oxygen atom and one ozone molecule:

$$O_2 + e \rightarrow 2O\ (O_2^*) \tag{7}$$

$$O_2 + O + M \rightarrow O_3 + M \tag{8}$$

$$O_2^* + O_2 \leftrightarrow O_3 + O \tag{9}$$

Technically, the synthesis of ozone is carried out in discharge tubes—ozonators. One of the most commonly used type ozonators are Siemens (Fig.10.3).

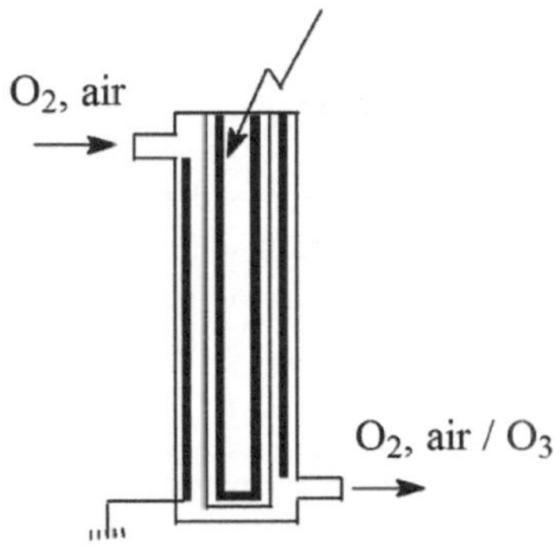

FIGURE 10.3 Principal scheme of ozonator Siemens.

This ozonator represents two brazed, one to another coaxial pipes with supported inside of the inner side and outside of the outer side electrically conductive coating of aluminum, silver, copper, and the like.A high voltage up to 20 kV is supplied to it, and dry and pure oxygen is left to pass through the ozonator. The ozone is synthesized on the other end of the ozonator.The ozone concentration depends on current parameters—voltage, frequency, and power, on the ozonator properties—thickness, length, and type of glass tube;on distance between the electrodes; and it alsodepends on temperature.

The analysis of the ozone concentration is developed by various physical and chemical methods discussed in details in monograph.[23] It should be noted the preference for the spectrophotometric method compared withthe iodometric method, which is determined by the lack of need of continuouspH monitoring during the ozone analysis and also by the possibility of direct observation of the inlet and outlet ozone concentrationsin the system, allowing precisefixingof the experimental time and its parameters.For determination of the ozone, amount can be used, not absolute values of concentration but the ratios of the proportional values of optical densities, which excludes the error influence atexpense of the inaccuracy in calibration of the instrument.

10.4 OZONE DECOMPOSITION

The reaction of ozone on the surface of solids is of interest from different points of view.Along with the known gas cycles depleting the atmospheric ozone, a definitive role playsits decomposition on aerosols that quantity already is growing at expense of the anthropogenic factor. The growth of ozone used in chemical industries set up the task for decomposition of the residual ozone on heterogeneous, environmental friendly catalysts.In humid environments and under certain temperatures and gas flow rates, this subject has not been yet entirely understood.

10.4.1 STRATOSPHERIC OZONE

The importance of the stratospheric ozone is determined by its optical properties—its ability to absorb the UV solar radiation with a wavelength of 220–300 nm. The absorption spectrum of ozone in the ultraviolet and visible region is given in Figure 10.4.

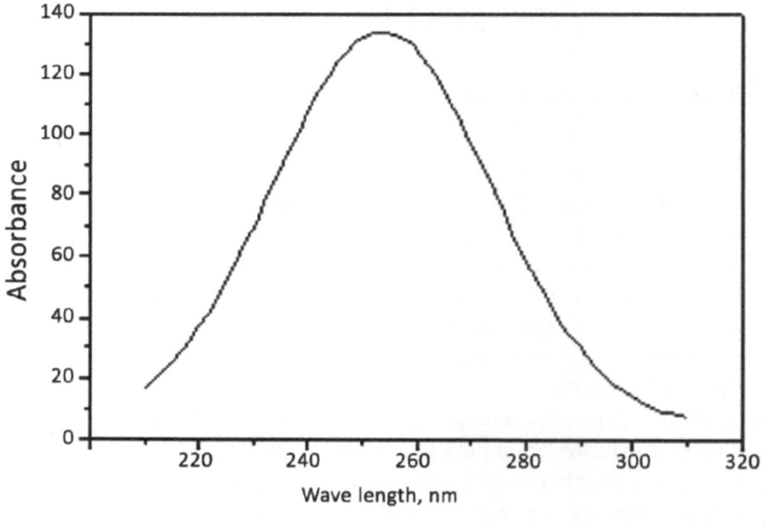

FIGURE 10.4 UV-absorption spectrum of ozone with maximum at 254 nm.

The main absorption band in the name of Hartley is in the range 200–300 nm with maximum at a wavelength of 254 nm. Stratospheric ozone layer has a thickness of 3 mm and its ability to absorb UV rays protects the earth's surface of biologically active solar radiation, which destroys the most important biological components, proteins, and nucleic acids.

The fundamentals of the photochemical theory of stratospheric ozone have been made by the English chemist Chapman,[24,25] according to whom the photochemical decomposition of ozone occurs to the following reactions:

$$O_3 + O \rightarrow O_2 + O_2 \quad (10)$$

$$O_3 + hv \rightarrow O_2 + O\,(^3P) \quad (11)$$

$$O_3 + hv \rightarrow O_2 + O\,(^1D) \quad (12)$$

The proposed mechanism leads to formation of oxygen atoms in ground and excited state. This cycle of decomposition is called the cycle of "residual oxygen." The photodissociation of ozone (11, 12) takes place under the action of solar radiation with wavelength less than 1134 nm. Decomposition of ozone following these reactions (11, 12) can be observed at any altitude near to ground level. The photochemical synthesis of ozone in the stratosphere requires its degradation in reaction (10). It was found that 20% of the stratospheric ozone decomposes following this reaction.[26]The effect of ozone layer on climate changes is related to the absorption of radiation, which occurs not only in the UV-region, but also in the IR-region of the spectrum. Absorbing IR rays from theEarth's surface, ozone enhances the greenhouse effect in the atmosphere.[27,28]The connection of this problem with climate is extremely complicated due to the variety of physical and chemical factors that influence the amount of atmospheric ozone.[29]

The destruction of ozone in the atmosphere is linked to catalytic cycles of the type:

$X + O_3 \rightarrow XO + O_2$	$X + O_3 \rightarrow XO + O_2$
$XO + O \rightarrow X + O_2$	$XO + O_3 \rightarrow X + 2O_2$
$O_3 + O \rightarrow O_2 + O_2$	$2O_3 \rightarrow 3O_2$

where: X is OH, NO, or Cl, formed by dissociation of freons in the atmosphere.[30]

The stratospheric cycles of ozone decomposition are discussed in a lot of works but fundamental contributions have most[26] and.[31] The effective action of the molecularcatalysts in the cycles of ozone decomposition is determined by their concentrations in the atmosphere, which depends on the rates of regeneration and exit from the respective cycle.[32] The ratio between the rate of decomposition of residual oxygen and the rate of catalyst outgoing from the cycle determines the length of the chain reaction and corresponds to the number of O_3 molecules destructed by one catalytic molecule. The number of stages of ozone decomposition in a single catalytic center can reach 106.[1] The reduction of the ozone amount over Antarctica, that is, the thinning of ozone layer is mainly due to the action of the chlorine cycle.[33–35] During this catalytic cycle, the presence of one chlorine atom in the stratosphere can cause the decomposition of 100,000 ozone molecules.

10.4.2 CATALYSTS FOR DECOMPOSITION OF OZONE

The use of ozone for industrial aims is related with the application of effective catalysts for its decomposition, as it was already mentioned, the release of ozone in the atmosphere near the ground level is dangerous and contaminates the air.[1,36–37] The ozone has a great number of advantages as oxidizing agent and, in that capacity, it has been used in different scientific investigationswith neutralization of the organic contaminants,[38,39] asthe ozonation efficiency increases in presence of the catalyst.[40,41] Due to its antibacterial and antiviral properties, ozone is one of most used agents in water treatment.[42] This facts result in considerable interest fromthe researchers in study of homogeneous and heterogeneous reactions of ozone decomposition, as well as in participationof ozone in multiple oxidation processes.

10.4.3 OZONE DECOMPOSITION ON THE SURFACE OF METALS AND METAL OXIDES

The first studies on the catalytic decomposition of gaseous ozone[43] showed that the degradation of the ozone is accelerated in the presence of Pt, Pd, Ru, Cu, W, and so on. Catalytic activity of metals in the decomposition of ozone is studied by Kashtanov et al.,[44] which highlight that the silver (Ag) show higher catalytic activity, compared to copper (Cu), palladium (Pd), and tin (Sn). It should be noted,from earlier works on decomposition of ozone, that the study of Schwab and Hartmann[45]has investigated catalysts based on metals from I to IV group and their oxides in different oxidation levels, and found out that the catalytic activity of these oxides increases with increase in the oxidation state of the metal. Below is presented the relationship between the catalytic activity degree of a number of elements and their oxides, in the reaction of the catalytic decomposition of ozone:

$Cu<Cu_2O<CuO$; $Ag<Ag_2O<AgO$; $Ni<Ni_2O_3$; $Fe<Fe_2O_3$; $Au<Au_2O_3$;

Pt<colloidal Pt

These results can be explained by the higher activity of the ions of the respective elements relative to the activity of the elements themselves, as well as the importance of ionic charge in the catalytic reaction. Other earlier studies have been presented in several articles by Emelyanov et al.,[46–48] which discussed catalytic decomposition of gaseous ozone at temperatures from −80 to +80°C and ozone concentration of 8.8 vol.%. For Catalysis,elements of the platinum group such as Pt, colloidal Pt, Pd, Ir and colloidal Ru, NiO, and Ni_2O_3have been used. Experiments have been carried out in a tube reactor at gas velocity of 5 l/h. Studies indicate identical activity of the nickel oxide and colloidal platinum at temperatures from 20 to 80°C. At −80°C, the activity of the nickel oxide falls to zero, while the platinum remains active. This is related to the low activation energy of decomposition of O_3 on the metal surface (~1 kcal/mol). It was also found out that the process of decomposition of the ozone in the presence of colloidal Rh and Ir at +3 and +20°C slowed down after 4 h, and the catalysts lost its activity. Sudak and Volfson's silver–manganese catalyst[49] for the decomposition of ozone was used at low temperatures in different gas mixtures. Commercial reactors operating with this catalyst have a long life of performance and maintain a constant catalytic activity. The preparation of other high-performance metal catalyst for the decomposition of ozone, including the development of special technology for synthesis, determination of the chemical composition, the experimental conditions, and the catalytic activity have been noted in a number of studies in Japanese authors published in the patent literature.[50–60] The main metals used are Pt, Pd, Rh, and Ce, as well as metals and metal oxides of Mn, Co, Fe, Ni, Zn, Ag, and Cu. The high price of precious metals stimulates the use of metal oxide catalyst supporters with a highly specific surface such as γ-Al_2O_3, SiO_2, TiO_2, ZrO_2, and charcoal. Hata et al.[50] have synthesized a catalyst containing 2% Pt as active component on a supporter composed of a mixture of 5% SiO_2 and 95% γ-Al_2O_3 having a specific surface area 120 m^2/g. They have impregnated chloroplatinic acid onto the carrier at 80°C then the sample has been dried successively at 120 and 400°C and atmospheric pressure. At operating temperature of 20°C, gas velocity of 20,000 h^{-1} and 10 ppm initial concentration of ozone in the fed gas, this catalyst shows 95% catalytic activity. Another catalyst comprising TiO_2, SiO_2, and Pt[52] degrades 94, 97, and 99% of the ozone in the air stream at temperatures of 20, 50, and 100°C respectively. Therui et al.[54] have deposited metals Mn, Co, Fe, Ni, Zn, Ag, or their oxides in the weight ratio in regard to carrier in the range of 0–60%, as well as Pt, Pd, and Rh of the 0–10 wt.%, and also mixed oxides such as TiO_2–SiO_2; TiO_2–ZrO_2, and TiO_2–SiO_2–ZrO_2, supported on colloidal polyurethane with 400 m^2/g specific surface. The catalysts have been placed in tube reactors and their catalytic activity measured in the decomposition of ozone at a concentration of 0.2 ppm has been 99%. In published patents,it may be noted that for catalysts with identical chemical composition various precursors and methods for synthesishave been used. For example, TiO_2 and MnO_2[56] are obtained from aqueous solutions of H_2SO_4 + $TiOSO_4$ and Mn $(NO_3).6H_2O$, while in another case,[57] TiO_2 was purchased from the manufacturer, and the MnO_2 is prepared by precipitation of aqueous solutions of $MnSO_4$ and NH_3 in an atmosphere of oxygen followed by calcination. The application of the proposed system conditions makes it possible to assess the effectiveness of the catalysts and may be used for development of new catalytic systems. It can be seen that many of the catalysts operate at ambient temperatures (293–323 K), high-space velocities (>20,000 h^{-1}), and exhibit high catalytic activity (conversion 95%). In recent years, more and more researchers are focused on the development, study, and application of catalysts for decomposition of ozone based on supported or native metal oxides, in regard with the already mentioned fact—the high price of metals of the platinum group. In this regard, the most widely used are the oxides of Mn, Co, Cu, Fe, Ni, Si, Ti, Zr, Ag, and Al.[61–69] It has been found that the high catalytic activity in the decomposition of ozone exhibit the oxides of transition metals,[70] and particularly the catalysts based on manganese oxide.[71] The manganese oxide is used as a catalyst for various chemical reactions including the decomposition of nitrous oxide[72–74] and isopropanol,[74,75] oxidation of methanol,[76] ethanol,[77] benzene,[78,79]

CO,[74,75,80]and propane[81] as well as for reduction of nitric oxide[82] and nitrobenzene.[83] In terms of technology for the control of air pollution, manganese oxides are used both in the decomposition of residual ozone, and the degradation of volatile organic compounds.[84,85] Oxides of Mn, Co, Ni, Cr, Ag, Cu, Ce, Fe, V, and Mo supported on γ-Al_2O_3 on cordierite foam (60 pores/cm^2 and geometry 5.1 × 5.1 × 1.3 cm) have been prepared and tested in the reaction of decomposition of ozone.[71] Experiments have been carried out at a temperature of 313 K, a linear velocity of 0.7 m/s, inlet ozone concentration of 2 ppm, relative humidity of 40%, and total gas flow rate of 1800 cm^3/s. It has been found out that the catalyst activity decreases over time and the measurements are made only when the speed of decomposition is stabilized. After comparison of the conversion degree of ozone, conclusions on their catalytic activity have been drawn, according to which they have been arranged as follows: MnO_2 (42%)>Co_3O_4 (39%)> NiO (35%)>Fe_2O_3 (24%)>Ag_2O (21%)>Cr_2O_3 (18%)>CeO_2 (11%)>MgO (8%)>V_2O_5 (8%)>CuO (5%)>MoO_3 (4%). The high dispersibility of the material is confirmed by X-ray diffraction analysis. TPR profile (spectrum) of supported and unsupported MnO_2 has been done. It has been found that the reduction temperature of the supported MnO_2, Co_3O_4, NiO, MoO_3, V_2O_5, and Fe_2O_3 were in the range of 611–735 K, while that of CeO_2 was above 1000 K. It has turned out that from all examined metal oxide samples, MnO_2 had the lowest reduction temperature because it exhibited a higher reduction ability in comparison with the others. On the basis of experiments, it has been proposed that the mechanism of decomposition of ozone on the catalytic surface which includes formation of intermediate ionic particles possessing either superoxide or peroxide features:

$$O_3 + 2^* \leftrightarrow {}^*O_2 + O^* \tag{13}$$

$$^*O_2 + {}^* \rightarrow 2O^* \tag{14}$$

$$2O^* \rightarrow O_2 + 2^* \tag{15}$$

where * denotes active center.

A distinguishing peculiarity of the proposed scheme is the assumption that the key intermediate particle $^*O_2$ is not desorbed immediately. This could happen if it has partial ionic character (O^{2-}, O_2^{2-}). The catalytic activity of iron oxide Fe_2O_3 in decomposition of ozone has alsobeen studied in earlier works of Rubashov et al.[86,87] It was found that Fe_2O_3 showed a catalytic activity only when it has been used in the form of particles with small diameter while catalyst that consisted aggregated particles which was not efficient. Furthermore, the stability of the catalyst is not high and its poisoning is due to the formation of oxygen directly associated with the surface. Reactivation of the catalyst is accomplished by its subjecting to the vacuum thermal treatment to remove oxygen. The same catalyst was also studied in terms of the fluidized bed, wherein the rate constant for the catalytic reaction is higher than the normal gas flow. By determining the constant of the decomposition of ozone on the surface of Fe_2O_3 at various temperatures,activation energy of the processcan be calculated: $E = (12.5 \pm 1)$ kcal/mole. This value differs from theactivation energy of ozone decomposition on compact sample of iron oxide, calculated by Schwab[45]that is in the order of 2–3 kcal/mol. The difference can be explained with the regime in which the kinetic process is carried out, in the first case, and diffusion in the second. Ellis et al.[88] have investigated the activity of 35 different materials with respect to decomposition of ozone as the best catalysts have turned out nickel oxide and charcoal. Measurements have been carried out at room temperature, the gas velocities between 250 and 2000 l/h, atmospheric pressure and ozone concentrations (1.1–25) $\times 10^{12}$ molecules/cm^3. The measured rate of decomposition of ozone corresponds to a first-order reaction. The authors ignoring the decomposition of ozone in the bulk have proposed a homogeneous decomposition on the surface:

$$O_3 + M \rightleftharpoons O_2 + O + M \quad (16)$$

$$O_3 + O \rightarrow 2O_2 \quad (17)$$

In this case, "M" is an active center on the surface, which reduces the energy of activation of the reaction and accelerates the decomposition. Assuming that the portion of the surface area occupied by ozone,θ is proportional to the partial pressure of ozone in the volume, the kinetic region of ozone leaving from the gas phase is equal to:

$$-dpo_3 / dt = k\theta = k' po_3 \quad (18)$$

It is obvious that the firstorder of reaction with respect to the decomposition of ozone have been established, which has been confirmed experimentally. The decomposition of ozone in gradient-free reactor is considered in work,[89]as thin film of nickel oxide coated on the walls of the reactor was used as a catalyst has used. Turbulence of gas flow in such type of reactor would avoid the physical transfer and will provide conditions for studying the kinetics only on the catalytic surface. In this case, the turbulent diffusion is superior to the rate of reaction and the process is conducted in the kinetic regime. It has been found that the rate constant in the range 10–40°C is calculated by the expression:

$$K = 10^{2.54} p^{-0.5} \exp(-1250/T) \text{ cm/s} \quad (19)$$

where$p$ is in atm.

The experiments that have been made with different pressures of the oxygen and ozone–helium mixtures have shown that the rate constant changes depending on the total pressure and not by the partial pressure of ozone. Hence, the authors have concluded that the catalyst operates in inner diffusion mode, that is, the reaction rate is limited by molecular diffusion in the pores of the catalyst. These and other assumptions made it possible to explain the resulting experimental dependence of the rate constant on the total pressure of the mixture in the gas phase. The kinetics of decomposition of ozone in gradient-free reactor was investigated in the work.[90]Activities of γ-Al_2O_3, hopkalite (MnO_2, CuO, bentonite), silver–manganese oxide, alumina–palladium, and alumina–platinum catalysts are measured. The experiments have been conducted at 60°C and initial ozone concentration of 2.7×10^{-5} mol/l. Silver–manganese oxide and hopkaliteare the most active of the tested catalysts. Kinetics of decomposition of ozone is determined to be of first order. It have also been defined the rate constants and energies of activation, as the latter being fluctuated in the range of 6–10 kcal/mol. Conclusions of authors on the active surface of the catalyst are of interest. It has been found that the activation energy does not depend on the dispersibility, but influences the catalytic activity. On the other hand, dispersibility does not increase the inner surface of the porous material. Hence, it has been concluded that the decomposition of ozone occurs in the outer kinetic region, that is, the heterogeneous reaction is limited mainly by the outer surface of the catalyst. During the study of the decomposition of ozone in the presence of nickel oxide[91] is established that the reaction to be first order. It has been proposed that the limiting stage appear to be adsorption of ozone to the catalytic surface. The EPR spectra revealed the presence of ozonide ion radical O^{3-}.[92,93] The formation of O^{3-} is explained by the process of electron transfer from the catalyst to the ozone, so that the surface Ni^{2+} ions are oxidized to the Ni^{3+}. Further studies of the authors have shown that approximately 5% of the total reaction takes place in this way. The remaining amount of the ozone has been decomposed by molecular mechanism:

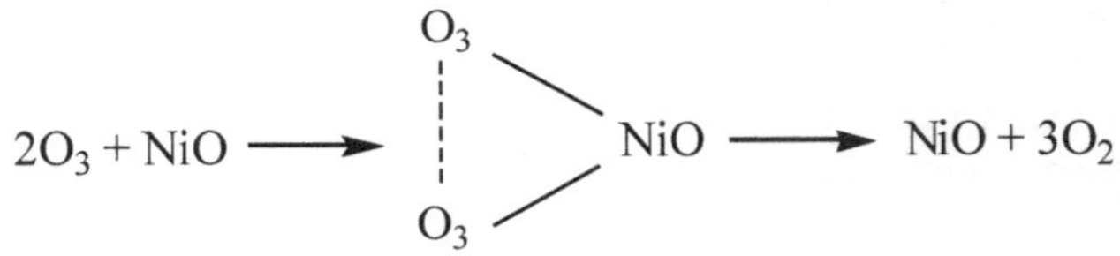

(20)

In some cases, the ozone does not remain in the molecular state and dissociate to atomic or diatomic oxygen species. Ozonide particles on catalytic surface are generated by the reaction:

$$O^- + O_2 \rightarrow O^{3-} \quad (21)$$

Figure 10.5 shows the characteristic bands of the EPR spectrum at $g_1 = 2.0147$, $g_2 = 2.0120$, and $g_3 = 2.0018$, due to the formation of the ionradical O^{3-}.

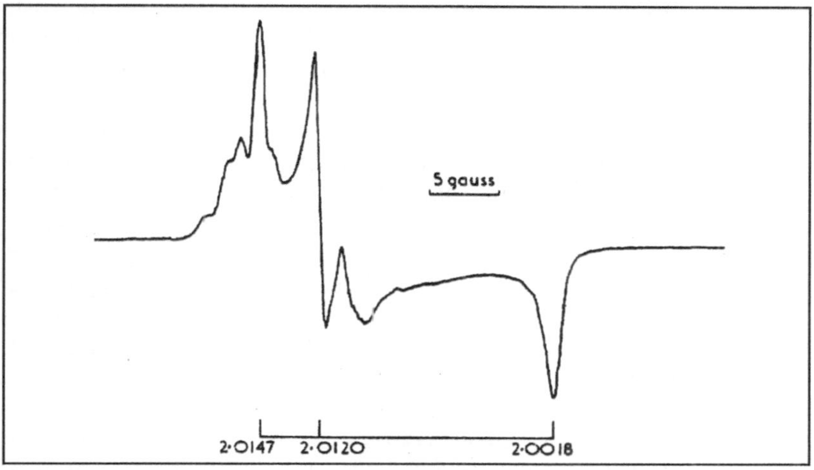

FIGURE 10.5 EPR spectrum of O^{3-} at temperature 77 K.

Martynov et al.[94] have studied the influence of the CuO, CoO, and NiO on the process of ozone decomposition. The catalysts have been prepared using aqueous solutions of the respective nitrates and have supported on the walls of the tubular reactor (l=25 cm, D=1.6 cm). The reactor has been heated to 370°C for 5 h, after which each of the catalysts was treated with ozone at a concentration of 0.5 vol.% for 6 h at a gas velocity of 100 l/h. After solving the diffusion-kinetic equation of the ozone decomposition reaction, the diffusion and the kinetic constants as well as the coefficients of ozone decompositioncan be defined. The recent ones characterize the decomposition rate of ozone molecules on the catalyst surface toward the rate of the hits of molecules with the surface. The results of these calculations for different samples are presented in Table 10.4 wherein ω is space velocity, γ is coefficient of ozone decomposition, and $k_{exp}k_{kin}$ are experimental and theoretical rate constants,respectively.

TABLE 10.4 Kinetic Parameters of the Reaction of Ozone Decomposition on Some Oxides.

Catalyst	ω, l/h	k_{exp}, s^{-1}	$\gamma.10^{-5}$	k_{kin}, s^{-1}
CuO	20	1.0	4.8	2.8
CoO	100	0.9	4.1	2.4
CoO	200	1.3	6.3	3.6
NiO	20	0.9	4.4	2.6
NiO	100	1.0	4.8	2.8
NiO	200	1.7	8.1	4.4

It is evident that the catalysts activity increases at ω =200 l/h. This is explained by the flow of catalytic reaction between kinetic and diffusion regions. Furthermore, the values of the rate constants of ozone decomposition in kinetic mode are almost three times higher than the experimental.

Radhakrishnan et al.[68] have used manganese oxide catalysts supported on Al_2O_3, ZrO_2, TiO_2, and SiO_2 for study of the support influence on the kinetics of decomposition of ozone. By using different physical methods of analysis such as "in situ" laser Raman spectroscopy, temperature-programmed desorption of oxygen, and measuring of specific surface area (BET), it is shown that the manganese oxide is highly dispersed on the support surface. The Raman spectra of the supported catalysts reveal the presence of Mn–O bands as a result from well-dispersed manganese oxide particles on the Al_2O_3- and SiO_2-supported catalysts. During the process of ozone decomposition on the catalytic surface, a signal from adsorbed particles appears at Raman spectra in the region of 876–880 cm^{-1}. These particles were identified as oxygen particles from peroxide type (O_2^{2-}), which disappear upon catalysts heating to 500 K. Using temperature-programmed desorptionof oxygen, the number of active manganese centers on the catalyst surface is calculated. After integration of the TPD peaks area corresponding to desorption of adsorbed oxygen, besides the density of active sites, the corresponding dispersion values of the catalysts are also identified. These resultsof the calculated specific surface areas of the catalytic samples are shown in Table 10.5. The catalysts were tested in reaction of ozone decomposition to determine their activity, and it has been found that the rate of decomposition increases with increasing the ozone partial pressure and the temperature. It has calculated the kinetic parameters of the reaction. It has been found that the activation energy is in the rangeof 3–15 kJ/mol, depending on the catalyst sample, as it is lower (3 kJ/mol) in the case of ozone decomposition on MnO_x/Al_2O_3. It was suggested that this is related to the structure of this catalyst, which is the only one of the tested samples that has mononuclear manganese center coordinated by five oxygen atoms. This was demonstrated using absorption fine-structure X-ray spectroscopy. Using this method, it was also found that the other three supported catalysts possess multi-core active manganese centers surrounded as well by five oxygen atoms.

TABLE 10.5 Densities of Active Centers and Specific Surfaces.

Catalyst	S_g, $m^2\ g^{-1}$	Density, mmol g^{-1} (O_2/O_3 TPD)	Dispersion, (%)
MnO_x/Al_2O_3	92	40	12
MnO_x/ZrO_2	45	163	47
MnO_x/TiO_2	47	31	9
MnO_x/SiO_2	88	13	4

A mechanism of ozone decomposition on MnO_x/Al_2O_3 catalyst has been proposed:

$$O_3 + Mn^{n+} \rightarrow O^{2-} + Mn^{(n+2)+} + O_2 \quad (22)$$

$$O_3 + O^{2-} + Mn^{(n+2)+} \rightarrow O_2^{2-} + Mn^{(n+2)+} + O_2 \quad (23)$$

$$O_2^{2-} + Mn^{(n+2)+} \rightarrow Mn^{n+} + O_2 \quad (24)$$

The presented mechanism consists of electron transfer from the manganese center to ozone, wherethe manganese is reduced by desorption of peroxide particles to form oxygen ($O_2^{2-} \rightarrow O_2 + 2e^-$).

Study of MnO_x/Al_2O_3 with absorption of fine-structure X-ray spectroscopy is also presented in publication.[95] The aim of the work is to receive important information on the catalytic properties of the supported manganese oxides in oxidation reactions using ozone. The structural changes in the manganese oxide supported on alumina have been detected in the process of catalytic ozone decomposition at room temperature. It has been found that during the ozone decomposition in presence of water vapor, the manganese is oxidized to higher oxidation state. At the same time, the water molecule combines to manganese active center due to cleavage of Mn–O–Al bond. The catalyst is completely regenerated after calcination in oxygen at 723 K.

In article,[96] the influence of nickel oxide in addition to the activity of cement-containing catalyst for ozone decomposition has been studied. The activity of the samples was measured by calculating the rate of decomposition γ that shows the degree of active interactions (leading to decomposition) of ozone molecules with catalytic surface. According toLunin, et al,[6] the expression for γ is:

$$\gamma = \frac{4\omega.\ln\left(C_o / C\right)}{V_t S} \quad (25)$$

where V_t—specific heat velocity of ozone molecules, S—geometric surface of the catalyst, ω—space velocity of the gas stream, C_0 and C—inlet and outlet ozone concentrations.

It has been found that the addition of nickel oxide in the catalyst composition improves its catalytic properties. Upon decomposition of wet ozone is observed; a decrease in the activity of all tested samples, as the calculated values for γ are 2–3 times lower compared with the value obtained after decomposition of dry ozone. On the basis of the measured values for energy of activation ($E_a = 5.9 \pm 0.3$ kJ/mol) of ozone decomposition in the region of high temperatures (300–400 K), and the corresponding values of E_a (15.2 ± 0.4 kJ/mol) in the region of low temperatures, it has been made a conclusion that in the first case, the process goes to the outer diffusion region, whereas in the second,into the inner diffusion region. The factor of diffusion suspension or the accessible part of the surface is estimated to make clear the role of the internal catalytic surface.This has permitted to perform calculation that at room temperature the molecules of ozone enter into the pores of the catalyst at a distance not more than ~10^{-4} cm.

Lin et al.[97] have studied the activity of a series oxide supports and supported metal catalysts with respect to decomposition of ozone in water. From the tested reaction conditions,the activated carbon has showed relatively high activity, while the zeolite support (HY and modernite), Al_2O_3, SiO_2, $SiO_2.Al_2O_3$,and TiO_2 have showed zero or negligible activity. From all supported metal catalysts that aresubmitted to ozone dissolved in water, the noble metals have highest activity in ozone decomposition excepting gold. The metals are deposited on four types of supports (Al_2O_3, SiO_2, $SiO_2.Al_2O_3$, and TiO_2). Highest activity was measured for the catalysts deposited on silica. It has been found that the catalyst containing 3% Pd/SiO_2 is most effective in the reaction of ozone decomposition from all tested samples. A comparison of some indicators for Pd catalysts deposited on different supports is presented in Table 10.6.

TABLE 10.6 Comparison of the Specific Surface Areas, the Size of Metal Particles and the Average Rates of Ozone Decomposition in Water Over Palladium-Containing Catalysts.

Catalyst	Specific surface area (m^2 g^{-1})	Size of metal particle (Å)	Average rate ($mg_{(O_3)}$ min^{-1} $g^{-1}_{(cat.)}$)
Pd/SiO_2	206	90	0.77
Pd/$SiO_2.Al_2O_3$	221	70	0.54
Pd/Al_2O_3	139	75	0.39
Pd/TiO_2	34	109	0.35

The first-order reaction of ozone decomposition on 3% Pd/$SiO_2$with respect toconcentration of ozonehas been determined. In the presence of the same catalyst, the calculated activation energy is about 3 kcal mol^{-1} and is assumed that the reaction occurs in the diffusion region. The proposed mechanism of catalytic decomposition of ozone in water is similar to the mechanism of decomposition of gaseous ozone. In Table 10.7, two possible reaction paths of ozone decomposition on metals or oxidesare shown, depending on the fact thatthe oxygen is adsorbed on the catalytic surface.

TABLE 10.7 Possible Mechanisms of Catalytic Ozone Decomposition in Water.[97]

Case of O_2 not adsorbed on metal	Case of O_2 adsorbed on metal
$O_3 \rightarrow O_{3(a)}$	$O_3 \rightarrow O_{3(a)}$
$O_{3(a)} \rightarrow O_{(a)} + O_2$	$O_{3(a)} \rightarrow O_{(a)} + O_{2(a)}$
$O_{(a)} + O_3 \rightarrow 2O_2$	$O_{(a)} + O_3 \rightarrow O_2 + O_{2(a)}$ $O_{2(a)} \rightarrow O_2$

In the literature,data are there too for the study of ozone decomposition in water in presence of aluminum (hydroxyl) oxide.[98] It has been suggested that the surface hydroxyl groups and the acid–base properties of aluminum (hydroxyl) oxides play important role in catalytic decomposition of ozone.

The environmental application of ozone in catalysis has been demonstrated in article,[99] devoted to the ozonation of naproxen and carbamazepine on titanium dioxide. The experiments were carried out in aqueous solution at $T = 25°C$ and in pH range of 3–7. The results have been indicated that naproxen and carbamazepine are completely destructed in the first few minutes of the reaction. The degree of mineralization during the non-catalytic reaction flow is measured up to about 50% and is formed primarily in the first 10–20 min. The presence of the catalyst is increased to more than 75% the degree of mineralization of the initial hydrocarbon. Furthermore, it has been found that the catalyst increases the mineralization in both acid and neutral solution, as the best results are obtained at slightly acidic media. This effect is related with possible adsorption of intermediate reaction products on Lewis acidic catalytic sites. It is also reflected that the titanium dioxide catalyzes the ozone decomposition in acidic media, whereas in neutral solution the ozone destruction is inhibited. This has precluded the flow of mechanism based on surface formation of hydroxyl radicals followed by their migration and complete reaction with the organic compounds. The variation of the total organic carbon is modeled as a function of the integral of the applied amount of ozone. On this basis, it is assumed that the reaction between organic compounds and ozone is of second order. For naproxen and carbamazepine,the calculated pseudo-homogeneous catalytic rate constants at pH 5 and catalyst amount of 1 g/l are $7.76 \times 10^{-3} \pm 3.9 \times 10^{-4}$ and $4.25 \times 10^{-3} \pm 9.7 \times 10^{-4}$l mmol^{-1} s^{-1}, respectively. The products of ozonation are investigated with a specific ultraviolet absorption at 254 nm. The wide application of metal oxide catalysts in ozone decomposition necessitated the use of different instrumental methods for analysis. Based on the results of X-ray diffraction, X-ray photoelectron spectroscopy, EPR, and TPD, it has been found that during the destruction of ozone on silver catalyst supported on silica, the silver is oxidized to complex mixture of Ag_2O_3 and AgO.[100] This investigation of catalytic ozone decomposition on Ag/SiO2 is carried out in the temperature range of–40°C to +25°C, it is determined as firstorder of reaction, and the calculated activation energy is 65 kJ/mol. X-ray diffraction has also been used to determine the phase composition of manganese oxide catalysts supported on γ-Al_2O_3 and SiO_2.[101] Moreover, the samples are characterized by Raman and IR spectroscopy. The supported catalysts are prepared by nitrate precursors using the impregnation method. In addition to the Raman spectral bands of β-MnO_2 and α-Mn_2O_3 phases, other signals are also registered and attributed to isolated Mn^{2+} ions present in tetraedrical vacations on the support surface, and in some epitaxial layerson γ–Mn_2O_3 and manganese silicate, respectively. The data from the IR spectra was not so much useful due to the fact that the supporter band overlaps the bands of manganese particles formed on the surface, and make it difficult to identify them.

Eynaga et al.[102] carried out catalytic oxidation of cyclohexane with ozone on manganese oxide supported on aluminum oxide at the temperature of 295 K. "In situ" IR studies have been performed for taking information on the intermediates formed on the catalytic surface at the time of oxidation. It has been found that the intermediates are partially oxidized to alcohols, ketones, acid anhydrides, and carboxylic acids. These compounds are subsequently decomposed by ozone. In the beginning, the activity of the catalyst gradually decreases whereit reaches a steady state with mole fractions of CO and CO_2,90 and 10%, respectively. It has been found that high-resistant particles, containing C=O, COO–,and CH groups, remain on the catalyst surface that caused the slow deactivation of the catalyst. The C=O groups are decomposed at relatively low temperatures (<473 K), while the COO–and CH groups are dissociated at temperatures>473 K.

The kinetics of gas-phase ozone decomposition was studied by heterogeneous interactions of ozone with aluminum thin films.[103] The ozone concentrations were monitored in real time using UV absorption spectroscopy at 254 nm. The films were prepared by dispersion of fine alumina powder in methanol, and their specific surface areas are determined by "in situ" adsorption of krypton at 77 K. It has been found that the reactivity of alumina decreases with increasing the ozone concentration. As consequence of multiple exposures to ozone of one film, it has been found that the number of active sites is greater than 1.4×10^{14} per cm^2 surface or it is comparable with the total number of active sites. The coefficients of ozone decomposition are calculated (γ) depending on the initial concentration of ozone in the reaction cell, using the expression:

$$\gamma = (4 \times k^{I} \times V)/(c \times SA) \tag{26}$$

where$c$ is the average velocity of gas-phase molecules of ozone, k^{I} is the observed initial rate constant of ozone decomposition from first order (for the first 10 s), SA is the total surface area of the aluminum film, and V is volume of the reactor. From the results, it was concluded that the coefficients of ozone decomposition on fresh films depend inversely on ozone concentration, ranging from 10^{-6} to 10^{14} molecules/cm^3 to $10^{-5}$to 10^{13} molecules/cm^3 ozone. It has also been observed that the coefficients of ozone decomposition do not depend on the relative humidity of gas stream. A mechanism of the reaction of catalytic ozone decomposition, based on detailed spectroscopic investigations of the catalytic surface,[104,105] where there is evidence for the formation of peroxide particles on the surfacewas discussed:

$$O_3 + O^* \longrightarrow O_2 + O^* \tag{27}$$

$$O_3 + O^* \longrightarrow O_2^* + O_2 \tag{28}$$

$$O_2^* \longrightarrow O_2 + * \tag{29}$$

It has been suggested that molecular oxygen could also be initiated by the reaction:

$$O^* + O^* \rightarrow 2^* + O_2 \quad (30)$$

It has been found that the reactivity of the oxidized aluminum film can be partially restored after being placed for certain period in a medium free of ozone, water vapor, and carbon dioxide.

10.4.3.1 DECOMPOSITION OF OZONE ON THE SURFACE OF THE CARBON FIBER AND INERT MATERIALS

The use of carbon material in adsorption and catalysis is related to their structural properties and surface chemical groups. Their structural properties are determined by the specific surface area and porosity, while the chemical groups on the surface of the catalyst are mainly composed of oxygen-containing functional groups. Ozone reacts with various carbon materials such as activated carbon, carbon black, graphite, carbon fiber, and so on. The character of these interactions depends both on the nature of the carbon surface and temperature.[106]Two types of interactions such as the complete oxidation leading to formation of gaseous carbon oxides and partial oxidation, producing surface oxygen-containing functional groupshas been tested.[106,107] With increase inthe temperature, the ratio between these two processes is displaced in the direction of complete oxidation. The latter is accompanied by an oxidative destruction on the surface with formation of gaseous carbon oxides. In addition, on the surface of the carbon,catalytic decomposition of ozonetakes place. The physicochemical properties of activated carbon treated with ozone, as well as the kinetics of the process related to the release of CO and CO_2have been studiedin a number of publications.[108–111] Subrahmanyam et al.[112] have examined the catalytic decomposition of ozone to molecular oxygen on active carbon in the form of granules and fibers at room temperature. It was found that the dynamics of the activity of the carbons is characterized by two distinct zones. The first one is the observed high activity with respect to the decomposition of ozone, which is mainly due to the chemical reaction of ozone with carbon. As a result of this interaction on the carbons, oxygen-containing surface groups are formed. Then sharp drop the conversion is registered and transition of the catalyst to a low active zone takes place. In this zone, the decomposition of the ozone to molecular oxygen takes place in a catalytic way. The activities of the carbons in dry environment on one hand and in presence of water vapor and NO_x on the other have been compared. The presence of water vapor has reduced catalytic activity, while the presence of the NO_x has improved activity, due to the change in the carbon surface functional groups. They can be modified in two ways: boiling in dilute nitric acid or thermal treatment at 1273 K in helium media. Positive results were obtained only in the first treatment. The decomposition of ozone with respect to the gasification of carbon with formation of CO_x proceeds with a selectivity of less than 25%. The catalysts were characterized by means of temperature-programmed decomposition of surface functional groups, IR, and X-ray photoelectron spectroscopy. Mechanism of decomposition of ozone on activated carbon has been proposed (Fig.10.6).

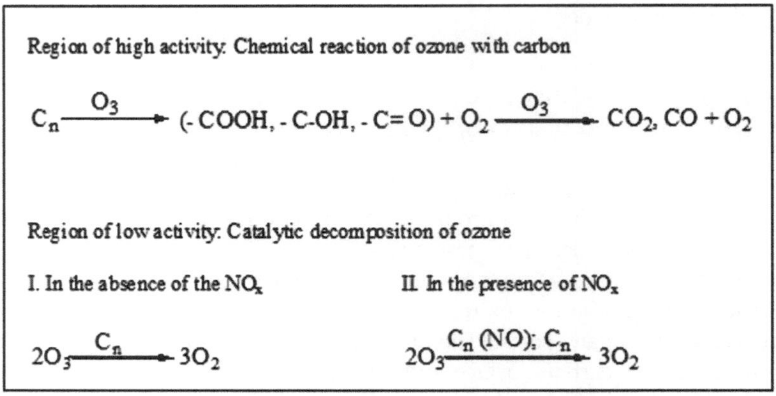

FIGURE 10.6 Simplified scheme of ozone decomposition on carbon.

In ozone decomposition process, it is difficult to draw a line between relatively inert materials and heterogeneous catalysts. However, the separation can be made on the basis of two evident signs: (1) the catalysts for ozone decomposition are specifically synthesized and (2) they have higher catalytic activity (higher values for the coefficient of decomposition γ) compared to the inert materials. In the literature, there are investigations on the reaction of ozone destruction on the surface of silica,[113] glass,[114] volcanic aerosols,[115,116] and ammonium hydrogen sulfate.[117] The published values for the coefficient of ozone decomposition γ for quartz and glass are in the order of $(1–2) \times 10^{-11}$, and those for aerosols and ammonium sulfatefrom 1.61×10^{-6} to 7.71×10^{-8}. It is interesting to mark the studies of ozone decomposition on the surface of Saharan dust[118] as well as the application of slurry-treated water as catalyst for ozone destruction in aqueous medium.[119]

The heterogeneous reaction between O_3 and authentic Saharan dust surfaces[118] was investigated in a Knudsen reactor at ~296 K. O_3 was destroyed on the dust surface and O_2 was formed with conversion efficiencies of 1.0 and 1.3 molecules; O_2 per O_3 molecule was destroyed for unheated and heated samples, respectively. No O_3 desorbed from exposed dust samples, showing that the uptake was irreversible. The uptake coefficients for the irreversible destruction of O_3 on (unheated) Saharan dust surfaces depended on the O_3 concentration and varied between $4.8 \times 10^{-5}$and

2.2×10^{-6} for the steady-state uptake coefficient. At very high O_3 concentrations, the surface was deactivated, and O_3 uptake ceased after a certain exposure period.

New, effective and stable ecological catalyst based on slurry[119] has been used in the process of ozone decomposition in water acidic medium. The catalyst was characterized by X-ray fluorescence, transmission electron microscopy, scanning electron microscopy, and X-ray diffraction. The sludge is essentially composed of different metallic and nonmetallic oxides. It has been investigated the effect of various experimental parameters such as catalyst amount, initial ozone concentration, and application of different metal oxide catalysts. The decomposition of dissolved ozone is significantly increased with the enhancement of the initial ozone concentration and the increment of catalyst amount from 125 to 750 mg. The order of activity of the tested catalysts such as$ZnO\approx$sludge$>TiO_2>SiO_2>Al_2O_3\approx Fe_2O_3$has been established. It has been found that ozone does not affect the catalyst morphology and its composition, and it is concluded that the sludge is promising catalyst for ozone decomposition in water.

In the end of the literature review, it can be concluded that except for the metals of platinum group, characterized by its high price, the metal oxide catalysts containing manganese oxide have the highest activity in decomposition of gaseous ozone and also in catalytic oxidation of pollutants. It is important to mention that unlike the inert materials for the oxide catalysts, there is nostrong dependence of catalytic activity from ozone concentration in gas phase. It should be emphasized that despite the great number of publications on the subject, the kinetics and the mechanism of ozone decomposition on the surface of heterogeneous metal oxide catalysts are not cleared up sufficiently.

KEYWORDS

- **catalysts**
- **synthesis**
- **kinetics**
- **mechanism**

REFERENCES

1. Rakovsky, S.K.; Zaikov, G.E.*Kinetic and Mechanism of Ozone Reactions with Organic and Polymeric Compounds in Liquid Phase;* monograph (2nded.), Nova Sci. Publ., Inc.: New York, 2007; pp 1–340.
2. Schonbein, C.F.*Pogg. Ann.*, **1840**,*49*,616 ; *Helb.Seances Acad. Sci.*, **1840**,*10*, 706.
3. *Ulmann's Encyclopedia of Industrial Chemistry*;A18, 1991; 349.
4. Razumovskii, S.D.; Rakovsky, S.K.; Shopov, D.M.; Zaikov, G.E. *Ozone and Its Reactions with Organic Compounds* (in Russian);Publ. House of Bulgarian Academy of Sciences:Sofia,1983.

5. Brown, T.L.;Eugene LeMay,H. Jr.;Bursten, B.E.;Burdge, J.R.*Chemistry: The Central Science*; 9th ed. (in English), Pearson Education, 2003; pp. 882–883.
6. Lunin, V.V.;Popovich, M.P.;Tkachenko, S.N.Physical Chemistry of Ozone (in Rus.);Publ. MSU, 1998; pp. 1–480.
7. Perry,R.H. and Green,D.*Perry's Chemical Engineer's Handbook;* McGraw-Hill: New York, 1989; pp. 3–147.
8. Thorp, C.*Bibliography of Ozone Technology*;1955;*2*(30).
9. Kutsuna, S.; Kasuda, M.; Ibusuki, T. Transformation and Decomposition of 1, 1, 1-Trechloroethane on Titanium Dioxide in the Dark and Under Photoillumination.*Atmosphere. Environ.*, **1994**, *28* (*9*), 1627–1631 .
10. Naydenov, A.; Stoyanova, R.; Mehandjiev, D. Ozone Decomposition and CO Oxidation on CeO_2.*J. Mol. Catal. A-Chem.*, **1995**, *98*(*1*), 9–14.
11. Rakitskaya, T.L.; Vasileva, E.K.; Bandurko, A.Yu.; Paina, V.Ya. Kinetics of Ozone Decomposition on Activated Carbons.*Kinet. Catal.*, **1994**, *35*(*1*), 90–92.
12. Tanaka, V.; Inn, E.C.J.; Watanabe, K.J.*Chem. Phys.***1953**, *21*(10), 1651 .
13. De More, W.B.; Paper, O.J.*Phys. Chem.***1964**, *68*, 412.
14. Beitker, K.H.; Schurath, U.; Seitz, N.*Int. J. Chem. Kinet.***1974**, *6,* 725.
15. Inn, E.C.J.; Tanaka, V.J.*Opt. Soc. Amer.***1953**, *43*(*10*), 870.
16. Griggs, M.J.*Chem. Phys.***1968**, *49*(*2*), 857.
17. Taube, H.*Trans. Faraday Soc.***1957**, *53,* 656.
18. Galimova, LG.;Komisarov,V.D.;Denisov,E.T.The Reports of Russian Academy of Sciences (in Rus.).*Ser. chem.***1973**, *2,* 307.
19. Alexandrov, Y.A.;Tarunin,B.I.;Perepletchikov, M.L.*J. phys. chem.*(*in Rus.*)**1983**, LVII(10), 2385–2397.
20. Hon, Y.S.; Yan, J.L.The Ozonolytic Cleavage of Cycloalkenes in the Presence of Methyl Pyruvate to Yield the Terminally Differentiated Compounds. *Tetrahedron, Lett.***1993**, *34*(41), 6591–6594.
21. Deninno, M.P.; McCarthy, K.E.The C-14 Radiolabelled Synthesis of the Cholesterol Absorption Inhibitor CP-148,623. A novel method for the Incorporation of a C-14 label in Enones. *Tetrahedron***1997**, *53*(32), 11007–11020.
22. Claudia, C.; Mincione, E.; Saladino, R.; Nicoletti, R. Oxidation of Substituted 2-Thiouracils and Pyramidine-2-Thione with Ozone and 3,3-Dimethyl-1,2-Dioxiran. *Tetrahedron,* **1994**, *50*(10), 3259–3272.
23. Razumovskii,S.D.;Zaikov,G.E.Ozone and its Reactions with Organic Compounds, Publ.Science (in Rus.): Moskow,1974.
24. Chapman, S.*Phil. Mag.***1930**, Ser. 7, 10, 369.
25. Chapman, S.*Met. Roy. Soc.***1930,***3*, 103.
26. Johnston, H.*Rev. Geoph. Space Phys.***1975**, *13,* 637.
27. NASA Reference Publication, 1162, *N. Y. Acad. Press***1986**.
28. Gerchenson,Yu.; Zvenigorodskii, S.; Rozenstein, V.*SuccessChemistry*(*in Rus.*), **1990**, *59,* 1601.
29. Johnston, H.*Photochemistry in the Stratosphere*; UCLA: Berkley, 1975; 20.
30. Crutzen, P.; Smalcel, R.*Planet Space Sci.***1983**, *31,* 1009.
31. Farmen, J.; Gardiner, B.; Shanklin, J.*Nature***1985**, *315,* 207.
32. Solomon, S.; Garcia, R.; Rowland, F.; Wueblles, P.*Ibid***1986**, *321,* 755.
33. Vupputuri, R.*Atm. Environ.***1988**, *22,* 2809.
34. Kondrat'ev, K.*MeteorologyandClimate*(*in Rus.*);1989;*19,* 212.
35. Kondrat'ev, K.*Success Chemistry* (*in Rus.*);1990;*59,* 1587.
36. Heisig, C.; Zhang, W.; Oyama, S.T. Decomposition of Ozone using Carbon Supported Metal Oxide Catalysts. *Appl. Catal. B: Environ.***1997,***14,*117 .

37. Rakitskaya, T.L.; Bandurko, A.Yu.; Ennan, A.A.; Paina, V.Ya.; Rakitskiy, A.S. Carbon-Fibrous-Material-Supported Base Catalysts of Ozone Decomposition. *Micro. Meso. Mater.***2001**, *43,* 153.
38. Skoumal, M.; Cabot, P. L.; Centellas, F.; Arias, C.; Rodriguez, R.M.; Garrido, J.A.; Brillas, E.*Appl. Catal. B Environ.***2006**,*66,* 228–240.
39. Bianchi, C.L.; Pirola, C.; Ragaini, V.; Selli, E.*Appl. Catal. B Environ.***2006**,*64,* 131–138.
40. Ma, J.; Sui, M.H.; Chen, Z.L.; Li, N.W.*Ozone Sci. Eng.***2004**, *26,* 3–10.
41. Zhao, L.; Ma, J.; Sun, Z.Z.*Appl. Catal. B Environ.***2008**,*79,* 244–253.
42. Von Gunten, U.*Water Res.***2003**, *37,* 1443–1463.
43. Monhot, W.; Kampschulte, W.*Ber.***1907**, *40,* 2891.
44. Kashtanov, L.; Ivanova, N.; Rizhov, B.;*J.Applied Chemistry* (*in Rus.*), **1936**, *9,* 2176.
45. Schwab, G.; Hartmanm,C.*J. Phys. Chem.*(*in Rus.*)**1964**, *6,* 72.
46. Emel'yanova, G.; Lebedev, V.; Kobozev, N.*J. Phys. Chem.* (*in Rus.*)**1964**, *38,* 170.
47. Emel'yanova, G.; Lebedev, V.; Kobozev, N.*J. Phys. Chem.* (*in Rus.*)**1965**, *39,* 540.
48. Emel'yanova, G.; Strakhov, B.*Advanced Problems Physical Chemistry* (*in Rus.*);1968;*2,* 149.
49. Sudak, A.; Vol'fson, V. Catalytic Ozone Purification of Air, *Scientific Notion*(*in Rus.*); Kiev, 1983;87.
50. Hata, K.; Horiuchi, M.; Takasaki, T. Jap. Pat.*CA*, 108, 61754u,1988.
51. Tchihara, S. Jap. Pat.*CA*, 108, 192035h,1988.
52. Kobayashi, M.; Mitsui, M.; Kiichiro, K. Jap. Pat.*CA*, 109, 175615a,1988.
53. Terui, S.; Sadao, K.; Sano, N.; Nichikawa, T. Jap. Pat.*CA*, 112, 20404p,1990.
54. Terui, S.; Sadao, K.; Sano, N.; Nichikawa, T. Jap. Pat.*CA*, 114, 108179b,1991.
55. Oohachi, K.; Fukutake, T.; Sunao, T. Jap. Pat.*CA*, 119, 119194g,1993.
56. Kobayashi, M. and Mitsui, K. Jap. Pat. 63,267,439, Nov 4, 1988, to Nippon Shokubal Kagaku Kogyo Co., Ltd.
57. Kuwabara, H. and Fujita, H. Jap. Pat. 3016640 Jan 24, 1991, to Mitsubishi Heavy Industries, Ltd.
58. Hata, K.; Horiuchi, M.; Takasaki, K. and Ichihara, S. Jap. Pat. 62,201,648, Sep 5, 1987, to Nippon Shokubai Kagaku Kogyo Co., Ltd.
59. Terui, S.; Miyoshi, K.; Yokota, Y. and Inoue, A. Jap. Pat. 02,63,552, Mar 2, 1990.
60. Yoshimoto, M.; Nakatsuji, T.; Nagano, K. and Yoshida, K. Eur. Pat. 90,302,545.0, Sep 19, 1990, to Sakai Chemical Industry Co., Ltd.
61. Oyama, S.T., Chemical and Catalytic Properties of Ozone. *Catal. Rev. Sci. Eng.***2000**, *42,* 279.
62. Einaga, H.; Futamura, S. Comparative Study on the Catalytic Activities of Alumina-supported Metal Oxides for Oxidation of Benzene and Cyclohexane with Ozone. *React. Kinet. Catal. Lett.***2004**, *81*, 121.
63. Tong, S.;Liu, W.;Leng, W.;Zhang, Q. Characteristics of MnO_2 Catalytic Ozonation of Sulfosalicylic Acid Propionic Acid in Water. *Chemosphere***2003**, *50,* 1359.
64. Konova, P.;Stoyanova, M.;Naydenov, A.;Christoskova,S.T.;Mehandjiev, D. Catalytic Oxidation of VOCs and CO by Ozone Over Alumina Supported Cobalt Oxide. *J. Appl. Catal. A: Gen.* **2006**,*298,*109 .
65. Stoyanova, M.;Konova, P.;Nikolov, P.;Naydenov, A.;Christoskova, S.T.;Mehandjiev, D. Alumina-Supported Nickel Oxide for Ozone Decomposition and Catalytic Ozonation of CO and VOCs. *Chem. Eng. J.***2006**,122, 41.
66. Popovich, M.; Smirnova, N.; Sabitova, L.; Filipov, Yu.*J. of Moskow Univerity*(*in Rus.*), *Ser. Chem.***1985**,*26,* 167.
67. Popovich, M.*J. of Moskow Univerity*(*in Rus.*), *Ser. Chem.***1988**,*29,* 29.
68. Radhakrishnan, R.; Oyama, S.T.; Chen, J.; Asakura, A. Electron Transfer Effects in Ozone Decomposition on Supported Manganese Oxide. *J. Phys. Chem. B***2001**, *105*(19), 4245.

69. Zavadskii, A.V.;Kireev, S.G.;Muhin, V. M.;Tkachenko, S. N.;Chebkin,V.V.;Klushin, V.N.;Teplyakov, D.E. Thermal Treatment Influence over Hopcalite Activity in Ozone Decomposition. *J. Phys. Chem.* (*in Rus.*)**2002**, *76*, 2278.
70. Imamura, S.; Ikebata, M.; Ito, T.; Ogita, T.*Ind. Eng. Chem. Res.***1991**,*30*, 217.
71. Dhandapani, B., Oyama, S.T.*J. Appl. Catal. B: Environ.***1997**, *11*, 129.
72. Lo Jacono, M.; Schiavello, M.TheInfluence of Preparation Methods on Structural and Catalytic Properties of Transition Metal Ions Supported on Alumina. *In Preparation of catalysts I*; DelmonB., JacobsP., Poncelet G.,Eds.;*Elsevier*: New York, 1976; 473.
73. Yamashita, T.; Vannice, A.*J. Catalysis***1996**,*161*, 254.
74. Ma, J.; Chuah, G.K.; Jaenicke, S.; Gopalakrishnan, R.; Tan, K.L.*Ber.Bunsenges. Phys. Chem***1995**, *100*, 585.
75. Ma, J.; Chuah, G.K.; Jaenicke, S.; Gopalakrishnan, R.; Tan, K.L.*Ber.Bunsenges. Phys. Chem***1995**, *99*, 184.
76. Baltanas, M. A.; Stiles, A.B.; and Katzer, J.R.*Appl. Catal.***1986**,28, 13.
77. Li, W.; Oyama, S.T. in *Heterogeneous Hydrocarbon Oxidation*;Warren B. K. and Oyama S. T.,Eds.;*ACS Symp. Ser.* 638, ACS: Washington, DC, 1996, 364.
78. Naydenov, A.; Mehandjiev, D.*Appl. Catal. A.: General.***1993**, *97*, 17 .
79. Einaga, H.; Ogata, A. Benzene Oxidation with Ozone OverSupported Manganese Oxide Catalysts: Effect of Catalyst Support and Reaction Conditions.*J. Hazard. Mater.***2008**, *164*, 1236.
80. Boreskov, G. K.*Adv. Catal.***1964**,*15*, 285.
81. Baldi, M.; Finochhio,E.; Pistarino, C.; Busca, G.*J. Appl.Catal. A: General***1998**, *173*, 61.
82. Kapteijn, F.; Singoredjo, L.; Andreini, A.; Moulijn, J.A.*J. Appl.Catal. B: Environ.***1994**, *3*, 173.
83. Maltha, A.; Favre, L.F.T.; Kist, H.F.; Zuur, A.P.; Ponec, V.*J. Catal.***1994**,*149*, 364.
84. Hunter, P.; Oyama, S.T.*Control of Volatile Organic Compound Emissions*; John Wiley & Sons, Inc.: New York,2000.
85. Subrahmanyam, Ch.; Renken, A.; Kiwi-Minsker, L. Novel Catalytic Non-Thermal Plasma Reactor for the Abatement of VOCs.*Chem. Eng. J.***2007**, *134*, 78.
86. Rubashov, A.M.; Pogorelov, V.V.; Strahov, B.V.*J. Phys. Chem.* (*in Rus.*)**1972**, *46* (9), 2283.
87. Rubashov, A.M.; Strahov, B.V.*J. Phys. Chem.* (*in Rus.*)**1973**,*47*(8), 2115.
88. Ellis, W.D.; Tomets, P.V.*Atmospheric Environment Pergamon Press.***1972**,*6*(10), 707.
89. Houzellot, J.Z.; Villermaux, J.*J. de Chemie Physique***1976**, *73*(7–8), 807.
90. Tarunin, B.I.; Perepletchikov, M.L.; Klimova, M.N.*Kinetic Catal.***1981**, *22*(2), 431.
91. Rakovsky, S.; Nenchev, L.; Cherneva, D.*Proc. 4th Symp. Heterogeneous Catalysis*; Varna, 1979, *2*, 231.
92. Che, M. and Tench A.J., *Adv. Catal.***1982**, 31, 77.
93. Tench, A.J. and Lawson, T.*Chem. Phys. Lett.***1970**, 459.
94. Martinov, I.; Demiduk, V.; Tkachenko, S.; Popovich, M.*J. Phys. Chem.* (*in Rus.*)**1994**, *68*, 1972.
95. Einaga, H.; Harada, M.; Futamura, S. Structural Changes in Alumina-supported Manganese Oxides during Ozone Decomposition, *Chem. Phys. Lett.***2005**, *408*, 377.
96. Martinov, I.V.; Tkachenko, S.N.; Demidyuk, V.I.; Egorova, G.V.; Lunin, V.V. NiO Addition Influence over Cement-containing Catalysts Activity in Ozone Decomposition. *J. of Moscow Univ.* (*in Rus.*)*Ser. Chem.***1999**, *40*, 355.
97. Lin, J.; Kawai, A.; Nakajima, T. Effective Catalysts for Decomposition of Aqueous Ozone. *Appl. Catal. B: Environ.***2002**,*39*, 157.
98. Qi, F.; Chen, Z.; Xu, B.; Shen, J.; Ma, J.; Joll, C.; Heitz, A. Influence of Surface Texture and Acid-Base Properties on Ozone Decomposition Catalyzed by Aluminum (Hydroxyl) Oxides. *Appl. Catal. B: Environ.***2008**,*84*, 684.

99. Rosal, R.; Rodriguez, A.; Gonzalo, M.S.; Garcia-Calvo, E. Catalytic Ozonation of Naproxen and Carbamazepin on Titanium Dioxide.*Appl. Catal. B: Environ.***2008**, *84,* 48.
100. Naydenov, A.; Konova, P.; Nikolov, P.; Klingstedt, F.; Kumar, N.; Kovacheva, D.; Stefanov, P.; Stoyanova, R.; Mehandjiev, D. Decomposition of Ozone on Ag/SiO_2Catalyst for Abatement of Waste Gases Emissions.*Catal. Today***2008,***137,* 471.
101. Buciuman, F.; Patcas, F.; Craciun, R.; Zhan, D.R.T. Vibrational Spectroscopy of Bulk and Supported Manganese Oxides.*Phys. Chem. Chem. Phys.***1998**, *1,* 185.
102. Einaga, H.; Futamura, S. Oxidation Behavior of Cyclohexane on Alumina-Supported Manganese Oxide with Ozone.*Appl. Catal. B: Environ.***2005**, *60,* 49.
103. Sullivan, R.C.; Thornberry, T.; Abbatt, J.P.D. Ozone Decomposition Knetics on Alumina: Effects of Ozone Partial Pressure, Relative Humidity and Repeated Oxidation Cycles.*Atmos. Chem. Phys.***2004,***4,* 1301.
104. Li, W.; Gibbs, G.V.; Oyama, S.T. Mechanism of Ozone Decomposition on Manganese Oxide: 1. In situ Laser Raman Spectroscopy and ab initio Molecular Orbital Calculations.*J. Am. Chem. Soc.***1998,***120,* 9041.
105. Li, W.; Oyama, S.T. The Mechanism of Ozone Decomposition on Manganese Oxide: 2. Steady-state and Transient Kinetic Studies. *J. Am. Chem. Soc.***1998**, *120,* 9047.
106. Atale, Hitoshi, Kaneko, Taraichi, Yano, Jap. Pat.*CA*, 123, 121871,1995.
107. Mori, Katsushiko, Hasimoto, Akira, Jap. Pat.*CA*, 118, 153488v,1993.
108. Kobayashi, Motonobu, Mitsui, Kiichiro, Jap. Pat.*CA*, 110, 120511d,1989.
109. Rakitskaya, T.L.; Vasileva, E.K.; Bandurko, A.Yu.; Paina, V.Ya.*Kinet. Catal.***1994,***35,* 103.
110. Aktyacheva, L.; Emel'yanova, G.*J. Phys. Chem.* (*in Rus.*)*Ser. Chem.***1990,***31,* 21.
111. Valdes, H.; Sanches-Polo, M.; Rivera-Utrilla, J.; Zaror, C.A. Effect of Ozone Treatment on Surface Properties of Activated Carbon.*Langmuir***2002**, *18,* 2111.
112. Subrahmanyam, C.; Bulushev, D.A.; Kiwi-Minsker, L. Dynamic Behaviour of Activated Carbon Catalysts DuringOzone Decomposition at Room Temperature.*Appl. Catal. B: Environ.***2005,***61,* 98.
113. Tkalich, V.S.; Klimovskii, A.O.; Lissachenko, A.A.*Kinet. Catal.***1984,***25*(5), 1109.
114. Olshina, K.; Cadle, R.D.; DePena, R.G.*J. Geoph. Res.***1979**, *84*(4), 1771.
115. Popovich, M.P.; Smirnova, N.N.; Sabitova, L.V.*J. Phys. Chem.* (*in Rus.*), *Ser. Chem.***1987,***28*(6), 548.
116. Popovich, M.P.*J. Phys. Chem.* (*in Rus.*)*Ser. Chem.***1988,***29*(5), 427.
117. Egorova, G.V.; Popovich, M.P.; Filipov, Yu.V.*J. Phys. Chem.* (*in Rus.*)*Ser. Chem.***1988,***29*(4), 406.
118. Hanisch, F.; Crowley, J.N. Ozone Decomposition on Saharan Dust: an Experimental Investigation.*Atmos. Chem. Phys. Discuss.***2002**, *2,* 1809.
119. Muruganadham, M.; Chen, S.H.; Wu, J.J. Evaluation of Water Treatment Sludge as a Catalyst for Aqueous Ozone Decomposition.*Catal. Commun.***2007,***8,* 1609.

CHAPTER 11

OZONE DECOMPOSITION ON THE SURFACE OF METAL OXIDE CATALYST

T. BATAKLIEV, V. GEORGIEV, M. ANACHKOV, S. RAKOVSKY, A. BERLIN[1], and G. E. ZAIKOV[2]

Institute of Catalysis, Bulgarian Academy of Sciences, Bonchev Str., bl.11, #218, 1113 Sofia, Bulgaria. E-mail: todor@ic.bas.bg

[1]N. N. Semenov Institute of Chemical Physics, Russian Academy of Sciences, 4 Kosygin str., Moscow 119991, Russia. E-mail: Berlin@chph.ras.ru

[2]N. M. Emanuel Institute of Biochemical Physics, Russian Academy of Sciences, 4, Kosygin str., Moscow 119334, Russian Federation. E-mail: Chembio@sky.chph.ras.ru

CONTENTS

ABSTRACT

The catalytic decomposition of ozone to molecular oxygen over catalytic mixture containing manganese, copper, and nickel oxides was investigated in the present work. The catalytic activity was evaluated on the basis of the decomposition coefficient γ, which is proportional to ozone decomposition rate and has been already used in other studies for catalytic activity estimation. The reaction was studied in the presence of thermally modified catalytic samples operating at different temperatures and ozone flow rates. The catalyst changes were followed by kinetic methods, surface measurements, temperature programmed reduction, and IR-spectroscopy. The phase composition of the metal oxide catalyst was determined by X-ray diffraction. The catalyst mixture has shown high activity in ozone decomposition at wet and dry O_3/O_2 gas mixtures.

11.1 INTRODUCTION

Ozone finds wide application in such important industrial processes like: purification of drinking water, bleaching of textiles, oxidation of sulfurous gas, complete oxidation of exhaust gases from production of nitric acid, and production of many organic compounds.[1] Ozone in the atmosphere protects the Earth's surface against UV-radiation, but on the ground level it is an air contaminant.[1–3] At this level, ozone can be removed by adsorption, absorption, thermal, and catalytic decomposition. The most effective catalysts for ozone decomposition are based on manganese oxide.[4–7] The main method for purification of waste gases containing residual ozone is the heterogeneous catalytic decomposition. Noble metals like Pt, Ag, Pd, and transition metal oxides including Co, Cu, and Ni supported on γ-Al_2O_3, SiO_2, and TiO_2 also are effective catalysts in this reaction,[8–14] as it can be mentioned for activated carbon fibers.[15]

The decomposition of ozone is a thermodynamically favored process with a heat of reaction of $\Delta H^0_{298} = -138$ kJ/mol and free energy of reaction of $\Delta G^0_{298} = -163$ kJ/mol.[16] The ozone structure is resonance stabilized, which is the reason for its relative stability. The coefficient of ozone decomposition γ was used in other studies for investigation of NiO addition influence over cement-containing catalysts activity,[17] and for the study of thermal treatment influence over oxide catalyst activity.[18]

The aim of present study is to apply mixed metal oxide catalyst for ozone decomposition to investigate its behavior at different conditions and to determine its composition and surface properties using different physical methods for analysis.

11.2 EXPERIMENTAL

The basic copper, manganese, nickel carbonates, and clay-bearing cement are milled in advance, then carefully mixed, crushed, and compressed under pressure

4 t/cm^2. The resulting tablets were treated hydrothermally at temperature of 80°C for 6 h, dried at 120°C for 6 h, and calcinated at 420°C for 6 h. The metal oxide catalyst based on the mixture of manganese oxide (20 wt%), copper oxide (10 wt%), nickel oxide (30 wt%), and clay-bearing cement (40 wt%) was thermally modified at 500°C for 2 h and finally was applied in our investigation as catalyst for ozone decomposition in dry and water-enriched gas flows. The catalyst was granulated and contained cylindrical grains with a diameter of about 5 mm and thickness of 3 mm.

The reactor for kinetic measurements was a glass tube (6 × 150 mm) filled in with 0.08–0.12 g of catalyst. Figure 11.1 shows the schematic of the experimental set-up for all kinetics.

The kinetic measurements of ozone degradation were performed at flow rates ranging from 6.0 to 24 l h^{-1} and ozone concentration from 1.0 to 1.2 mM. Ozone was generated by passing dry oxygen through a high-voltage silent-discharge ozone generator. About 1 mM ozone concentration was achieved at 15–20 kV. The inlet and outlet ozone concentrations were monitored using an UV absorption-type ozone analyzer at 300 nm.

The specific surface area of the catalyst (72 m^2/g) was measured by N_2 adsorption–desorption isotherms at 77 K using BET method in a FlowSorb 2300 instrument (Micromeritics Instrument Corporation). IR studies were performed in the transmittance mode using a Nicolet 6700 FT-IR spectrometer (Thermo Electron Corporation). A mixture of KBr and manganese oxide catalyst (100:1) was milled in an agate mortar manually before the preparation of pellets. The spectra were obtained by averaging 50 scans with 0.4 cm^{-1} resolution.

A typical TPR experiment is done by passing a H_2 stream over a catalyst while it is heated linearly and monitoring the consumption of H_2 with a thermal conductivity detector or mass spectrometer. In our study, a 10% H_2/Ar mixture was used and the consumption of H_2 was monitored using a thermal conductivity detector. A linear heating rate of 0.17 K s^{-1} was used for the experiment. X-ray diffraction (XRD) analysis was used to determine the crystalline metal oxide phases for the supported catalyst. A Bruker D8 Advance powder diffractometer with Cu Kα radiation source and SolX detector was used. The samples were scanned from 2θ angles of 10–80° at a rate of 0.04° s^{-1}. The X-ray power operated with a current of 40 mA and a voltage of 45 kV.

11.3 RESULTS

The catalytic activity was evaluated on the basis of the coefficient [19] that is proportional to ozone decomposition rate and to catalyst efficiency. It has been already used in other studies.[17,18]

$$\gamma = \frac{4\omega}{V_t S} \ln \frac{[O_3]_0}{[O_3]},$$

where ω is the flow rate, V_t—specific heat rate of ozone molecules, S—geometrical surface of catalyst sample, and $[O_3]_0$ and $[O_3]$—inlet and outlet ozone concentrations, respectively.

In general, there exists no precise estimation of γ by solving the diffusion-kinetic equation. This is possible in some special cases, for example, to find γ using the approximate method of Frank-Kamenetsky (method of equally accessible surface).[20] Equally accessible surface is that surface where, in each section, the molecules fall with equal probability. The rate of the chemical reaction on the surface is expressed by concentration of reacting molecules in the volume near the surface. For reactions of first order:

$$W_s = K_s C_s = KC'.$$

The parameters dimension is as follows: w_s—(molecules/cm^2.s), κ_s и κ—(s^{-1}) и (cm/s), c_s и c'—(molecules/cm^2) и (molecules/cm^3). It has been suggested that the molecular flow from volume to surface does not depend on the reaction rate, and with approximation it is defined of the equation:

$$J = \beta(C - C'),$$

where β—coefficient of mass transfer, having dimension as the rate constant *k*, equal to cm s^{-1}, *c* and *c'*—concentration in the regions of the volume, where the flow is passing through.

The distance between the surface and the region with concentration c', the ozone molecules pass without collisions with average specific heat rate v_T. The number of hits on unit of surface per unit of time $z = v_T c'$ and taking into account the definition of coefficient of ozone decomposition, it has been found that:

$$\gamma = \frac{kc'}{z} = \frac{4k}{v_T}.$$

Thus, the coefficient γ is related to the rate constant *k*. Now we could consider the case when the surface, where the reaction takes place, is located in an unlimited volume of gas. In stationary conditions, the molecular flow toward the surface is equal to the chemical reaction rate:

$$w_s = \frac{\beta kc}{k + \beta} = k_{\text{eff}} c,$$

where *c*—concentration of actives molecules standing to great distance from the catalytic surface.

Thus, the rate of reaction on the surface is expressed by the concentration in the volume and the effective rate constant that depends on the rate constant *k* and the coefficient of mass transfer β, obviously:

$$\frac{1}{k_{\text{eff}}} = \frac{1}{k} + \frac{1}{\beta}.$$

If $\beta >> k$, then $c' = c$ and $k_{\text{eff}} = k$: the total reaction rate is limited by the no hits stage with constant *k*. In this case, the reaction proceeds in the kinetic region. If $\beta << k$, $c' << c$, and $k_{\text{eff}} = \beta$, the reaction rate is determined by the rate of mass transfer and the reaction occurs in the diffusion region.

On the other hand,[19] when operating in stationary conditions, the relationship between the rate constant of the reaction of heterogeneous ozone decomposition and the coefficient of ozone decomposition γ is given by the formula:

$$\gamma = \frac{4\omega}{Sv_T} \ln \frac{[O_3]_0}{[O_3]},$$

where *V*—volume of reactor, *S*—geometric surface of catalyst, and v_T—specific heat rate of ozone molecules.

Taking into account, the dependence of the gas flow rate from the reactor radius, and considering the concentration of active ozone molecules and the reaction time of ozone decomposition, after complex mathematical transformations, the latter formula passes into the expression:

$$\gamma = \frac{4\omega}{Sv_T} \ln \frac{[O_3]_0}{[O_3]}.$$

where ω—gas flow rate, $[O_3]_0$ и $[O_3]$—inlet and outlet ozone concentrations.

This formula is applicable for tubular type reactor, when the catalyst is supported as thin layer on the walls in the inner side of the tube. The expression is also convenient for calculation of catalytic activity in the case when the reactor is filled with granulated catalytic samples having specific geometric surface. Therefore, this method has been used by us for calculation of the catalysts activity in the process of ozone decomposition.

11.4 DISCUSSION

Figure 11.1 shows the changes occurring in catalytic activity of the cement-containing catalyst, when the calcination temperature of the samples is different.

All the experiments were made in dry conditions and ozone flow rate of 6 l h^{-1}. The $MnO_x/CuO/NiO$ catalyst has catalytic activity in ozone decomposition that does not change dramatically with an increase in calcination temperature. However, we can see that the catalyst is more active when the calcination temperature is in the range of 400–500°C. In Figure 11.2, changes of γ are shown at two different flow rates in temperature range of 258–323 K. The calculated activation energy is 5 kJ/mol. In these experiments, the reactor was kept up at constant temperature for enough time to get the necessary value. The duration of the decomposition reaction and the reaction time were much smaller than the time of cooling or heating the reactor. This means that the temperature inside the reactor was maintained constant during the measurement of ozone decomposition rate. The difference between the values of γ at different flow rates is due to the low loading of catalyst. The low values of E_a are directly connected with the fact that limited stage of reaction is the adsorption of ozone on catalytic surface.

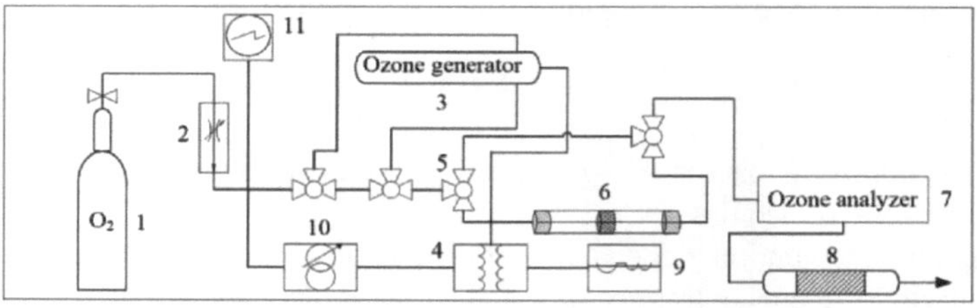

FIGURE 11.1 Experimental set-up of reaction system for catalytic decomposition of ozone: 1—oxygen; 2—flow controller; 3—ozone generator; 4—transformer; 5—three way turn cock; 6—reactor charged with catalyst sample; 7—ozone analyzer; 8—reactor for decomposition of residual ozone; 9—current stabilizer; 10—autotransformer; 11—voltmeter.

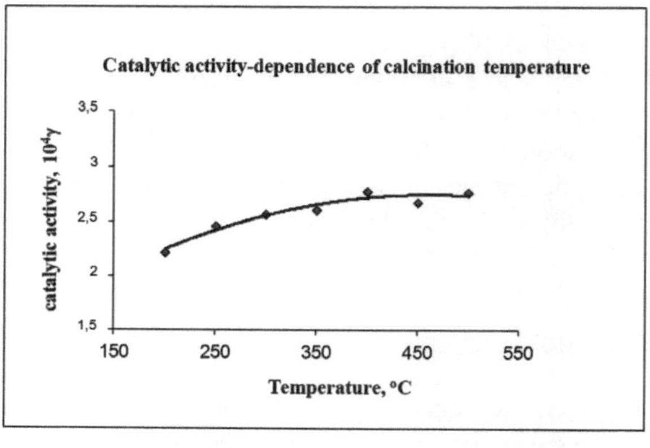

FIGURE 11.2 Dependence of the catalytic activity in ozone decomposition on calcination temperature.

In Figure 11.3, changes of γ are presented depending on temperature and humidity of the gas flow. The measured activation energy was also 5 kJ/mol. It was found out that the humidity of the gas flow decreases the catalytic activity by 10%, but the catalyst does not disturb its stability in the time of ozonation.

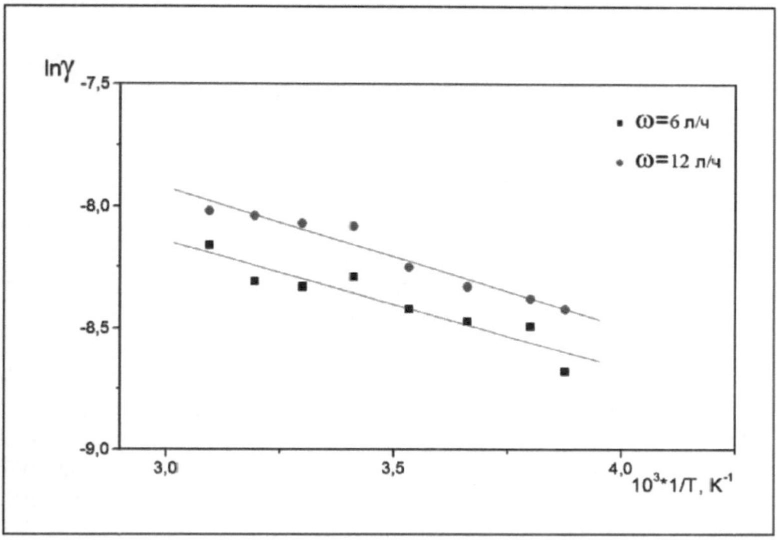

FIGURE 11.3 Dependence of the catalytic activity in ozone decomposition on reaction temperature at two different flow rates.

Figure 11.4 presents the changes of γ at different flow rates at two temperatures, −258 and 298 K. It can be seen that γ is proportional to the flow rate at both temperatures. The coefficient of decomposition depends on temperature, but at 258 K, the steady state of the curve is reached faster. The obtained values of γ are close to the coefficients reported in literature.[17] Therefore, it could be concluded that at low values of gas flow rate, a reason for the dependence of the catalytic activity from gas flow rate is the influence of the external diffusion over the kinetics of the process of heterogeneous catalytic ozone decomposition on catalyst surface, that is, the process takes place mainly in the outer diffusion region or in the transition diffusion-kinetic region.

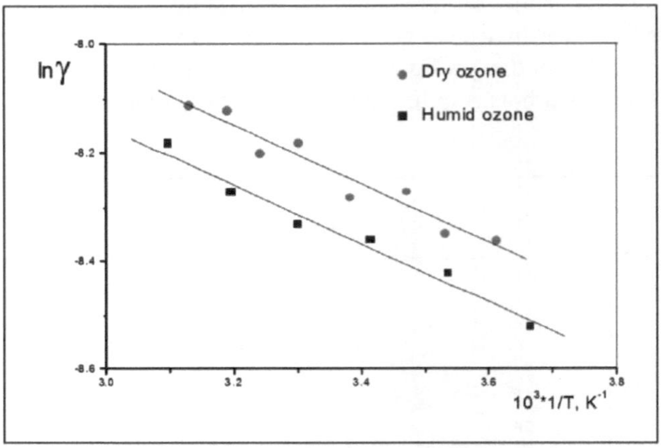

FIGURE 11.4 Temperature dependence of catalytic activity at dry and humid conditions, temperature range 273–323 K, ozone flow rate 8 l h^{-1}.

Figure 11.5 shows the dependence of γ on gas flow humidity at 298 K. The values of γ decrease with increase in the humidity, but nevertheless these values remain relatively high. The effect of water vapor may be result of thin film formation on the catalytic surface that makes the diffusion of ozone to catalytic centers more difficult. The humidity of O_3/O_2 gas flow was measured to be 50%.

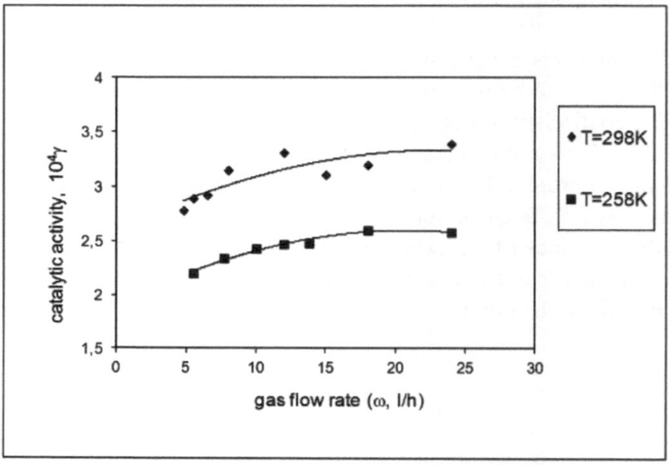

FIGURE 11.5 Flow rate dependence of catalytic activity at 258 and 298 K, ozone flow rate range 5–25 l h^{-1}.

A popular technique used to characterize manganese oxide containing catalyst and to determine the identity of the manganese oxide phase at high loadings (>6%) has been X-ray diffraction (XRD).[21,22] The X-ray diffraction results for the cement-containing metal oxide catalyst are presented in Figure 11.6. The diffractogram for the catalyst sample showed peaks with a certain number of large intensities at different 2θ values. The peaks at 39 and 35.5° correspond to copper oxide (CuO). The diffraction features for the catalyst at 33 and 55.1° are indicative of bixbyite-o (Mn_2O_3). The metal oxide catalyst sample peaks at 43, 37.2, and 62.9° are due to nickel oxide (NiO). The catalyst diffraction peaks at 36.1, 32.4, and 59.9° correspond to hausmannite (Mn_3O_4). Finally, the cement diffraction peaks at 20, 25.5° and 29.5, 47.5° are due to grossite ($CaAl_4O_7$) and calcite ($CaCO_3$), respectively. In conclusion, the information that can be deduced from the X-ray diffractogram for this catalyst is that there are three metal oxides as the manganese oxide is present in two forms—Mn_2O_3 and Mn_3O_4. It could be also seen that the cement support of the catalyst is built mainly by two components—$CaAl_4O_7$ and $CaCO_3$.

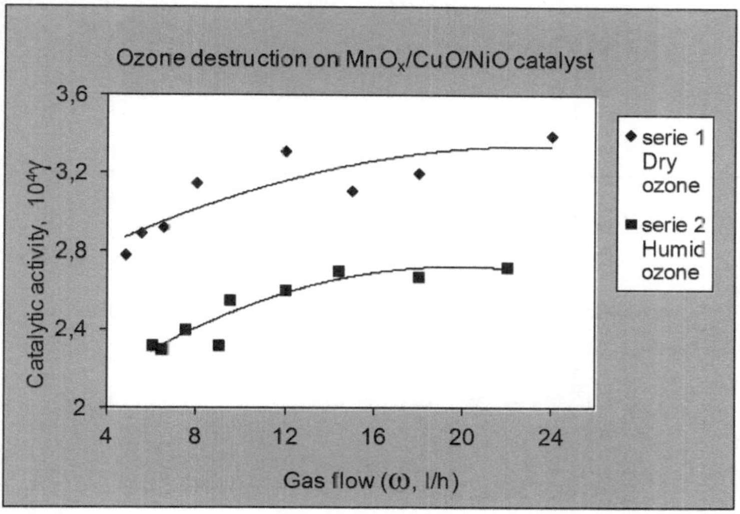

FIGURE 11.6 Flow rate dependence of catalytic activity at dry and humid conditions, ozone flow rate range 4–26 l h^{-1}.

The TPR experiment was carried out for the supported cement metal oxide catalyst (Fig. 11.7). The H_2 consumption was monitored by thermal conductivity detector in the course of time. Manganese-containing catalyst was already studied using TPR.[23] The peak temperatures of reduction in Figure 11.7 are 527, 596, 643, and 976 K or the reduction temperature of the catalyst was in the range of 527–976 K. The bulk reduction peaks at 596 and 643 K can be identified for the manganese oxide in the cement-containing metal oxide mixture while the peak at 976 K can be related with reduction of the nickel oxide.[8,23]

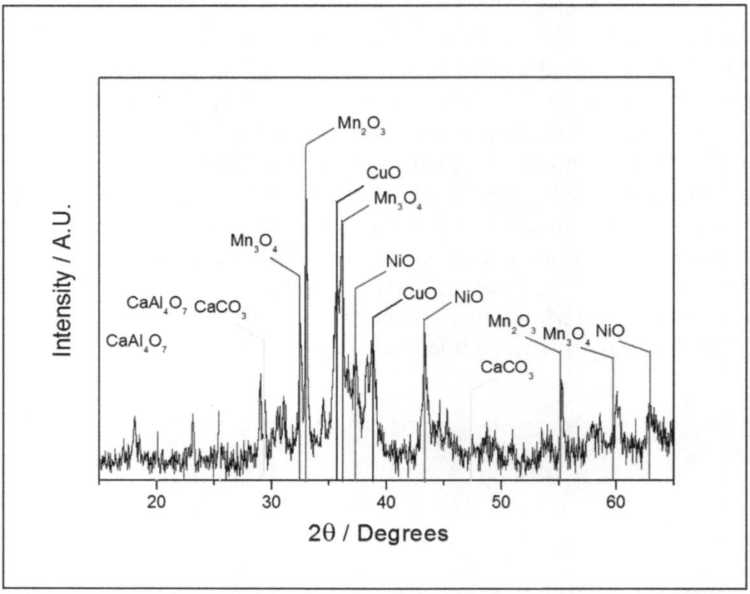

FIGURE 11.7 XRD of MnO_x/CuO/NiO catalyst supported on cement.

The most possible mechanism[6,7] of catalytic ozone decomposition can be presented as follows:

$$O_3 + * \longrightarrow O_2 + O* \quad (1)$$

$$O_3 + O* \longrightarrow O_2* + O_2 \quad (2)$$

$$O_2* \longrightarrow O_2 + * \quad (3)$$

where the symbol * was used to denote surface sites. In step (1), ozone decays and the finding that the adsorbed ozone that does not desorb, ascertains the irreversibility of steps (1) and (2). Further, peroxide particles are formed in accordance with step (2) and then oxygen is desorbed from the catalytic surface in step (3). The finding that the peroxide species could not be formed from molecular oxygen at any conditions shows the irreversibility of step (3).

The FT-IR spectra of the catalyst before and after ozone decomposition are shown in Figure 11.8. The two similar spectra indicate that the catalyst does not change practically during the reaction. A broad band at 3415–3425 cm^{-1} and also the band at 1410–1430 cm^{-1} are associated with the vibrations of water molecules.[15,24] The intensive bands at 515–530 cm^{-1} in accordance with literature[25] were assigned to the stretching vibration of the surface metal–oxygen bond.

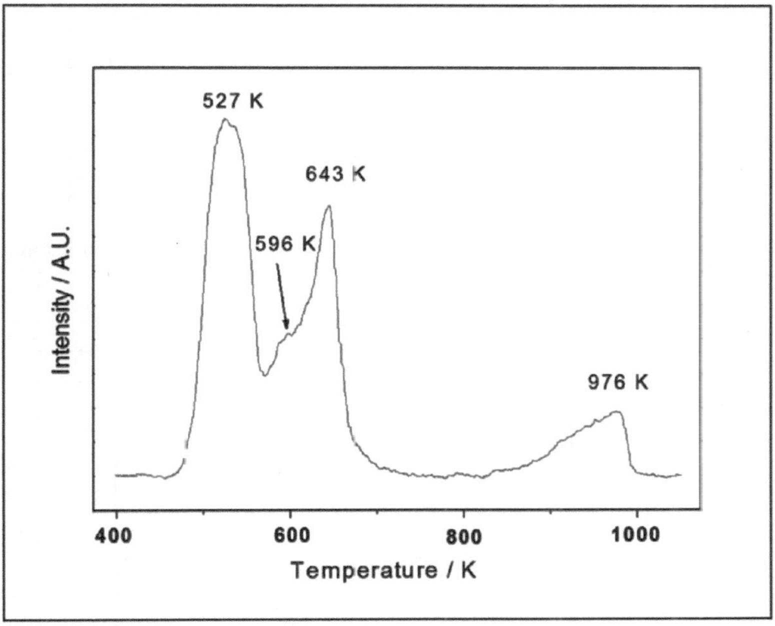

FIGURE 11.8 TPR of $MnO_x/CuO/NiO$ catalyst supported on cement.

The FT-IR spectra of cement–oxide catalyst after dry ozone decomposition (a) and after humid ozone decomposition (b) are presented in Figures 11.9 and 11.10. The spectra are almost identical, showing that the catalyst structure is not altered during the humid catalytic reaction. The broad adsorption band at 3430 cm^{-1} appears from the stretching vibration of hydrogen-bonded hydroxyl groups.[15] The adsorption band at 1635 cm^{-1} is due to vibrations of water molecules.[14] The intensive band at 520–530 cm^{-1} appears at higher manganese concentrations and, in accordance with literature, can be attributed to well-defined Mn_2O_3 phase.[22]

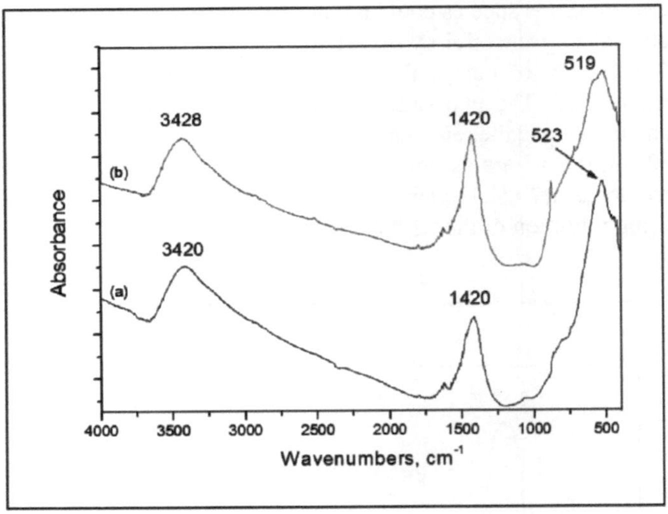

FIGURE 11.9 FT-IR spectra of MnO_x/CuO/NiO catalytic samples obtained before ozone decomposition (a) and after ozone decomposition for 8 h (b).

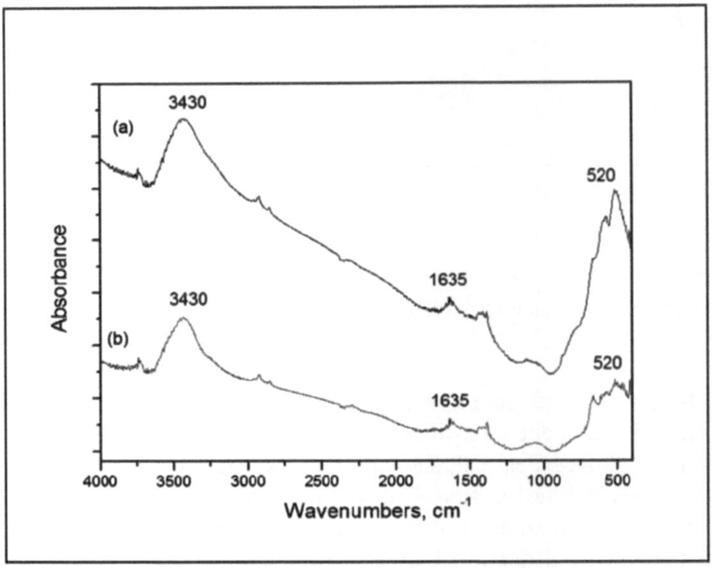

FIGURE 11.10 FT-IR spectra of cement-oxide catalyst after dry ozone decomposition (a) and after humid ozone decomposition (b).

11.5 CONCLUSION

1. The catalyst based on MnO_x/CuO/NiO has high efficiency in the reaction of ozone decomposition both at dry and wet conditions.
2. The catalyst work is stable and its activity does not change dramatically by varying temperature, O_3/O_2 flow rate, and humidity.
3. XRD analysis has proven to be a useful tool for the identification of all metal oxide phases in catalyst mixture. The temperature-programmed reduction of the catalyst denotes its high reducibility.
4. The FT-IR spectral band at 526 cm^{-1} shows that the catalyst stays unchangeable during the ozone decomposition for about 8 h.
5. The FT-IR spectra after humid ozone decomposition indicates that even after decomposing ozone in humid conditions, the catalyst structure does not change practically during the reaction.

KEYWORDS

- **ozone**
- **catalyst**
- **decomposition**
- **synthesis**
- **kinetics**
- **mechanism**

REFERENCES

1. Rakovsky, S.; Zaikov, G. *Kinetic and Mechanism of Ozone Reactions with Organic and Polymeric Compounds in Liquid Phase*; monograph (2nd ed.), Nova Sci. Publ., Inc.: New York; 2007; pp 1–340.
2. Heisig, C.; Zhang, W.; Oyama ,S. T. Decomposition of Ozone using Carbon Supported Metal Oxide Catalysts. *Appl. Catal. B: Environ.* **1997,** *14,* 117.
3. Rakitskaya, T. L.; Bandurko, A. Yu.; Ennan, A. A.; Paina, V. Ya; Rakitskiy, A.S. Carbon-Fibrous-Material-Supported Base Catalysts of Ozone Decomposition. *Micro. Meso. Mater.* **2001,** *43,* 153.
4. Radhakrishnan, R.; Oyama, S. T.; Ohminami, Y.; Asakura, K. Structure of MnO_x/Al_2O_3 Catalyst: A Study Using EXAFS, In Situ Laser Raman spectroscopy and Ab Initio Calculations. *J. Phys. Chem.* **2001,** *105*, p 9067.
5. Radhakrishnan, R.; Oyama, S. T.; Chen, J.; Asakura, A. Electron Transfer Effects in Ozone Decomposition on Supported Manganese Oxide. *J. Phys. Chem. B.* **2001,** *105* (19), 4245.
6. Li, W.; Gibbs, G. V.; Oyama, S. T. Mechanism of Ozone Decomposition on Manganese Oxide: 1. In situ Laser Raman Spectroscopy and ab initio Molecular Orbital Calculations. *J. Am Chem. Soc.* **1998,** *120,* 9041.

7. Li, W.; Oyama, S. T. The Mechanism of Ozone Decomposition on Manganese Oxide: 2. Steady-state and Transient Kinetic Studies. *J. Am. Chem. Soc.* **1998,** *120,* 9047.
8. Dhandapani, B.; Oyama, S. T. Gas Phase Ozone Decomposition Catalysts. *J. Appl. Catal. B: Environ.* **1997,** *11,* 129.
9. Oyama, S. T. Chemical and Catalytic Properties of Ozone. *Catal. Rev. Sci. Eng.* **2000,** *42,* 279.
10. Einaga, H.; Futamura, C. S. Comparative Study on the Catalytic Activities of Alumina-supported Metal Oxides for Oxidation of Benzene and Cyclohexane with Ozone. *React. Kinet. Catal. Lett.* **2004,** *81,* 121.
11. Tong, S.; Liu, W.; Leng, W.; Zhang, Q. Characteristics of MnO_2 Catalytic Ozonation of Sulfosalicylic Acid Propionic Acid in Water. *Chemosphere* **2003,** *50,* 1359.
12. Li, W.; Oyama, S. T.; Absolute Determination of Reaction Mechanisms by In Situ Measurements of Reaction Intermediates. *Top. Catal.* **1999,** *8,* 75.
13. Konova, P.; Stoyanova, M.; Naydenov, A.; Christoskova, S. T.; Mehandjiev, D. Catalytic Oxidation of VOCs and CO by Ozone Over Alumina Supported Cobalt Oxide. *J. Appl. Catal. A: Gen.* **2006,** *298,* 109.
14. Stoyanova, M.; Konova, P.; Nikolov, P.; Naydenov, A.; Christoskova, S. T.; Mehandjiev, D. Alumina-Supported Nickel Oxide for Ozone Decomposition and Catalytic Ozonation of CO and VOCs. *Chem. Eng. J.* **2006,** *122,* 41.
15. Subrahmanyam, C.; Bulushev, D. A.; Kiwi-Minsker, L. Dynamic Behaviour of Activated Carbon Catalysts during Ozone Decomposition at Room Temperature. *J. Appl. Catal. B: Environ.* **2005,** *61,* 98.
16. Perry, R. H.; Green, D. *Perry's Chemical Engineer's Handbook*; McGraw-Hill: New York, 1989; pp 3–147.
17. Martinov, I. V.; Tkachenko, S. N.; Demidyuk, V. I.; Egorova, G. V.; Lunin, V. V. NiO Addition Influence Over Cement-containing Catalysts Activity in Ozone Decomposition. *J. of Moscow Univ.* (*in Rus.*), Ser. 2, Chemistry, **1999,** *40,* 355 .
18. Zavadskii, A.V.; Kireev, S. G.; Muhin, V.M.; Tkachenko, S. N.; Chebkin, V. V., Klushin, V. N.; Teplyakov D.E. Thermal Treatment Influence over Hopcalite Activity in Ozone Decomposition. *J. of Phys. Chem.* (*in Rus.*) **2002,** 76, 2278.
19. Lunin, V. V.; Popovich, M. P.; Tkachenko, S. N. *Physical Chemistry of Ozone* (*in Rus.*); Moscow University Publ. House: Moscow, RU, 1998; pp 377–444.
20. Frank-Kamenetskii, D.A. *Diffusion and heat transfer in chemical kinetics* (*in Rus.*), M., (1987).
21. Jacono, M. Lo; Schiavello, M. The Influence of Preparation Methods on Structural and Catalytic Properties of Transition Metal Ions Supported on Alumina. In *Preparation of catalysts I*; Delmon, B., Jacobs, P., Poncelet, G., Eds.; New York, 1976; 473.
22. Buciuman, F.; Patcas, F; Craciun, R.; Zhan, D. R. T., Vibrational Spectroscopy of Bulk and Supported Manganese Oxides. *Phys. Chem. Chem. Phys.* **1998,** *1,* 185.
23. Kapteijn, F.; Van Langeveld, A. D.; Moulijn, J. A.; Andreini, A.; Vuurman, M. A.; Turek, M. A.; Jehng, J. M.; Wachs, I. E. Alumina-Supported Manganese Oxide Catalysts. *J. Catal.* **1994,** *150,* 94.
24. Gomez-Serrano, V.; Alvarez, P. M.; Jaramillo, J.; Beltran, F.J. Formation of Oxygen Structures by Ozonation of Carbonaceous Materials Prepared from Cherry Stones-II, Kinetic Study. *Carbon,* **2002,** *40,* 513.
25. Bielanski, A.; Haber, J. Oxygen in Catalysis. Marcel Dekker Inc.: New York, 1991.

CHAPTER 12

SYNTHESIS OF O-INCLUDING COMPOUNDS BY CATALYTIC CONVERSATION OF OLEFINS

G. Z. RASKILDINA[1], N. G. GRIGOR'EVA[2], B. I. KUTEPOV[2], S. S. ZLOTSKY[1], and G. E. ZAIKOV[3]

[1]Ufa State Petroleum Technological University, 1 Kosmonavtov Str.,450062 Ufa, Russia; Phone (347) 2420854. E-mail: nocturne@mail.ru

[2]Institute of Petrochemistry and Catalysis of RAS, 141 pr. Oktyabria, 450075 Ufa, Russia

[3]N. M. Emanuel Institute of Biochemical Physics, Russian Academy of Sciences, 4, Kosygin str., Moscow 119334, Russian Federation. E-mail: Chembio@sky.chph.ras.ru

CONTENTS

ABSTARCT

By studying the reactions of hydratation of norbornene and 2-vinyl-2-methil-gem-dichlorocyclopropane and the reactions of styrene and norbornene with different alcohols and carbonic acids in presence of heterogenic catalyst, it was found that the selected zeolite H-Beta is active and is selective catalyst for these reactions. Alcohol, ethers, esters, and diesters of norbornene have exo-configuration. It has been established that reaction of norbornene with diols, catalyzing by zeolite Beta, leads to the formation of esters, which did not find before.

12.1 INTRODUCTION

We have been investigating the reactions of commercially available alkenes (norbornene and 2-vinyl-2-methil-*gem*-dichlorocyclopropane) with water in the presence of zeolite catalyst H-Beta, which is successfully used in petrochemical processes such as alkylation, isomerization, and so on.[1,2] The results of this reactions are formation of corresponding alcohols. It was determined that norbornene is hydrated easier (T = 10°C, under atmospheric pressure) than 2-vinyl-2-methil-*gem*-dichlorocyclopropane (T= 150°C, 6 bar gauge autoclave pressure). The results have reported that the heterogeneous catalytic joining of water to olefins is of interest, being an effective and cheap method for obtaining corresponding alcohols. Furthermore, we investigated additional reaction of alcohols and ethers to norbornene and styrene. Addition of *O*-including compounds to the multiple carbon–carbon bonds in presence of homogeneous and heterogeneous catalysts finds a wide application in synthesis of ethers and esters.[3] The use of homogeneous (mineral acids) and heterogeneous (cationite) catalysts has some drawbacks and do not provide required high yields and selectivity of products.

12.2 RESULTS AND DISCUSSION

The alcohols obtained by hydratation of olefins is widespread in petrochemical synthesis.[4–6] Recently, hydratation over mineral (sulfuric, phosphoric) and organic (toluene sulfonic) acids and over a cationite KU-2-8 in H^+-form was shown.[7–10]

We investigated the hydratation of model alkenes (norbornene **2** and 2-vinyl-2-methil-*gem*-dichlorocyclopropane **3** over zeolite catalyst H-Beta, which is widely used on industrial scale.[8–10]

The corresponding alcohols (**4, 5**) are obtained at 10°C (under atmospheric pressure) in water solvent. Conversion of more active norbornene **2** is 90% for 2 h (yield of alcohol is 50%), while less active 2-vinyl-2-methil-*gem*-dichlorocyclopropane **3** gives alcohol **5** (yield 24%) in an autoclave at 150°C, 6 bar gauge autoclave pressure for 4 h is shown in (Table 12.1) (Scheme 12.1).

TABLE 12.1 The Interaction of Olefins 2, 3 with Water 1 (Molar Ratio Olefins:Water = 1:24; 20 wt% Catalyst).

Reagents		Temperature (°C)	Time (h)	Кa, %	Selectivity (%)
	2	10	2	90	4 (50)
1	3	80	24	26	5 (24)
		150*	2	48	5 (25)

[a]*K*[a]—conversion of olefin.
[b]*Autoclave.

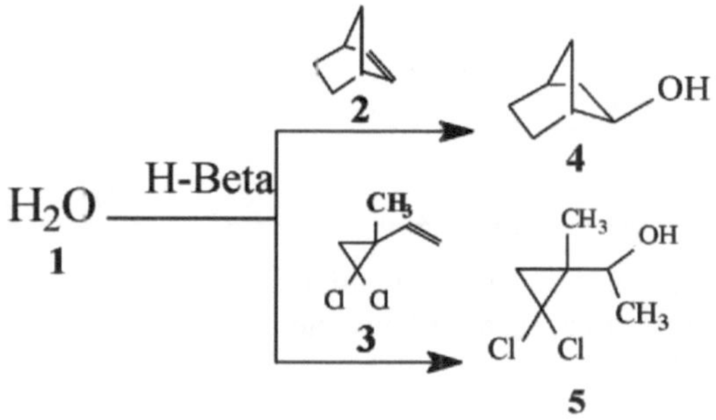

SCHEME 12.1 Hydratation of olefins.

In the present work, we found (Scheme 12.2) that monohydric alcohols (**7–9**) of various structures (butyl, allylic, benzyl) are connected to styrene (**6**) and norbornene (**2**) selectively with formation of corresponding ethers (**10–15**).

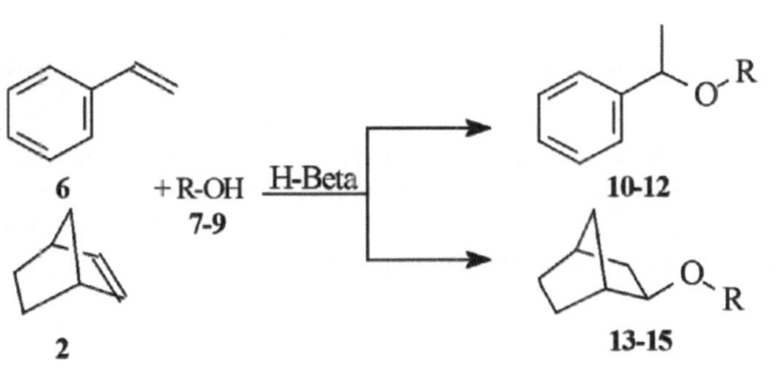

R=n-Bu (**7; 10, 13**); All (**8; 11, 14**); Bn (**9; 12, 15**)

SCHEME 12.2 Formation of ethers (10–15) by reaction of alcohols and olefins 2, 6.

In the studied conditions (Table 12.2), there is total conversion of olefin; and selectivity of formation of target ethers is 85% more, which is weakly dependent on the structure of reagents.

TABLE 12.2 Reaction of Olefins 6, 2 with Alcohols 7–9 in Presence Zeolite H-Beta (Molar Ratio Olefin: Alcohol = 1:3, 20 wt% Catalyst, T=80°C, 5 h).

Olefin	Alcohol	Selectivity, (%)
6	7	10 (92)
2	7	13 (93)
6	8	11 (89)
2	8	14 (90)
6	9	12 (92)
2	9	15 (95)

In the case reaction of diols (**16, 17**) with norbornene, the reaction proceeds by consistent formation of mono- (**18, 19**) and diethers (**20, 21**) in presence of zeolite H-Beta (Scheme 12.3).

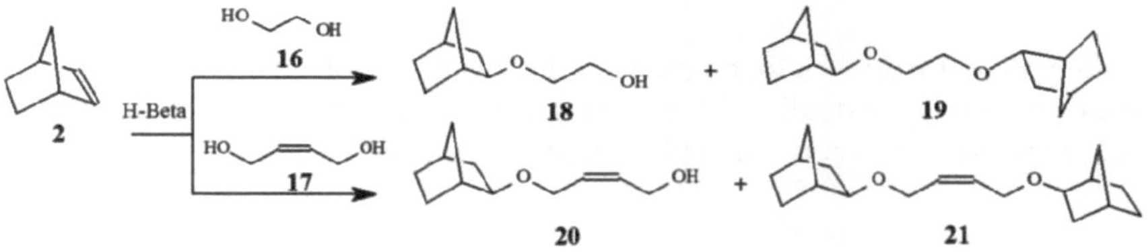

SCHEME 12.3 Formation of mono- and diethers of norbornene over zeolite H-Beta.

By increasing the temperature from 50 to 80°C, conversion of norbornene **2** changes slightly. However, selectivity of diethers' formation increases from 4 to 32%. Moreover, increasing the molar ratio olefin:diol from 1:3 to 3:1, selectivity of diethers' formation increases more than twice (Table 12.3).

TABLE 12.3 Reaction of Norbornene with Diols 16, 17 (Molar Ratio 2:diol = A:B, 20 wt.% Catalyst H-Beta, 5 h).

Diol	A:B	T, °C	Selectivity, (%)	
			Monoether	Diether
		50	18 (92)	19 (4)
16	1:3	60	18 (84)	19 (12)
		80	18 (58)	19 (32)
	3:1		20 (25)	21 (68)
17	1:1	80	20 (46)	21 (41)
	1:3		20 (60)	21 (30)

In the studied conditions, activities of ethylene glycol (**16**) and *cis*-2-butene-1,4-diol (**17**) are similar. Bicyclic olefin **2** reacts quantitatively with monohydric acids (**22–25**) producing appropriate esters. These yields are 80–99% (a four-fold molar excess of acid) and depend on identity of acids (Scheme 12.4).

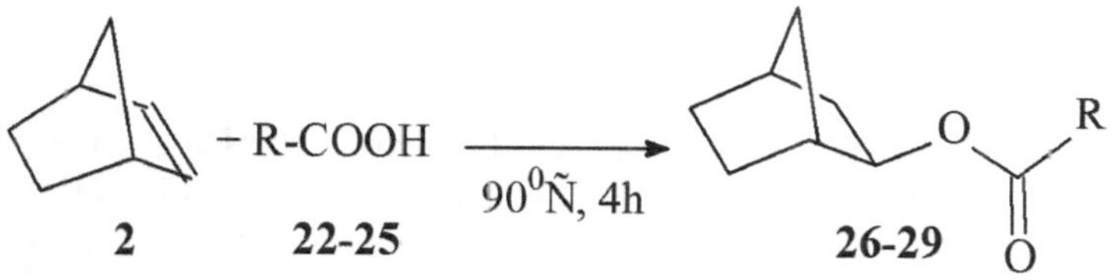

R= -CH_3 (**22, 28**); -n-C_3H_7 (**23, 27**); -CH_2Cl (**24, 28**); -C_3H_5(**25, 29**)

SCHEME 12.4 Formation of esters (26–29) by reacting monobasic carboxylic acids (22–25) with norbornene 2.

At the same time, judging from yield of monoethers, monochloracetic **24,** and methacrylic **25** acids are six times lesser active than acetic acid **22**.

The reaction of olefin **2** and dicarbonic acids is going with the formation of mono- and diethers appropriately (Scheme 12.5).

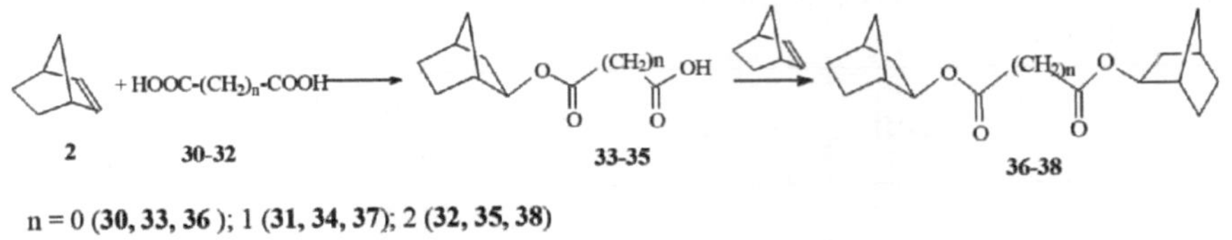

SCHEME 12.5 Formation esters (33–35) by react monocarbonic acids (30–32) with norbornene 2.

Spatial structure of diesters of norbornene **(36–38)** identified by methods of homo- (COSY, NOESY) and heteronuclear (HSQC, HMBC) two-dimensional ^{1}H, ^{13}C NMR spectroscopy. So, in ^{13}C NMR spectra of target esters **(36–38)**, signals of atoms C-7 of bicyclic fragment are situated in the range of 35.6–36.2 ppm area, indicating that *exo*-configuration is because signals *endo*-isomers are located in the weaker field (~40 ppm area).

Exo–exo-configuration of dibicyclo[2.2.1]hept-2-yl diester of malonic acid **37** is confirmed by interaction of protons of atoms C-2 and C-12 with protons of atoms C-6 and C-16. In case of *endo– endo* isomer, protons of atoms C-2 and C-12 correlate with protons of atoms C-7 and C-17, that is not found in our case. Esters by *exo*-, *endo*-configuration are not have in the products of reactions from NMR-spectrums.

12.3 EXPERIMENTAL

An HRGS 5300 Mega Series "Carlo Erba" chromatograph with a flame ionization detector was used for the qualitative and quantitative analysis of starting material and reaction products. The chromatograph was equipped with a thermo-conductivity were registered using the «Bruker AVANCE-400» spectrometer (400.13 and 100.62 MHz, respectively) in $CDCl_3$ solvent, where benzene-d_6, toluene-d_8 were used as internal standards. High-resolution mass spectra were measured on a Fisons Trio 1000 instrument, whose chromatograph was equipped with a DB-560 quarts column (50 m); the temperature of the column was increased from 50 to 320°C with a programmed heating rate of 4°C min^{-1}; the electron impact (70 eV).

In order to carry out the reactions, zeolite BEA (Beta) (mole ratio SiO_2/Al_2O_3 = 18.0), synthesized in the JSC "Angarsk Catalysts and Organic Synthesis" in NH_4-form, was used as catalyst. Zeolite Beta was converted H-form by heated in air at 540°C for 3 h. Before experiments, the catalyst sample was dried in air for 4 h at 350°C.

12.3.1 CATALYST

Zeolite NH_4-Beta was produced by the public corporation Angarsk Factory of Catalysts and Organic Synthesis. Zeolite NH_4-Beta was transferred into H-Beta form by calcinations at 540°C for 4 h before all experiments.

12.3.2 THE INTERACTION OF NORBORNENE 2 WITH WATER OVER H-BETA ZEOLITE

The zeolite H-Beta (0.28 g) was added to a mixture of norbornene **2** (1.22 g, 0.013 mmol) and water **1** (50 g, 3.13 mmol) at 10°C and stirred for 2 h. After the catalyst was filtered and the reaction mass was extracted with ether, the latter was removed at the reduced pressure. The alcohol **4** (72°C/20 mm Hg) was isolated under vacuum. The mixture of *exo*-2-norborneol **4** and di-norbornyl ether of this mixture was (% w/w): 50/50% accordingly. Compounds were identified by NMR spectroscopy.

EXO-2-NORBORNEOL (4)

^{1}H NMR, δ: 1.02–1.05 (m, 3H, C^6H_a, C^7H_a, C^5H_a), 1.12–1.18 (m, 2 H, C^3H_a, C^5H_b), 1.36–1.51 (m, 2H, C^7H_b, C^6H_b), 1.62–1.67 (m, 1H, C^3H_b), 2.12–2.35 (m, 2H, C^4H, C^1H), 3.68 (d, 1H, C^2H). ^{13}C NMR, δ: 24.39 C^6, 29.27 C^5, 34.38 C^7, 35.40 C^4, 42.37 C^3, 44.34 C^1, 74.95 C^2. IR spectrum: 2851–2954 (C–H, CH_2), 3437 (–OH); m/z: 122 M^+ (30), 107 (100), 79 (93), 77 (52), 43 (30), 51 (21), 105 (10), 50 (10), 80 (7); Kovaè index I_k 1065.

12.3.3 THE INTERACTION OF 2-VINYL-2-METHIL-GEM-DICHLOROCYCLOPROPANE 3 WITH WATER OVER H-BETA ZEOLITE

The zeolite H-Beta (0.43 g) was added to a mixture of 2-vinyl-2-methil-*gem*-dichlorocyclopropane **3** (1.96 g, 0.013 mmol) and water **1** (50 g, 3.13 mmol) at 80°C/150°C* and stirred for 24/2*h. After the catalyst was filtered and the reaction mass was extracted with ether, the latter was removed at the reduced pressure. The alcohol **5** (95°C/25 mm Hg) was isolated under vacuum. The mixture of 1,1-dichloro-2-methyl-2-(hydroxyethyl-1)cyclopropane **5** of this mixture was 24%. Compound was identified by NMR spectroscopy.

After the completion of the reaction, mass was separated from the catalyst by filtration. The conversion of initial olefins and the quantitative composition of the alcohol fraction were determined using gas–liquid chromatography (GLC). The chemical stricter of alcohols (**4, 5**) was established by means of GC/MS spectrometry and NMR spectroscopy.

1,1-DICHLORO-2-METHYL-2-(HYDROXYETHYL-1)CYCLOPROPANE (5)

^{1}H NMR, δ: 1.23–1.24 2H (d, cyclC^{3}H$_2$), 1.42 3H (s, C^{5}H$_3$), 1.70 1H (s, OH), 1.92 3H (s, C^{6}H$_3$), 4.01–4.06 1H (m, C^{4}H ^2J = 12.8, ^3J = 6). ^{13}C NMR, δ: 20.71 C^6, 23.39 C^5, 34.66 C^3, 44.70 C^2, 66.39 C^4, 77.26 C^1. IR spectrum: 2877-2972 (C–H, CH$_2$), 3418 (-OH); m/z: 169 M$^+$ (0.6), 45 (100), 124/126/128 (35/22/4), 87/89/91 (6/35/11), 53 (14), 53 (14), 43 (13), 51 (11). Kovaè index I_k = 1123.

* Autoclave

12.3.4 METHOD OF REACTION OF OLEFINS (2) WITH ALCOHOLS (7–9)

A mixture of 0.255 M alcohol **7** (or 0.255 M alcohol **8,** or 0.255 M alcohols **9**) and 0.085 M norbornene **2,** 20 wt.% catalyst H-Beta was carried out at 80°C and mixed intensively for 5 h. The reaction mass was separated from the catalyst by filtering the reaction termination and unreacted alcohol was removed at a low pressure. Ethers were isolated by vacuum distillation for calibration.

1-n-butyl-1-phenylethan (10): b.p. 95–96°C (10 mm Hg). ^{1}H-NMR (CDCl$_3$, δ ppm, *J*Hz): 0.86 (t, 3H, CH$_3$), 1.33 (m, 2H, CH$_2$), 1.35 (m, 2H, CH$_2$), 1.45 (d, 3H, CH$_3$), 3.42 (q, 2H, OCH$_2$), 4.30 (q, 1H, CH), 7.3 (m, 5H, ArH). ^{13}C-NMR (CDCl$_3$, δ ppm): 13.78 C^6, 18.90 C^5, 21.15 C^2, 32.65 C^4, 69.75 C^3, 72.86 C^1, 124.97–127.04 Ar, 143.03 C$^{1'}$.

1-allyloxy-1-phenylethan (11): b.p. 68°C (5 mm Hg): ^{1}H-NMR (CDCl$_3$, δ ppm, *J*Hz): 1.5 (d, 3H, CH$_3$), 3.85 (dddd., 1H, CH$_a$, ^2J 12.8, ^3J 5.8), 3.94 (dddd., 1H, CH$_b$, ^2J 12.8, ^3J 5.2), 4.51 (dd., 1H, CH, ^3J 6.4), 5.19 (dd., 1H, CH$_a$, ^2J 1.6, ^3J 10.4), 5.29 (dd., 1H, CH$_b$, ^2J 1.6, ^3J 17.2), 5.95 (dddd., 1H, CH, ^3J 5.2, ^3J 5.8, ^3J 10.4, ^3J 17.2), 7.28–7.40 (m., 5H, Ar).

1-cyclohexyloxy-1-phenylethan (12): b.p. 91–92°C (9 mm Hg): ^{1}H-NMR (CDCl$_3$, δ ppm, *J*Hz): 1.46 (d, 3H, CH$_3$), 1.12–1.20 (m, 6H, CH$_2$), 1.78–1.94 (m, 4H, CH$_2$), 3.40 (m, 1H, CH), 4.64 (q, 1H, ArCH), 7.23–7.33 (m, 5H, ArH).

Exo-2-(butoxy)bicyclo[2.2.1]heptane (13): b.p. 77°C (27 mm Hg). ^{1}H-NMR (CDCl$_3$, δ ppm, *J*Hz): 0.91 (m, 3H, CH$_3$), 0.96–1.07 (m, 3H, CH$_2$), 1.32–1.41 (m, 3H, CH$_2$), 1.46–1.55 (m, 4H, CH$_2$), 2.20 (s, 1H, CH), 2.29 (d, 1H, CH, ^2J = 4), 3.24–3.40 (m, 1H, CH), 3.24–3.40 (m, 2H, CH$_2$). MS (70eV), m/z (*J.*, %): 168 [M-1]$^+$ (≤1), 94 (100); 66 (74); 79 (65); 67 (56), 41 (49), 95 (39), 57 (28), 83 (19), 55 (17), 68 (16), 56 (13), 112 (12).

2-(alliloxy)bicyclo[2.2.1]heptane (14): b.p. 78°C (10 mm Hg). ^{1}H-NMR (CDCl$_3$, δ ppm, *J*Hz): 0.95–1.12 (m, 3H, CH$_2$), 1.37–1.60 (m, 5H, CH$_2$), 2.25 (s, 1H, CH), 2.33 (d, 1H, CH), 3.89–4.00 (m, 2H, CH$_2$), 5.10–5.18 (d, 1H, CH$_a$, ^2J =

1.6, ^3J = 10.4), 5.22–5.32 (d, 1H, CH_b, ^2J = 1.6, ^3J = 17.2), 5.83–5.99 (m, 1H, CH). MS (70eV), m/z (*J*., %): 152 [M-1]$^+$ (1), 67 (100); 41 (56); 94 (51); 95 (41), 66 (34), 79 (29), 55 (27), 93 (23), 91 (11), 81 (11), 77 (11).

2-(benzyloxy)bicyclo[2.2.1]heptane (15): b.p. 70°C (20 mm Hg). ^{1}H-NMR ($CDCl_3$, δ ppm, *J*Hz): 0.92–1.05 (m, 3H, CH_2), 1.25–1.40 (m, 2H, CH_2). 1.51–1.56 (m, 3H, CH_2), 2.19 (s, 1H, CH), 2.28 (s, 1H, CH), 3.64–4.51 (m, 1H, CH), 3.64–4.51 (m, 2H, CH_2), 7.00–7.05 (m, 5H, Ar).

2-(bicyclo[2.2.1]heptyl-2-oxy)ethanol (18): b.p. 130°C (10 mm Hg). ^{1}H-NMR ($CDCl_3$, δ ppm, *J*Hz): 0.95–1.10 (m, 3H, CH_2), 1.38–1.59 (m, 5H, CH_2), 2.23 (m, 1H, CH), 2.33 (m, 1H, CH), 2.55 (1H, OH), 3.39 (d, 1H, CH), 3.49–3.57 (m, 2H, CH_2), 3.70–3.77 (m, 2H, CH_2). ^{13}C–NMR ($CDCl_3$, δ ppm): 24.61 C^6, 28.55 C^5, 34.77 C^7, 35.17 C^4, 39.56 C^3, 40.36 C^1, 63.71 C^9, 67.66 C^8, 83.0 C^2. MS (70eV), m/z (*J*., %): 156 [M-1]$^+$ (2), 95 (100); 67 (49); 155 (19), 111 (18), 94 (16), 94 (15), 66 (15), 41 (15).

2,2'-[ethane-1,2-diylbis(oxy)]bicyclo[2.2.1]heptane (19): b.p. 158°C (5 mm Hg). ^{1}H-NMR ($CDCl_3$, δ ppm, *J*Hz): 0.86–1.26 (m, 8H, CH_2), 1.32–1.40 (m, 8H, CH_2), 1.55–1.58 (m, 2H, CH_2), 1.63 (m, 1H, CH), 1.73 (m, 2H, CH_2), 2.04 (m, 2H, CH, CH), 3.20 (m, 2H, CH_2), 3.42 (m, 2H, CH_2), 3.57 (d, 1H, CH). MS (70eV), m/z (*J*., %): 250 [M-1]$^+$ (1), 95 (100); 94 (95); 66 (87), 79 (77), 67 (52), 45 (47), 41 (33), 55 (19), 57 (18); 83 (17); 44 (16); 77 (13); 65 (13); 43 (14); 53 (11).

Exo-4-(bicyclo[2.2.1]hept-2-yloxy)but-2-en-1-ol (20): b.p. 141°C (2 mm Hg). ^{1}H-NMR ($CDCl_3$, δ ppm, *J*Hz): 0.95–1.03 (m, 2H, C^6H_b, C^3H_a), 1.03–1.14 (m, 2H, C^6H_a, C^3H_b), 1.37–1.48 (m, 1H, C^5H_b), 1.50–1.60 (m, 1H, C^5H_a), 2.17 (s, 1H, C^1H), 2.25 (s, 1H, –OH), 2.34 (d, 1H, C^4H), 3.40 (d, 1H, C^2H), 3.95–4.09 (m, 1H, C^8H_a), 4.19(m, 1H, C^8H_b, ^2J = 6.4, ^3J = 18.8), 4.22 (d, 2H, $C^{11}H_a$, $C^{11}H_b$, ^2J = 4.4 ^3J = 16.8), 5.67–5.75 (m, 1H, C^9H), 5.76–5.85 (m, 1H, $C^{10}H$). ^{13}C-NMR ($CDCl_3$, δ ppm): 24.58 C^6, 28.40 C^5, 35.14 C^4, 39.51 C^3, 40.28 C^1, 58.53 C^{11}, 64.03 C^8, 82.67 C^2, 128.82 C^{10}, 131.82 C^9. MS (70eV), m/z (*J*., %): 182 [M-1]$^+$ (2), 164 (7), 138 (12), 109 (12), 95 (100), 94 (22), 81(18), 79 (46), 77 (10), 71 (17), 70 (27), 67 (90), 66 (40), 57 (10), 55 (27), 53 (18), 43 (40).

Exo-exo-[(2Z)-but-2-en-1,4-diylbis(oxy)]bisbicyclo[2.2.1]heptane (21): b.p. 183°C (2 mm Hg). ^{1}H-NMR ($CDCl_3$, δ ppm, *J*Hz): 0.97–1.12 (m, 6H, C^6H_a, $C^{14}H_a$, C^5H_a, $C^{17}H_a$, C^7H_a, $C^{18}H_a$), 1.34–1.48 (m, 6H, C^3H_a, $C^{14}H_a$, C^5H_b, $C^{16}H_b$, C^6H_b, $C^{17}H_b$), 1.49–1.58 (m, 4H, C^7H_b, $C^{18}H_b$, C^3H_b, $C^{14}H_b$), 2.23 (s, 2H, C^4H, $C^{15}H$). 2.30–2.34 (d, 2H, C^1H, $C^{12}H$), 3.35–3.40 (dd, 2H, C^2H, $C^{13}H$), 3.95–4.05 (m, 2H, C^8H_a, C^8H_b), 4.09 (dd, 1H, $C^{11}H_a$), 4.19 (dd, 1H, $C^{11}H_b$), 5.63–5.74 (m, 1H, $C^{10}H$), 5.75–5.88 (m, 1H, C^9H). ^{13}C-NMR ($CDCl_3$, δ ppm): 23.75 C^6, 27.62 C^5, 33.93 C^7, 34.27 C^4, 38.71 C^3, 39.42 C^1, 57.54 C^{11}, 63.04 C^8, 81.25 C^2, 129.41 C^{10}, 131.96 C^9. MS (70eV), m/z (*J*., %): 276 [M-1]$^+$ (<1), 164 (6), 96 (10), 95 (100), 93 (6), 79 (6), 70 (10), 67 (42), 41 (13).

12.3.5 METHOD OF REACTION OF NORBORNENE (2) WITH MONOCARBOXYLIC ACIDS (22-25)

A mixture of 0.34 M acetic acid **22** (or 0.34 M *n*-butyric acid **23**, or 0.34 M chloracetic acid **24**, or 0.34 M methacrylic acid **25**), 0.085 M of norbornene, 20 wt.% catalyst H-Beta was carried out at 90°C and mixed intensively for 4 h. For homogenization of initial compounds (**24, 25**), nonane was used as a solvent. The reaction mass was separated from the catalyst by filtering after the reaction termination and unreacted acid was removed at a low pressure. Esters were isolated by vacuum distillation for calibration.

Exo-bicyclo[2.2.1]hept-2-yl ester of acetic acid (26): b.p. 95°C (20 mm Hg). ^{1}H-NMR ($CDCl_3$, δ ppm, *J*Hz): 1.08–1.18 (m, 4H, C^3H_a, C^6H_a, C^6H_b, C^3H_b), 1.42–1.54 (m, 3H, C^4H_b, C^7H_a, C^7H_b), 1.73 (m, 1H, C^4H_a), 2.02 (s, 3H, C^9H_3), 2.30 (m, 2H, C^2H, C^5H), 4.61 (d, 1H, C^1H). ^{13}C-NMR ($CDCl_3$, δ ppm): 21.42 C^9, 24.33 C^3, 28.13 C^4, 35.24 C^5, 35.37 C^7, 39.60 C^6, 41.40 C^2, 77.60 C^1, 170.82 C^8. MS (70eV), m/z (*J.*, %): $[M-1]^+$ 154 (6), 43 (100), 66/67 (70/68), 94/95 (68/52), 79 (65), 111/112 (64/51), 41 (53), 71 (52).

Exo-bicyclo[2.2.1]hept-2-yl ester of n-butyric acid (27): b.p. 106°C (10 mm Hg). ^{1}H-NMR ($CDCl_3$, δ ppm, *J*Hz): 0.96 3 (t, 3H, $C^{11}H_3$), 1.17–1.20 (m, 3H, C^3H_a, C^7H_a, C^4H_a), 1.39–1.49 (m, 2 H, C^6H_a, C^4H_b), 1.50–1.58 (m, 2H, C^7H_b, C^3H_b), 1.65 (m, 2H, $C^{10}H_2$), 1.69–1.78 (m, 1H, C^6H_b), 2.25 (t, 2H, C^9H_a, C^9H_b), 2.27–2.32 (m, 2H, C^5H, C^2H), 4.62 (d, 1H, C^1H). ^{13}C-NMR ($CDCl_3$, δ ppm): 13.66 C^{11}, 18.54 C^{10}, 24.31 C^3, 28.85 C^4, 35.26 C^7, 35.38 C^9, 36.59 C^5, 39.65 C^6, 41.45 C^2, 77.29 C^1, 173.40 C^8. MS (70eV), m/z (*J.*, %): $[M-1]^+$ 182 (2), 71 (100), 95 (52), 43 (40), 139 (31), 111 (30), 154 (15), 79 (12).

Exo-bicyclo[2.2.1]hept-2-yl ester of chloracetic acid (28): b.p. 95°C (6 mm Hg). ^{1}H-NMR ($CDCl_3$, δ ppm, *J*Hz): 1.09–1.22 (m, 3H, C^3H_b, C^7H_b, C^4H_b), 1.44–1.57 (m, 4H, C^3H_a, C^4H_a, C^7H_a, C^6H_a), 1.75–1.80 (m, 1H, C^6H_b), 2.32 (m, 1H, C^5H), 2.36 (d, 1H, C^2H), 4.04 (s, 2H, C^9H_a, C^9H_b), 4.72 (d, 2H, C^1H). ^{13}C-NMR ($CDCl_3$, δ ppm): 24.13 C^6, 28.03 C^5, 35.23 C^7, 35.35 C^4, 39.38 C^3, 41.20 C^9, 41.39 C^1, 79.71 C^2, 167.01 C^8. MS (70eV), m/z (*J.*, %): $[M-1]^+$ 188 (<1), 66/67/68 (100/78/29), 94/95 (56/64), 77/79 (38/87), 41 (40), 49 (20), 55 (17), 42 (13), 53 (10).

Exo-bicyclo[2.2.1]hept-2-yl ester of methacrylic acid (29): b.p. 90°C (6 mm Hg). ^{1}H-NMR ($CDCl_3$, δ ppm, *J*Hz): 1.12–1.20 (m, 3H, C^3H_a, C^4H_a, C^7H_a), 1.43–1.57 (m, 4H, C^3H_b, C^4H_b, C^6H_a, C^7H_a), 1.74–1.79 (m, 1H, C^6H_b), 1.93 (s, 3H, $C^{11}H_3$), 2.31 (m, 1H, C^5H), 2.35 (m, 1H, C^2H), 4.68 (m, 1H, C^1H), 5.52 (s, 1H, $C^{10}H_a$), 6.07 (s, 1H, $C^{10}H_b$). ^{13}C-NMR ($CDCl_3$, δ ppm): 18.26 C^{11}, 24.24 C^3, 28.17 C^4, 35.33 C^7, 35.37 C^5, 39.57 C^6, 41.45 C^2, 77.74 C^1, 124.81 C^{10}, 136.90 C^9, 167.09 C^8. MS (70eV), m/z (*J.*, %): $[M-1]^+$ 180 (<1), 69 (100), 41 (84), 66 (71), 94 (56), 95 (40), 79 (19), 70 (19), 109 (11), 97 (11), 55 (10), 124 (10), 137 (10).

12.3.6 METHOD OF REACTION OF NORBORNENE (2) WITH DICARBOXYLIC ACIDS (30–32)

A mixture of 0.34 M oxalic acid **30** (or 0.34 M malonic acid **31**, or 0.34 M succinic acid **32**), 0.085 M of norbornene, 20 wt.% catalyst H-Beta was carried out 90°C and mixed intensively for 4 h. For homogenization of initial compounds (**30–32**) nonane was used as a solvent. The reaction mass was separated from the catalyst by filtering after the reaction termination and unreacted acid was removed at a low pressure. Esters were isolated by vacuum distillation for calibration.

The resulting physicochemical properties, NMR-spectra, and mass-spectra of compounds **33–35** correspond to literature data.[11]

Exo-dibicyclo[2.2.1]hept-2-yl ester of oxalic acid (36): b.p. 122°C (7 mm Hg). ^{1}H-NMR ($CDCl_3$, δ ppm, *J*Hz): 1.12–1.23 (m, 6H, C^6H_a, C^5H_a, C^7H_a, $C^{15}H_a$, $C^{14}H_a$, $C^{16}H_a$), 1.44–1.62 (m, 8H, C^6H_b, C^5H_b, C^3H_a, C^7H_b, $C^{15}H_b$, $C^{14}H_b$, $C^{12}H_a$, $C^{16}H_b$), 1.76–1.82 (dddd, 2H, C^3H_b, $C^{12}H_b$), 2.33 (s, 2H, C^4H, $C^{13}H$), 2.43 (d, 2H, C^1H, $C^{10}H$), 4.75 (d, 2H, C^2H, $C^{11}H$). ^{13}C-NMR ($CDCl_3$, δ ppm): 24.11 C^3, C^{15}, 28.03 C^5, C^{14}, 35.29 C^7, C^{16}, 35.37 C^4, C^{13}, 39.26 C^3,C^{12}, 41.30 C^1, C^{10}, 80.46 C^2, C^{11}, 158.05 C^8, C^9. MS (70eV), m/z (*J.*, %): $[M-1]^+$ 278 (0.1), 95/96 (100/8), 66/67/68 (8/24/2), 41 (9), 77/79 (3/4), 93 (4), 53/55 (2/4), 65 (3).

Exo-dibicyclo[2.2.1]hept-2-yl ester of malonic acid (37): b.p. 172°C (2 mm Hg). ^{1}H-NMR ($CDCl_3$, δ ppm, *J*Hz): 1.08–1.19 (m, 6H, C^5H_a, $C^{15}H_a$, C^5H_a, $C^{16}H_a$, C^7H_a, $C^{17}H_a$), 1.43–1.59 (m, 8H, C^3H_a, $C^{13}H_a$, C^7H_b, $C^{17}H_b$, C^5H_b, $C^{15}H_b$, C^6H_b, $C^{16}H_b$), 1.72–1.77 (m, 2H, C^3H_b, $C^{13}H_b$), 2.30 (s, 2H, C^4H, $C^{14}H$), 2.35 (s, 2H, C^1H, $C^{11}H$), 3.29 (s, 2H, C^9H_a, C^9H_b), 4.67 (d, 2H, C^2H, $C^{12}H$). ^{13}C-NMR ($CDCl_3$, δ ppm): 24.18 C^6, C^{16}, 28.10 C^5, C^{15}, 35.24 C^7, C^{17}, 35.34 C^4, C^{14}, 39.34 C^3, C^{13}, 41.33 C^1, C^{11}, 42.26 C^9, 78.79 C^2, C^{12}, 166.34 C^8, C^{10}. MS (70eV), m/z (*J.*, %): $[M-1]^+$ 292 (0.2), 95 (100), 67 (13), 111 (10), 199 (8).

Exo-dibicyclo[2.2.1]hept-2-yl ester of succinic acid (38)L: b.p. 180°C (1 mm Hg). ^{1}H-NMR ($CDCl_3$, δ ppm, *J*Hz): 1.12–1.24 (m, 6H, C^7H_a, C^5H_a, $C^{16}H_a$, C^6H_a, $C^{17}H_a$, $C^{18}H_a$), 1.42–1.53 (m, 8H, $C^{14}H_a$, C^3H_a, C^5H_b, $C^{16}H_b$, C^7H_a, $C^{18}H_b$, C^6H_b, $C^{17}H_b$), 1.70–1.75 (m, 2H, C^3H_b, $C^{14}H_b$), 2.29 (m, 4H, C^4H, $C^{15}H$, C^1H, $C^{12}H$), 2.57 (m, 4H, C^9H_a, C^9H_b, $C^{10}H_a$, $C^{10}H_b$), 4.63 (d, 2H, C^2H, $C^{13}H$). ^{13}C-NMR ($CDCl_3$, δ ppm): 24.25 C^6, C^{17}, 28.13 C^5, C^{16}, 29.58 C^9, C^{10}, 35.25 C^7, C^{18}, 35.36 C^4,C^{15}, 39.52 C^3, C^{14}, 41.39 C^1, C^{12}, 77.89 C^2, C^{13}, 171.93 C^8, C^{11}. MS (70eV), m/z (*J.*, %): $[M-1]^+$ 306 (0.1), 95 (100), 67 (24), 195 (18), 55 (9), 79 (9), 111 (9), 213 (9), 41 (7), 162 (7).

12.4 CONCLUSION

The obtained results indicate that the zeolite H-Beta is active and selective catalyst for synthesis of alcohols, ethers, and esters from olefins with acids and alcohols.

It should be noted that offered methods are simple compared to methods based on the use of traditional acid catalysts. In this case, products of reactions are separated from the catalyst by filtering and zeolite H-Beta can be recovered for use later.

Besides high activity and selectivity, zeolite catalyst H-Beta makes it possible to obtain new structure compounds, which were not obtained based on homogeneous acidic catalysts.[12]

KEYWORDS

- **heterogenic catalyst**
- **norbornene**
- **styrene**
- **2-vinyl-2-methil-gem-dichlorocyclopropane**
- **zeolite H-Beta**
- **alcohols**
- **ethers**
- **esters**
- **diesters**

REFERENCES

1. Valencia, S.; Corma, A.; Cambor, M. *Micropor. Mesopor. Mat.* **1998,** *25,* 59–74.
2. Minchayev, C. M., Kondratyev, D. A. *Uspehi Khimii* («*Successes of Chemistry*», in Rus.), 1983; *52(12)*, 21–73.
3. Reutov, O. A.; Kurts, A. L.; Butin, K. P. *Organic Chemistry (in Rus.)*. Part 2, 1999; 624.
4. Patent 2,345,573, US Herman A. Bruson http://www.freepatentsonline.com/2345573.html. 1944.- #4.
5. Yang, X.; Chatterjee, S.; Zhang, Z.; Zhu, X.; and Charles U. P. Jr. *Ind. Eng. Chem. Res.* **2010,** *49,* 2003.
6. Brown, H.; Kawakami, J. *J. Amer. Chem. Soc.* **1990,** *92,* 1970.
7. Butlerov, A. M. *Selected Works in Organic Chemistry*; Publishing House of the Academy of Sciences of the USSR: Moscow, RU, 1951; p 333.
8. Azinger, Ph. *Chemistry and Technology of Monoolefins*; Gostoptehidat: Moscow, RU, 1960; p.467.
9. Sumio, A. *Chem. Eng.* **1973,** 56, 80.
10. Menjajlo, A. T. *Synthesis of Alcohols and Organic Compounds from Oils*. Tr. NIISa. Goschimizdat: Moscow, Ru, 1960; pp 226.
11. Mamedov, M. K. *J. Org. Chem.* (in Rus.), **2006**, *42* (8), 1159–1162.
12. Gasanov, A. G.; Nagiev, A. V. *J. Org. Chem.* (in Rus.), **1994**, *30* (5), 707–709.

CHAPTER 13

STRESS BIREFRINGENCE OF HIGHLY CROSS-LINKED POLYMERS OF OPTOELECTRONIC APPLICATION IN THE CONTEXT OF HEREDITY THEORY

N. V. ULITIN[1], I. I. NASYROV[1], R. R. NABIEV[1], D. A. SHIYAN[1], N. K. NURIEV[1], G. E. ZAIKOV[2]

[1]Kazan National Research Technological University, 68 Karl Marx Street, 420015 Kazan, Republic of Tatarstan, Russian Federation. Fax: +7 (843) 231-41-56. E-mail: n.v.ulitin@mail.ru

[2]N. M. Emanuel Institute of Biochemical Physics, Russian Academy of Sciences, 4, Kosygina st., Moscow 119334, Russian Federation. Fax: +7(499)137-41-01. E-mail: Chembio@sky.chph.ras.ru

CONTENTS

ABSTRACT

Aiming to control by stress birefringence, the radiotransparent fiberglass plastic products based on highly cross-linked polymer matrices, theoretical regularities for mathematical description of this property were developed. Computer physical modeling of topological structure of experimental objects was carried out on epoxy–amine polymers with different cross-link density taken as an example. And constants of this model were specified. The adequacy of the model was demonstrated by comparison of the model-calculated against experimental approach to thermal polarization curves.

13.1 INTRODUCTION

To manufacture protective domes for radar installations, radiotransparent fiberglass plastics are used, that is, polymer composite materials consisting of highly cross-linked polymer matrix reinforced by glass fiber. When in use, radiotransparent products are subject to static (in particular, by its own weight) or dynamic loads, and highly cross-linked polymer matrix demonstrates an effect of stress birefringence. The actual challenge that arises during application of radiotransparent fiberglass plastics is reduction of stress birefringence. Therefore, the aim of this work is to develop a mathematical model for describing stress birefringence of highly cross-linked polymer matrices in all physical states of the ones (glassy state, rubbery state, and transition state between them).

13.2 MATHEMATICAL MODEL

We can define relative deformation of polymer solid body of free shape in the form of tensor as follows[1]:

$$u_{ik} = \frac{1}{2} J\tau_{ik} - \frac{1}{3} B_{\infty} p\delta_{ik}, \qquad (13.1)$$

where B_{∞} is balanced bulk creep compliance, (MPa^{-1}); $\boldsymbol{J}$ is relaxation operator of shear compliance (MPa^{-1}), τ_{ik} is tensor of shear stress (MPa); p is pressure, which is compressing any volume element without changing its shape (MPa); δ_{ik} is Kronecker symbol.

Contributions B_{∞} in u_{ik} for highly cross-linked polymer matrices are very small and are basically ignored.[1] If the deformation of cross-linked polymer matrix is not accompanied by destruction of its chemical structure, then:

$$J = J_{\infty} J_{\mathrm{N}}, \qquad (13.2)$$

where J_∞ is balanced shear compliance at the given temperature (MPa^{-1}); $\boldsymbol{J}_N$ is a normalized to 1 relaxation Volterra operator.

In this paper, all the discussions were made for highly cross-linked polymer matrices, whose topological structure is spatially uniform. Highly cross-linked polymer matrices on supramolecular level of the structure is characterized by micro-heterophasicity: apart from gel fraction formed by globules and their aggregates, micro-dispersed formations (sol fraction) are also present here, which are formed of linear and/or branched macromolecules of low molecular weight. Since different topologies of cross-linked macromolecules forming microgel supramolecular formations are found in the ones equally often over the entire volume, they are considered statistically equivalent, and highly cross-linked polymer matrices are considered spatially uniform, owing to their topological structure. An important experimental proof of that is only inversely proportional dependence of balanced shear compliance J_∞ (MPa^{-1}) verses temperature (T, K):

$$J_\infty = A_\infty / T, \tag{13.3}$$

where A_∞ is independent of T constant of rubbery state, K/MPa.

Since relaxation spectrum of shear compliance consists of β- and α-branches[1], therefore, for highly cross-linked polymer matrices with spatially uniform topological structure, operator $\boldsymbol{J}_N$ is as following:

$$J_N = w_{J,\beta} + (1 - w_{J,\beta}) J_{N,\alpha}, \tag{13.4}$$

where $w_{J,\infty}$ is weighting coefficient, independent of T, and reflecting contribution of β-transitions in J_∞; $\boldsymbol{J}_{N,\alpha}$ is fractional exponential operator connected to distribution of α-relaxation time $L_{J,\alpha}(\theta)$:

$$J_{N,\alpha} = \int_{-\infty}^{\infty} L_{J,\alpha}(\theta)[1 - \exp(-t/\theta)]\,d\ln\theta \tag{13.5}$$

In eq13.5, θ is relaxation time, t is current time.

Normalized to 1 α-mode has been described using distribution developed by Yu. N. Rabotnov[2]:

$$L_{J,\alpha}(\theta) = \sin[\pi(1 - \Xi_{J,\alpha})] / [2\pi\{ch[(1 - \Xi_{J,\alpha})\ln(\theta/\Theta_{J,\alpha})] + \cos[\pi(1 - \Xi_{J,\alpha})]\}] \tag{13.6}$$

where $\Theta_{J,\alpha}$ is average α-relaxation time; $\Xi_{J,\alpha}$ is independent of T distribution width ($0 \le \Xi_{J,\alpha} \le 1$).

Operator $\boldsymbol{J}_{N,\alpha}$ in glassy state takes on the value equal to 0, in rubbery state equal to 1, and in transition state between these physical states it is equal to 0–1. That is why eqs 13.1 through 13.6 will describe $\boldsymbol{J}$ of highly cross-linked polymer matrices with spatially uniform topological structure in all their physical states.

Stress birefringence was reduced to ***J***, considering that ordered orientation of solid body molecules[2] is occurring by deformational shear. The point of departure for our discussions was the equation linking dielectric permittivity of the deformed polymer dielectric with independent components of relative deformation tensor[3]:

$$\varepsilon_{ik} = \varepsilon_0 \delta_{ik} + a_1 \gamma_{ik} + \frac{1}{3}(a_1 + 3a_2) u_{ll} \delta_{ik}, \tag{13.7}$$

here ε_0 is dielectric permittivity of the nondeformed solid body; γ_{ik} is deformation shear tensor; a_1, a_2 are polarization coefficients; and u_{ll} is volumetric compression deformation tensor.

[1] α-branch reflects cooperative mobility of network nodes, which are not participating in local movements; β-branch is connected with local conformation mobility;

[2] this orientation is the purpose of polarization anisotropy.

From eq 13.7, equation of Brewster-Wertheim law can be obtained:

$$\Delta n = \xi_\infty \Delta \gamma = C_\infty \Delta \tau, \tag{13.8}$$

where Δn is stress birefringence; $\Delta\gamma$ and $\Delta\tau$ are differences of main shear deformations and stresses in the given point; C_∞ is balanced electromagnetic susceptibility (MPa^{-1}) connected with J_∞ equation:

$$C_\infty = 0.5 \xi_\infty J_\infty, \tag{13.9}$$

here ξ_∞ is balanced elastic coefficient of electromagnetic susceptibility.

Let us introduce the relaxation operators for electromagnetic susceptibility ***C*** (MPa^{-1}) and for elastic coefficient of electromagnetic susceptibility ***ξ***, assuming that they are connected with ***J*** in the form of equation below:

$$C = 0.5 \boldsymbol{\xi} J \tag{13.10}$$

The importance of eq 13.10 lies in the fact that we get a result of coincidence of relaxation spectra ***ξ*** and shear module. Operator ***ξ*** is as follows:

$$\xi = \xi_\infty (w_{\xi,\beta} + (1 - w_{\xi,\beta}) G_{N,\alpha}), \tag{13.11}$$

where $w_{\xi,\infty}$ is weighting coefficient, independent of *T*, and reflecting contribution of β-transitions in ξ_∞; $\boldsymbol{G}_{N,\alpha}$ is fractional exponential operator, reverse to $\boldsymbol{J}_{N,\alpha}$.

Based on the rule of multiplication of the fractional exponential operators[2] from eq 13.10, we get:

$$C = C_\infty (w_{C,\beta} + (1 - w_{C,\beta}) J_{N,\alpha}), \tag{13.12}$$

where $w_{C,\beta}$ is weighting coefficient independent of T and reflecting contribution of β-transitions in C_∞.

The result of eq 13.12 shows coincidence of relaxation spectra $\boldsymbol{J}$ and $\boldsymbol{C}$, hereby $\boldsymbol{C}$, as well as $\boldsymbol{J}$, will cover all physical states of highly cross-linked polymer matrices. To apply the obtained regularities in practice, we must know the temperature dependence of α-relaxation time:

$$\lg(\Theta_{J,\alpha}(T)/\Theta_{J,\alpha}(T_g)) = 40\left((f_g/(f_g+\alpha(1-f_g)(T-T_g))-1\right),\ \alpha=\begin{cases}\alpha_g,\ T<T_g\\ \alpha_\infty,\ T>T_g,\end{cases} \quad (13.13)$$

where T_g is glass transition temperature, K; α_g, α_∞ are coefficients of thermal expansion in glassy and rubbery states, K^{-1}; f_g is fractional free volume at T_g.

Equation 13.13 is obtained analytically from equation below:

$$\lg(\Theta_{J,\alpha}(T)/\Theta_{J,\alpha}(T_g)) = (f_g^{/}+\left[0.025\alpha_\infty(1-f_g)/f_g\right](T-T_g))^{-1}-(f_g^{/})^{-1},$$

suggested by Ferry[4] to describe relaxation of linear, branched, and lightly cross-linked polymers at higher than T_g temperatures, taking into account the known temperature function f_g of highly cross-linked polymer matrices.[5] It is to be noted that value $f_g^{/} = 0.025$, which Ferry identified with fractional free volume at T_g, is not the same in reality, and for highly cross-linked polymer matrices, it is included in coefficient 40.

So, theoretical regularities of stress birefringence of highly cross-linked polymer matrices, forming under the influence of temperature fields and mechanical stresses, are fully described by an aggregate of eqs 13.2 through 13.6, 13.8, 13.10 through 13.13. According to these equations, stress birefringence can be assessed based on parameters such as: T_g, α_g, α_∞, A_∞, ξ_∞, $w_{C,\infty}$, $w_{J,\infty}$, $\Theta_{J,\alpha}$, and $\Xi_{J,\alpha}$. In this connection, theoretical and experimental assessment of these values were carried out to demonstrate adequacy of the introduced theoretical representations and their operability for evaluation of stress birefringence for highly cross-linked epoxy–amine polymers at different conditions.

13.3 EXPERIMENTAL PART

13.3.1 EXPERIMENTAL OBJECTS

Experimental objects became highly cross-linked polymer matrices with a various cross-link density on the basis of diglycidyl ether of bisphenol-A (DGEBA—Fig. 13.1(a)), cured by mixtures of hexylamine (HA—Fig. 13.1(b)) and hexamethylenediamine (HMDA—Fig. 13.1(c)) at a variation of a molar ratio of the ones $x = n$ (HA)/n (HMDA) from 0 to 2 (step 0.5) taking into account a stoichiometry of epoxy groups and hydrogen of the amine group.

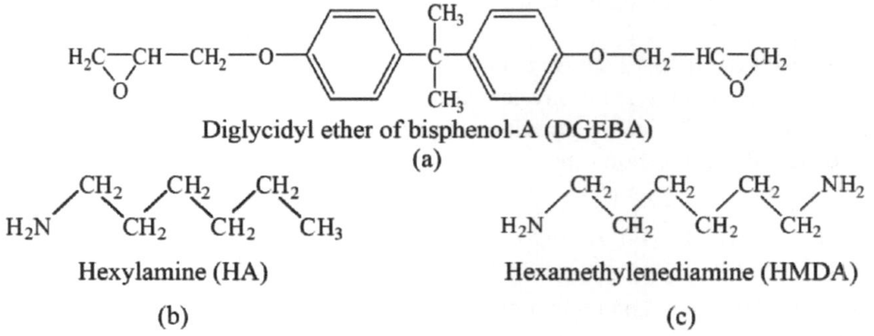

FIGURE 13.1 Initial substances.

Preparation of epoxy–amine compositions: DGEBA and HMDA weighted on scales METTLER TOLEDO AB304-S/FACT (up to 0.0001 g) and heated to 315 K, then molt of HMDA and required amount of HA were mixed with DGEBA; then the mixture was stirred to form a homogeneous mass and was poured into an ampoule and vacuumized for 1 h in the "freeze–thaw" conditions; ampoule filled with argon and sealed. Curing conditions: at 293 K for 72 h, at 323 K for 72 h, at 353 K for 72 h, and at 393 K for 72 h (chosen on the basis of the representations set forth in[6]).

13.3.2 EXPERIMENTS BY PHOTOELASTIC METHOD

One of the methods of experimental determination of stress birefringence is a photoelastic method. The tests of the experimental objects were conducted on the test facilities[7] designed to measure the relative stress birefringence in the center of the disk and the horizontal diameter relative deformation of the disk. Disc is made of the test material and 18 mm in diameter and 3 mm in thickness. At the test, the disc was compressed by the concentrated forces on the vertical diameter. Measurement error of the relative deformation and the stress birefringence does not exceed 3 and 1%, respectively. To determine the T_g, α_g, and α_∞, test facilities were used as a dilatometer. Dilatometric curves $u(T)$ are temperature dependences of the relative deformation of unloaded samples, when cooling of the ones with a constant average speed of 0.4 K/min were averaged in the results over four measurements. The T_g, α_g, and α_∞ were determined by the method of ordinary least squares according to eq 13.9:

$$u(T)=\frac{1}{3}\left[\alpha_\infty\left(T_g-T_0\right)+\alpha\left(T-T_g\right)\right],\quad \alpha=\begin{cases}\alpha_g, & T<T_g\\ \alpha_\infty, & T>T_g\end{cases},$$

where T_0 is initial temperature of the sample (rubbery state), K.

J_∞ and C_∞ are dependent of T, and therefore, were determined on the basis of measurements of the relative deformation and stress birefringence with four loads for a series of temperatures above $T_g + 30$ K followed by averaging the results of four tests for each load. Substituting the obtained values of J_∞ and C_∞ in eqs 13.3 and 13.9, allows determining the experimental A_∞ and ξ_∞.

C and ***J*** in glassy state has been calculated in terms of measuring results of relative deformations and stress birefringence using four loads at 298 K with subsequent averaging of results of four tests for each load. Experimental values $w_{J,\beta}$ and $w_{C,\beta}$ were defined by substitution of ***J***, ***C*** values (in glassy state), and A_∞, ξ_∞ in eqs 13.2, 13.4, and 13.12, on condition that in the glassy state $\boldsymbol{J}_{N,\alpha} = 0$. Creep and photo creep curves are the development of relative deformation and stress birefringence into time at constant T under the influence of constant stress. They were taken for several T of transit state between glassy and rubbery states, $T_g \pm 15$ K, and were averaged over the results of four measurements. Empirical values of creep and photo creep functions were calculated by eq 13.8.[8]

13.4 MODELING OF TOPOLOGICAL STRUCTURE OF EXPERIMENTAL OBJECTS

Modeling of topological structure of experimental objects was carried out in two ways: using computer modeling and graphs theory.

13.4.1 COMPUTER MODELING

The topological structure of highly cross-linked polymer matrices in the length interval of 0.25–2 nm is a random fractal.[9] Models of topological structure (Fig. 13.2) of experimental objects were arranged in the Bullet Physics Library (www.bullet-physics.org) with use of known values of Van-der-Waals volume of the atoms and link lengths between them,[5] plus fractal dimension (d_f) (Table 13.1) defined by the Bullet Physics Library from the following expression[9]:

$$N_{st} \infty R_W^{d_f},$$

where N_{st} is a number of random segments within the sphere radius R_W.

TABLE 13.1 The Theoretical and Experimental Values of Glass Transition Temperature, the Constant of Rubbery State, the Balanced Elastic Coefficient of Electromagnetic Susceptibility and Weighting Coefficients.

x	<l>	n_{3f}	d_f	T_g, K			A_∞, K/MПa			ξ_∞			$w_{J,\beta}$			$w_{C,\beta}$		
				theor.	exp.	ε,%	theor.	exp.	ε,%	theor.	exp.	ε,%	theor.	exp.	ε,%	theor.	exp.	ε,%
0.0	1.00	0.4000	2.63	380	382	2	33.5	35.0	4	0.0240	0.0263	9	0.0231	0.0180	28	0.0407	0.0280	45
0.5	1.25	0.3333	2.65	373	372	1	47.3	53.0	11	0.0230	0.0235	2	0.0164	0.0150	9	0.0298	0.0260	15
1.0	1.50	0.2857	2.69	360	361	1	62.1	69.3	10	0.0220	0.0224	2	0.0125	0.0140	11	0.0232	0.0230	1
1.5	1.75	0.2500	2.72	350	350	0	77.7	81.6	5	0.0220	0.0207	6	0.0100	0.0120	17	0.0188	0.0200	6
2.0	2.0	0.2222	2.74	344	344	0	93.9	93.5	4	0.0220	0.0192	15	0.0083	0.0080	4	0.0160	0.0180	11

* The relative difference of the theoretical value versus the experimental value was calculated by formula $\varepsilon = |(\text{exp.value} - \text{theor.value})/\text{exp.value}| \cdot 100$, %.

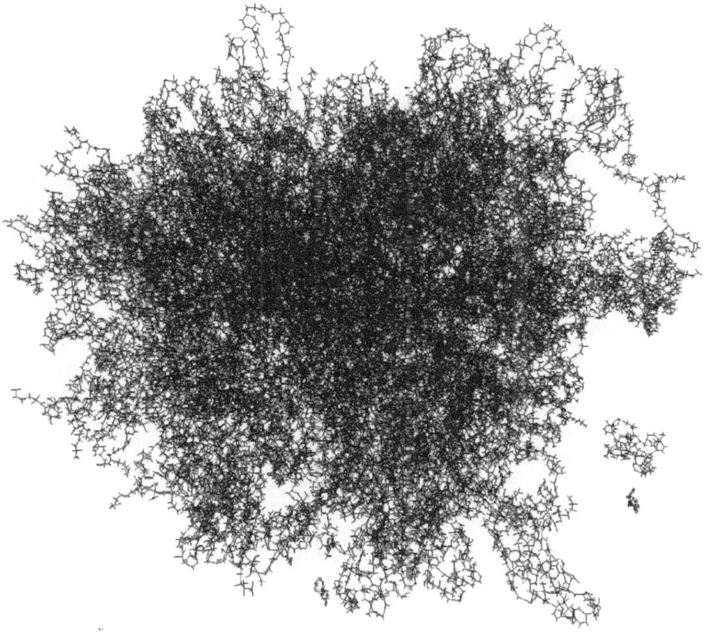

FIGURE 13.2 The model of topological structure of the experimental object composed of the composition $x = 2.0$.

The nature of modeling was in polyaddition imitation. For each experimental object, the model was arranged in such a way that the topological structure was spatially uniform and the total number of elastically effective nodes comprised minimum 10,000.

13.4.2 MODELING ON THE BASIS OF GRAPHS THEORY

Figure 13.3 shows a structure of the repeating fragment common to all polymer series. If we designate the numbers of HA fragments with two methylene groups attached to nitrogen atom as N_{2f}, the elastically effective nodes as N_{3f}, the tetramethylene fragments as N_{σ}, and DGEBA fragments as N_{π}, and N_{tot} being the total number of links, then, according to stoichiometry, we work out as follows:

$$N_{2f} = xn(HMDA), N_{3f} = 2n(HMDA), N_{\sigma} = n(HMDA),$$

$$N_{\pi} = (2 + x)n(HMDA),\ N_{tot} = (5 + 2x)n(HMDA)\ .$$

$$n_{3f} = N_{3f} / N_{tot} = 2/(5+2x), \quad <l> = N_\pi / N_{3f} = 1+0.5x.$$

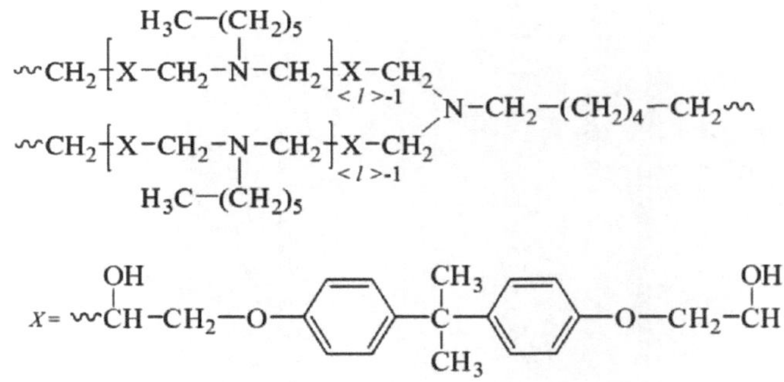

FIGURE 13.3 The repeating fragment of the topological structure covering all experimental objects.

In this case, random parameter representing the number of elastically effective nodes n_{3f} (Table 13.1), and the number average degree of polymerization of the intermodal chain $< l >$ (Table 13.1) are as follows:

$$n_{3f} = N_{3f} / N_{tot} = 2/(5+2x), \quad <l> = N_\pi / N_{3f} = 1+0.5x$$

13.5 THEORETICAL ASSESSMENT OF CONSTANTS FOR STRESS BIREFRINGENCE MODEL

13.5.1 GLASS TRANSITION TEMPERATURE

Glass transition temperature[9] (Table 13.1):

$$T_g = Cl_{st}^{d_f - d}, \text{ K}$$

where $C = 270$ K is a constant, d is Euclidean dimension, and l_{st} is an average size of a random segment, nm.

The difference between theoretical and experimental values T_g is maximum 2%.

13.5.2 COEFFICIENTS OF THERMAL EXPANSION

Calculated frombelow equations,[9] values of α_g and α_∞ for all experimental objects turned out to be approximately equal, and in average were $3.2 \cdot 10^{-4}$ K^{-1} and $6.1 \cdot 10^{-4}$ K^{-1}, respectively. These values agree with experimental values: $4.3 \cdot 10^{-4}$ K^{-1} and

$7.0 \cdot 10^{-4}$ K^{-1}, the relative difference between the theory and the experiment is 26 and 13%, respectively.

$$\alpha_g = \frac{-3(T_g \times 10^{-4} - 2d_f + 5) - \sqrt{9(T_g \times 10^{-4} - 2d_f + 5)^2 - 12T_g^2 \times 10^{-8}}}{2T_g^2 \times 10^{-4}}, \text{K}^{-1},$$

$$\alpha_\infty = \frac{0.106 + \alpha_g T_g}{T_g} = \frac{0.106}{T_g} + \alpha_g, \text{K}^{-1}.$$

13.5.3 CONSTANT OF RUBBERY STATE AND BALANCED ELASTIC COEFFICIENT OF ELECTROMAGNETIC SUSCEPTIBILITY

Constant of rubbery state is as follows[5]:

$$A_\infty = (0.5 f\, C_{\text{tot}} n_{3\text{f}} RF)^{-1}, \tag{13.14}$$

where f is functionality of the network nodes; F: front-coefficient; R: gas constant (8.314 J/(mol×K)); C_{tot}: concentration of elemental links of the topological structure at T_g, mol/cm^3.

By definition, we have:

$$C_{\text{tot}} = N_{\text{tot}} / V_{\text{tot}}(T_g),$$

where $V_{\text{tot}}(T_g)$ is volume of elemental links at T_g (cm^3):

$$V_{tot}(T_g) = N_\sigma V_\sigma(T_g) + N_{3f} V_{3f}(T_g) + N_{2f} V_{2f}(T_g) + N_\pi V_\pi(T_g),$$

here, $V_\sigma(T_g)$, $V_{3f}(T_g)$, $V_{2f}(T_g)$, and $V_\pi(T_g)$ are molar volumes of elemental links at T_g, cm^3/mol. Then:

$$C_{tot} = (5 + 2x) d(T_g) / (M_\sigma + 2M_{3f} + xM_{2f} + (2 + x)M_\pi), \tag{13.15}$$

where M_σ, M_{3f}, M_{2f}, and M_π are molar mass of elemental links, g/mol; $d(T_g)$ is polymer density at T_g, g/cm^3.

For highly cross-linked polymer matrices temperature dependence of density is as follows[5]:

$$d(T) = k_g M_{r.f.} \Big/ \left(10^{-24} \left[1 + \alpha(T - T_g)\right] N_A \left(\sum_i \Delta V_i \right)_{r.f.} \right), \alpha = \begin{cases} \alpha_g, T < T_g \\ \alpha_\infty, T > T_g \end{cases},$$

here, $d(T)$ is polymer density at T, g/cm^3; k_g is molecular packing coefficient at T_g; $M_{\text{r.f.}}$ is molar mass of the repeating fragment of network, g/mol; 10^{-24} is conversion coefficient, from Å^3 to cm^3; N_A is Avogadro constant; and $\left(\sum_i \Delta V_i\right)_{r.f.}$ is Van-der-Waals volume of the repeating fragment of network.

The molar mass of the repeating fragment of network is $M_{r.f.} = 441 <l> - 43$ g/mol.

Taking into account that for highly cross-linked polymer matrices at T_g, the $k_g \approx 0.681$,[5] we work out:

$$d(T) = 1.13(441<l> - 43)/\left(\left[1+\alpha\left(T - T_g\right)\right](458.9<l> - 57.4)\right), \alpha = \begin{cases} \alpha_g, T < T_g \\ \alpha_\infty, T > T_g \end{cases}$$

Thus, substitution of T_g, α_g, and α_∞ allows finding the $d(T)$ values at any T values. The molar masses of the elemental links are as follows: $M_\sigma = 56$ g/mol, $M_{3f} = 56$ g/mol, $M_{2f} = 127$ g/mol, and $M_\pi = 314$ g/mol.

Substitution of resultant expression for C_{tot}

$$C_{tot} = \left(1.13\left(5+2x\right)\left(441<l> - 43\right)\right)/\left(\left(796+441x\right)\left(458.9<l> - 57.4\right)\right)$$

in eq 13.14 yields:

$$A_\infty = \left(796+441x\right)\left(458.9<l> - 57.4\right)/\left(1.695FR\left(5+2x\right)\left(441<l> - 43\right)n_{3f}\right).$$

The F values for highly cross-linked polymer matrices with spatially uniform topological structure are within the range of 0.65–0.85 and should increase in a linear fashion as the network density grows up.[10] On this basis we receive:

$$F = 1.125n_{3f} + 0.4$$

$$A_\infty = \left(796+441x\right)\left(458.9<l> - 57.4\right)/\left(1.695\left(1.125n_{3f}+0.4\right)R\left(5+2x\right)\left(441<l> - 43\right)n_{3f}\right) \quad (13.16)$$

Experimental values of A_∞ (Table 13.1) can be determined by way of approximation of empirical J_∞ values (2). Small difference between theory and experiment for A_∞ (4–11%) prove the adequacy of the eq 13.16. Experimental values of F can be determined from A_∞ values, for that, on the basis of experimental polymer density values, preliminary calculation of concentration of elemental links should be found by eq 13.15.

To assess ξ_∞, the following equation was established:

$$\xi_\infty = -3K(\partial\delta\varepsilon/\partial T)/(\alpha_\infty\sqrt{\varepsilon_0}), \quad (13.17)$$

where ε_0 is dielectric permittivity of the unstressed polymer at T_g, K is constant, depending on polymer topological structure; $\partial\delta\varepsilon/\partial T$ is one of the components of the derivative ε_0 in T, which relates to electromagnetic anisotropy in stressed polymer, K^{-1}; ε_0 and $\partial\delta\varepsilon/\partial T$ were calculated incrementally according to standard methods.[5] Constant K was determined by the method described in the work,[3] its value for experimental objects is 15.7 (Figure 13.4).

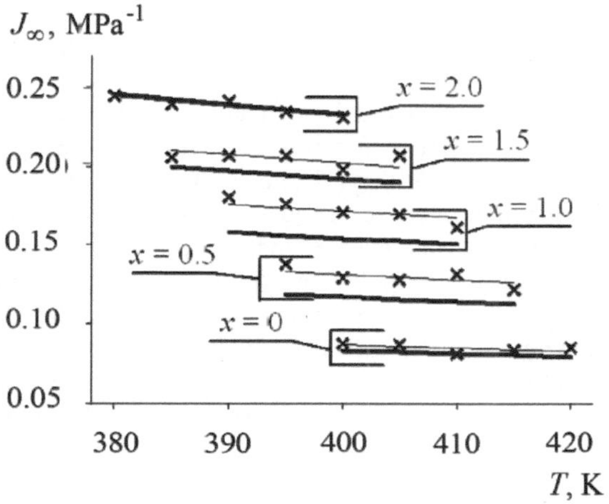

FIGURE 13.4 Temperature dependence of balanced shear compliance (cross: experiment, line: approximation).

Finally (Table 13.1):

$$\xi_\infty = 0.07\left[\frac{212.7362 < l > +5.309}{458.9 < l > -57.4}\right]\sqrt{\frac{186.860314 < l > -24.388346}{455.465782 < l > -54.939368}}\cdot$$

The difference between theoretical and experimental values, ξ_∞, was maximum 15%.

13.5.4 WEIGHTING COEFFICIENTS

Weighting coefficients (Table 13.1):

$$w_{J,\beta} = \frac{J_{\beta,\infty}}{A_\infty}T, \quad w_{C,\beta} = \frac{C_{\beta,\infty}}{0.5\xi_\infty A_\infty}T,$$

where $C_{\beta,\infty} = \dfrac{J_{\beta,\infty} a_1}{4n_0}$; $J_{\beta,\infty} = \dfrac{2}{3} \times \dfrac{1+\mu_{\beta,\infty}}{1-2\mu_{\beta,\infty}} B_\infty$; $\mu_{\beta,\infty} = \dfrac{d_f}{d-1} - 1$.

Here a_1 is coefficient (refer to eq 13.7); n_0 is a refractive index of the nondeformed polymer dielectric, which is in glassy state. The a_1 and n_0 are calculated incrementally according to standard methods at 298 K.[5] Table 13.1 shows a comparison of experimental and theoretical determination of the weighting coefficients.

13.5.5 PARAMETERS OF THE RELAXATION SPECTRUM

Parameters of the relaxation spectrum theoretically cannot be assessed neither within the framework of the increments method, nor within the framework of the fractal approach. Empirical values of $\Theta_{J,\alpha}$ (T) and $\Xi_{J,\alpha}$ were determined from the creep and photo creep curves (method of least squares with regularization of solutions by singular decomposition[11]). It was revealed that for the highly cross-linked polymer matrices with spatially uniform topological structure $\Xi_{J,\alpha}$ is determined by network topology and is independent of temperature (no splitting of α-transition). The experimentally found value $\Xi_{J,\alpha}$ increases linearly from 0.4 to 0.65 with an increase $< l >$ (for a priori prediction, the value of 0.5 can be used). Example of $\Theta_{J,\alpha}$ (T) dependence of temperature is shown in Figure 13.5: the eq 13.13 describes both branches of the experimental curve lg$\Theta_{J,\alpha}$ (T) with high accuracy. The obtained value of the share of the fluctuation free volume was averaged over all experimental objects, the average value was $f_g = 0.095$. This result is consistent with the currently accepted value for the highly cross-linked polymer matrices –0.09.

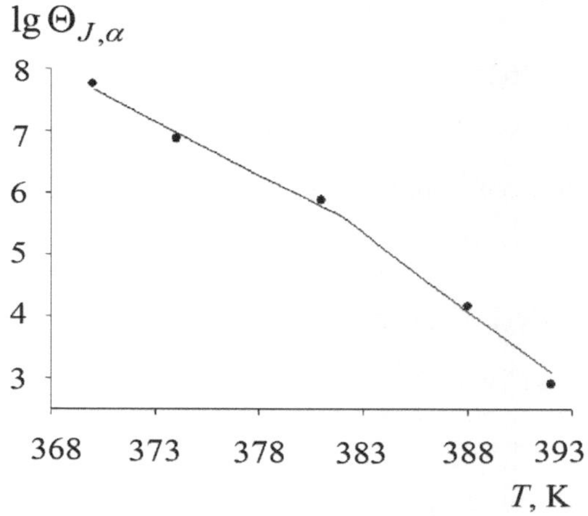

FIGURE 13.5 The dependence of lg$\Theta_{J,\alpha}$ on T for the object of the composition $x = 0$ (dots: experiment, line: approximation).

13.5.6 ADEQUACY OF THE MODEL

The adequacy of the model of stress birefringence of highly cross-linked polymer matrices was experimentally demonstrated by comparing of the predicted and the actual course of the thermal polarization curves (Fig. 13.6).

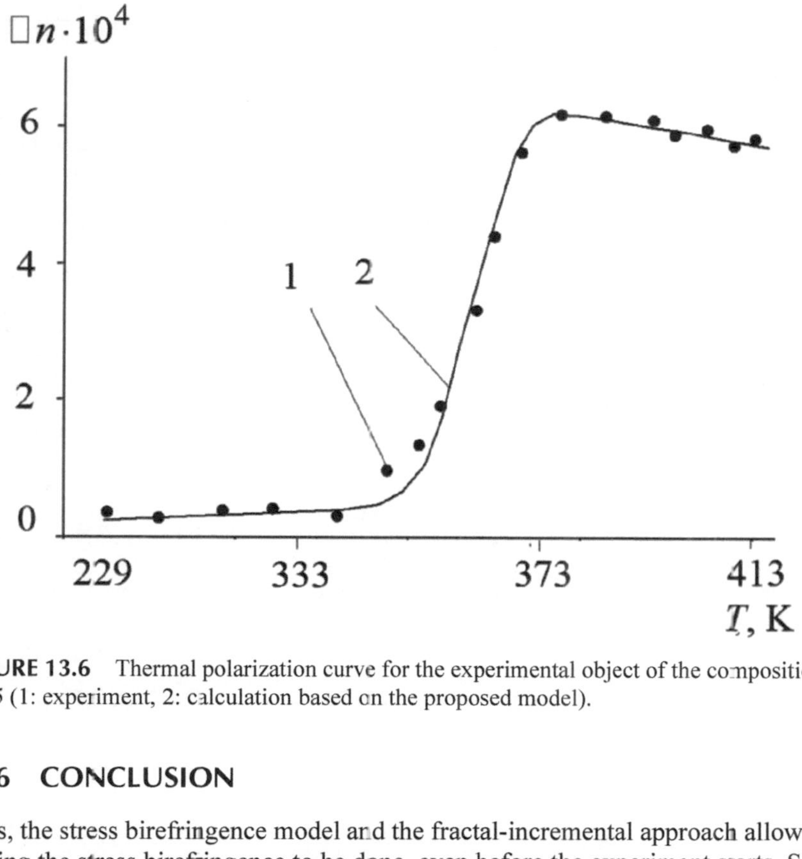

FIGURE 13.6 Thermal polarization curve for the experimental object of the composition $x = 1.5$ (1: experiment, 2: calculation based on the proposed model).

13.6 CONCLUSION

Thus, the stress birefringence model and the fractal-incremental approach allow assessing the stress birefringence to be done, even before the experiment starts. So, a priori, we can assess the maximum ultimate value of stress birefringence for which highly cross-linked polymer matrix is capable in the given operating conditions of radiotransparent fiberglass. Knowing the value of stress birefringence of highly cross-linked polymer matrices, one can assess the change of fiberglass radiotransparency coefficient applicationwise. Hence, the possibility of highly cross-linked polymer matrix application for fiberglass is being justified.

KEYWORDS

- **highly cross-linked epoxy–amine polymers**
- **modeling**
- **heredity theory**
- **fractal analysis of macromolecules**

REFERENCES

1. Davide, S. A.; De Focatiis, C. P. B.. Prediction of Frozen-In Birefringence in Oriented Glassy Polymers Using a Molecularly Aware Constitutive Model Allowing for Finite Molecular Extensibility. *Macromolecules* **2011**, *44* (8), 3085–3095.
2. Rabotnov, Y. N. *Mechanics of a Deformable Solid Body*; Nauka: Moscow, RU, 1979; p 744 (in rus).
3. Blythe, T.; Bloor, D. *Electrical Properties of Polymers;* Cambridge University Press: Cambridge, UK, 2005; p 492.
4. Ferry, J. D. *Viscoelastic Properties of Polymers*, 3rded.; John Wiley & Sons, Inc.: New York, Chichester, Brisbane, Toronto, Singapore, 1980; p 641.
5. Askadskii, A. A. *Computational Materials Science of Polymers;* Cambridge International Science Publish: Cambridge, UK, 2003; p 650.
6. Irzhak, V. I.; Mezhikovskii, S. M. Structural Aspects of Polymer Network Formation Upon Curing of Oligomer Systems. *Russ. Chem. Rev.* **2009**, *78* (2), 165–194.
7. Zuev, B. M.; Arkhireev, O. S. The Initial Stage in the Fracture of Stressed Dense-Cross-Linked Polymer Systems. *Polym. Sci. U.S.S.R.* **1990**, *32* (5), 941–947.
8. *Handbook of Thermal Analysis and Calorimetry;* Cheng, S. Z. D., Ed.; Applications to Polymers and Plastics, Vol. 3; Elsevier Science B.V.: Amsterdam, 2002; pp 1–45.
9. Novikov, V. U.; Kozlov, G. V. Structure and Properties of Polymers in Terms of the Radical Approach. *Rus. Chem. Rev.* **2000**, *69* (6), 523–549.
10. Irzhak, V. I. Topological Structure and Relaxation Properties of Polymers, *Russ. Chem. Rev.* **2005**, *74*, 937.
11. Tihonov, A. N.; Arsenin, V. Ja. *Methods of Decission of Incorrect Problems*, 3rd ed.; Nauka: Moscow, RU, 1986; p 287 (in rus).

CHAPTER 14

INVESTIGATION OF VARIOUS METHOD FOR STEEL SURFACE MODIFICATION

I. NOVÁK[1], I. MICHALEC[1], M. VALENTIN[1], M. MARÔNEK[2], L. ŠOLTÉS[3], J. MATYAŠOVSKÝ[4], P. JURKOVIČ[4]

[1]Department of Welding and Foundry, Faculty of Materials Science and Technology in Trnava, 917 24 Trnava, Slovakia

[2]Slovak Academy of Sciences, Polymer Institute of the Slovak Academy of Sciences, 845 41 Bratislava, Slovakia

[3]Institute of Experimental Pharmacology of the Slovak Academy of Sciences, 845 41 Bratislava, Slovakia

[4]VIPO, Partizánske, Slovakia. E-mail: upolnovi@savba.sk

CONTENTS

14.1 INTRODUCTION

The surface treatment of steel surface is often used, especially in an automotive industry, which creates the motive power for research, design, and production. New methods of surface treatment are also developed having major influence on improvement of the surface properties of steel sheets while keeping the price at reasonable level.[1–3]

The nitrooxidation is one of the nonconventional surface treatment methods, which combine the advantages of nitridation and oxidation processes. The improvement of the mechanical properties (tensile strength, yield strength) together with the corrosion resistance (up to level 10) can be achieved.[4–7] The fatigue characteristics of the nitrooxidized material can be also raised.[5]

Steel sheets with surface treatment are more often used, especially in an automotive industry that creates the motive power for research, design, and production. New methods of surface treatment are also developed having major influence on improvement of the surface properties of steel sheets while keeping the price at reasonable level.

Previous outcomes[1,3,8,4,10] dealt with the welding of steel sheets treated by the process of nitrooxidation by various arc and beam welding methods. Due to high oxygen and nitrogen content in the surface layer, problems with high level of porosity had occurred in every method. The best results were achieved by the solid-state laser beam welding, by which the defect-free joints were created. Due to high initial cost of the laser equipment, further research was directed to the joining method that has not been tested. Therefore, the adhesive bonding was chosen, because the joints are not thermally affected, they have uniform stress distribution and good corrosion resistance.

The goal of the paper is to review the adhesive bonding of steel sheets treated by nitrooxidation and to compare the acquired results to the nontreated steel.

14.2 EXPERIMENTAL

For the experiments, low-carbon deep-drawing steel DC 01 EN 10130/91 of 1 mm in thickness was used. The chemical composition of steel DC 01 is documented in Table 14.1.

TABLE 14.1 Chemical Composition of Steel DC 01 EN 10130/91.

EN designation	C (%)	Mn (%)	P (%)	S (%)	Si (%)	Al (%)
DC 01 10130/91	0.10	0.45	0.03	0.03	0.01	–

14.2.1 CHEMICAL MODIFICATION

The base material was consequently treated by the process of nitrooxidation in fluidized bed. The nitridation fluid environment consisted of the Al_2O_3 with granularity of 120 μm. The fluid environment was wafted by the gaseous ammonia. After the process of nitridation, the oxidation process started immediately. The oxidation itself was performed in the vapors of distilled water. Processes parameters are referred in Table 14.2.

TABLE 14.2 Process of Nitrooxidation Parameters.

	Nitridation	Oxidation
Time (min)	45	5
Temperature (°C)	580	380

14.2.2 ADHESIVES

In the experiments, four types of two-component epoxy adhesives made by Loctite Company (Hysol 9466, Hysol 9455, Hysol 9492, and Hysol 9497) were used. The properties of the adhesives are documented in Table 14.3.

TABLE 14.3 The Characterization of the Adhesives.

	Hysol 9466	Hysol 9455	Hysol 9492	Hysol 9497
Resin type	Epoxy	Epoxy	Epoxy	Epoxy
Hardener type	Amin	Methanethiol	Modified amin	
Mixing ratio (resin:hardener)	2:1	1:1	2:1	2:1
Elongation (%)	3	80	0.8	2.9
Shore hardness	60	50	80	83

14.2.3 METHODS

The experiments were done at the Faculty of Materials Science and Technology, Department of Welding and Foundry in Trnava. The adhesive bonding was applied on the grinded as well as non-grinded surfaces of the material to determine the grinding effect on total adhesion of the material so as on ultimate shear strength of the joints. The grinded material was prepared by grinding with silicone carbide paper up to 240 grit.

Before the adhesive bonding, the bonding surfaces (both grinded as well as non-grinded) were decreased with aerosol cleaner. The overlap area was 30 mm. To ensure the maximum strength of the joints, the continuous layer of the adhesive was coated on the overlap area of both bonded materials. The thickness of the adhesive layer was 0.1 mm and was measured by a calliper. The joints were cured under fixed stress for 48 h at the room temperature. The dimensions of the joints are referred in Figure 14.1.

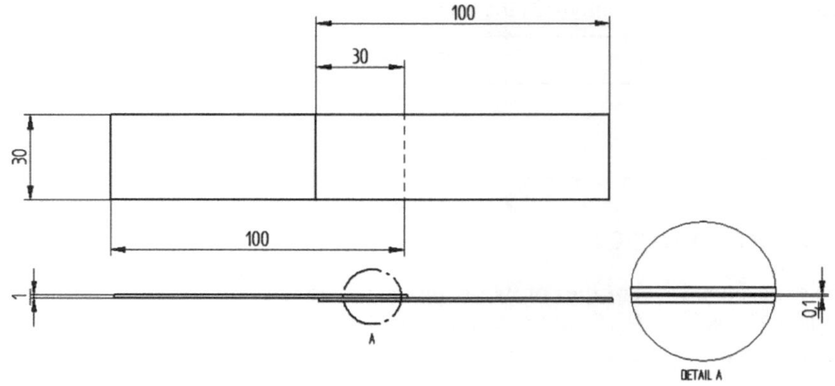

FIGURE 14.1 The dimension of the bonded joints.

The mechanical properties of the joints were examined by the static shear tests. As a device, the LaborTech LabTest SP1 was used. The conditions of the tests were set in accordance with STN EN 10002-1. The static shear tests were repeated on three separate samples and an average value was calculated.
The fracture areas were observed in order to obtain the fracture character of the joints. The JEOL JSM-7600F scanning electron microscope was used as a measuring device.
The differential scanning calorimetry (DSC) was performed on Netzsch STA 409 C/CD equipment. As the shielding gas, Helium with purity of 99.999% was used. The heating process starts at the room temperature and continued up to 400°C with heating rate 10°C/min. The DSC analysis on Hysol 9455 was done on Diamond DSC Perkin Elmer, capable of doing analyses from −70°C.

14.3 RESULTS AND DISCUSSION

The material analysis represents the first step of evaluation. There are many factors having an influence on the joint quality. The properties of the nitrooxidized material depends on the treatment process parameters. For the adhesive bonding, the surface layer properties are important because of that, the high adhesion is needed to ensure the high strength of the joint.

The overall view on the microstructure of the nitrooxidized material surface layer is referred in Figure 14.2a. On the top of the surface, the oxide layer (Fig. 14.2b) was created. This layer had a thickness of ~700 μm. Beneath the oxide layer, the continous layer of ε-phase, consisting of nitrides $Fe_{2\text{-}3}N$ and with the thickness of 8–10 μm was observed.

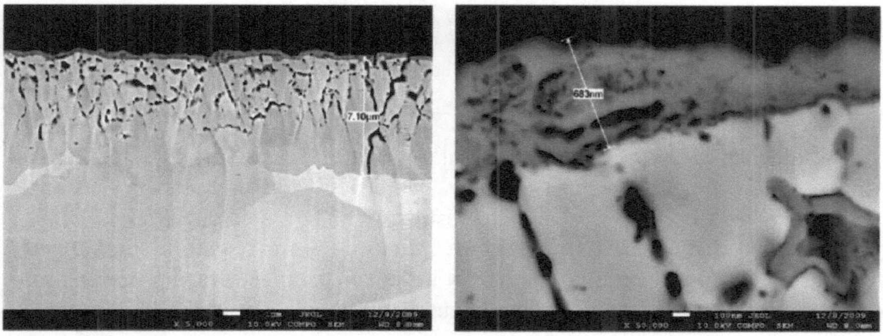

FIGURE 14.2 The microstructure of the surface layer. (a) Overall view and (b) detail view on the oxide layer.

The surface energy measurements were performed due to obtain the properties of the material, which are important for adhesive bonding. For observing the grinding effect on the total surface energy, the measurements were done on the base as well as on the grinded material.

To determine the surface energy, the portable computer-based instrument See-System was used. Four different liquids (distilled water, formamide, diiodomethan, and ethylene glycol) were instilled on the material surface and contact angle was measured. The Owens–Wendt regression model was used for the surface energy calculation. The total amount of six droplets were analyzed of each liquid. The results (Table 14.4) proved that the nitrooxidation treatment had a strong effect on material surface energy, where the decrease by 28% in comparison to non-nitrooxidized material had occurred. The surface energies of grinded and non-grinded material without nitrooxidation were very similar, while the increase of surface energy of grinded nitrooxidized material by 35% in comparison to non-grinded material was observed. In the case of barrier plasma-modified steel, the surface energy is higher than that of unmodified material and namely its polar component is significantly higher than polar component of surface energy for unmodified sample as well as for sample modified by nitrooxidation.

TABLE 14.4 The Surface Energy of Materials.

Material type	Total surface energy (mJ/m^2)	Dispersion component (mJ/m^2)	Acid–base component (mJ/m^2)
DC 01	38.20	35.62	2.58
DC 01 grinded	38.80	33.66	5.14
Nitrooxidized	27.60	25.79	1.80
Nitrooxidized grinded	37.31	32.54	4.77
Barrier plasma treated	39.99	33.69	6.30

The mechanical properties of the material were obtained by the static tensile test. Total amount of three measurements were done and the average values are documented in Table 14.5. Based on the results, it can be stated that after the process of nitrooxidation, the increase of yield strength by 55% and tensile strength by 40% were observed. The barrier plasma did not influence the mechanical properties after surface modification of steel, which remained the same as for unmodified sample.

TABLE 14.5 Mechanical Properties of the Base Materials.

	Yield strength (MPa)	Tensile strength (MPa)
DC 01	200	270
Nitrooxidized	310	380
Barrier plasma Treated	200	270

The tensile test of the adhesives were carried out on the specimens, which were created by curing of the adhesives in special designed polyethylene forms for 48 h. The results are shown in Table 14.6. In three of the adhesives (Hysol 9466, Hysol 9492, and Hysol 9497), very similar values were observed while in case of Hysol 9455, only tensile strength of 1 MPa was observed.

TABLE 14.6 The Mechanical Properties of the Adhesives.

	Hysol 9466	Hysol 9455	Hysol 9492	Hysol 9497
Tensile strength (MPa)	60	1	58	65

The differential scanning calorimetry was performed due to obtain the glass transition temperature as well as the melting points of the adhesives. The results are shown in Table 14.7. To measure the glass transition temperature of Hysol 9455, the measurements had to be started from the cryogenic temperatures. The results of such a low glass transition temperature explained the low tensile strength of the Hysol 9455, where at the room temperature, the mechanical behavior changed from rigid to rubbery state. The results of DSC analyses are given in Figure 14.3–14.6.

TABLE 14.7 The Glass Transition Temperatures of Adhesives.

	Hysol 9466	Hysol 9455	Hysol 9492	Hysol 9497
Glass transition temperature (°C)	52.6	5.0	61.2	62.4
Melting point (°C)	315.7	337.0	351.8	327.3

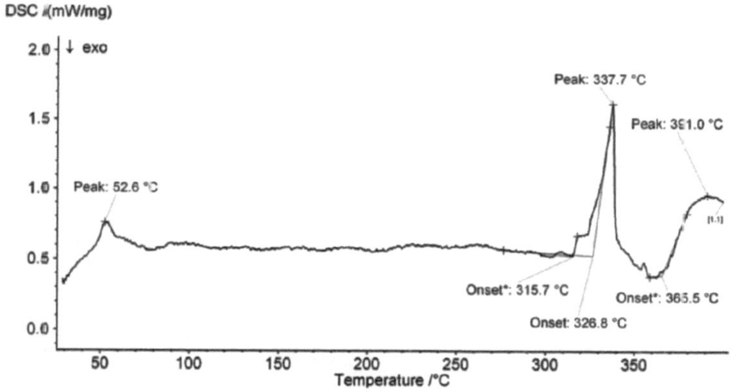

FIGURE 14.3 The DSC analysis of the Hysol 9466.

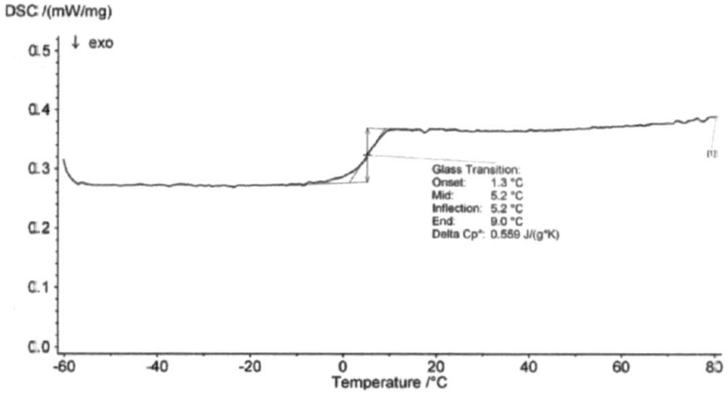

FIGURE 14.4 The DSC analysis of the Hysol 9455.

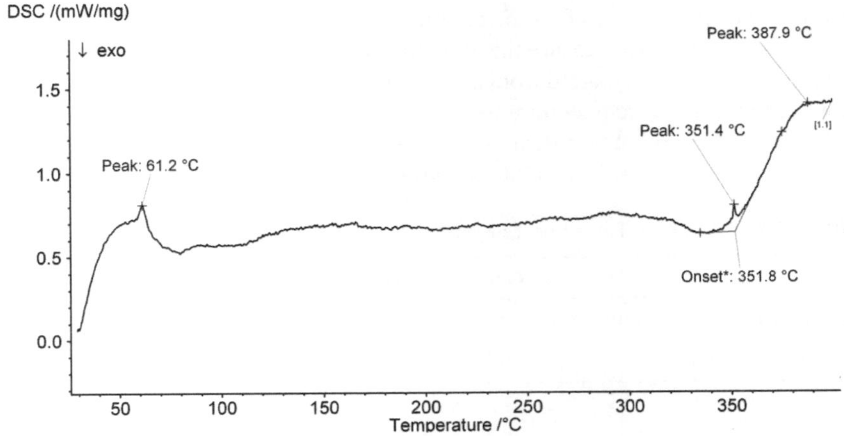

FIGURE 14.5 The DSC analysis of the Hysol 9492.

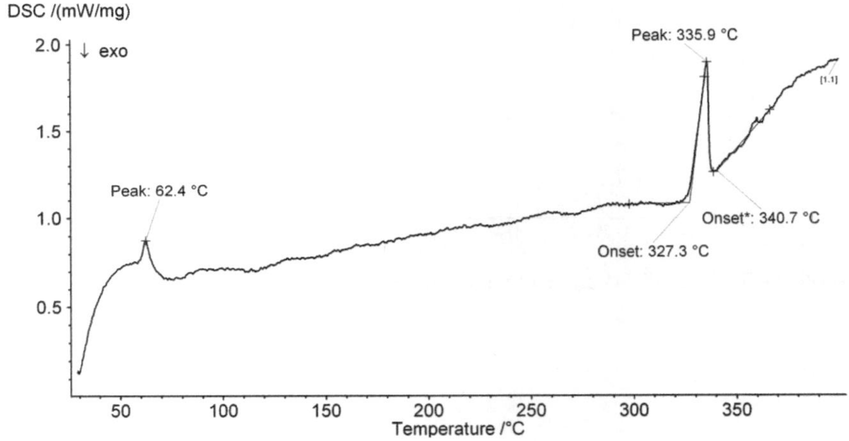

FIGURE 14.6 The DSC analysis of the Hysol 9497.

The adhesive joint evaluation consisted of observing the mechanical properties and fracture surface, respectively.

In order to obtain mechanical properties of the joints, the static shear tests were carried out. Results (Table 14.8) showed that the highest shear strength was observed in grinded nitrooxidized joints. The Hysol 9466 provided joints with the highest shear strength

TABLE 14.8 The Results of Shear Test of the Joints.

Material	Shear strength (MPa)			
	Hysol 9466	Hysol 9455	Hysol 9492	Hysol 9497
DC 01	9.0	2.8	7.1	6.0
DC 01 grinded	8.9	2.9	7.2	6.1
Nitrooxidized	12.9	3.1	7.8	5.4
Nitrooxidized grinded	12.9	5.9	12.7	7.0
Barier plasma treated	13.8	6.1	14.2	7.8

The mechanical properties of the joints made on non-nitrooxidized material did not depend on the surface grinding. The mechanical properties of the joints made on nitrooxidized material, in comparison to DC 01, were higher by 43% in case of Hysol 9466, 11% in case of Hysol 9455, and 10% in case of Hysol 9492. In the case of adhesive Hysol 9497, the decrease of the shear strength had occurred. The joints produced on grinded nitrooxidized material, in comparison to DC 01, had a higher shear strength by 43% in case of Hysol 9466, 110% in case of Hysol 9455, 79% in case of Hysol 9492, and 15% in case of Hysol 9497. The shear strength of adhesive joint was for barrier plasma-modified steel for all kinds of adhesive Hysol higher than unmodified and nitrooxidized steel.

Results of fracture morphology of the joints made of non-nitrooxidized material are shown in Figure 14.7a. Only the adhesive type of fracture morphology (Fig. 14.7b) was observed in every type of the adhesive. Cohesive and combined fracture type were not observed. It can be stated that the adhesion forces were not strong enough, hence, the joints were fractured between the material and adhesive.

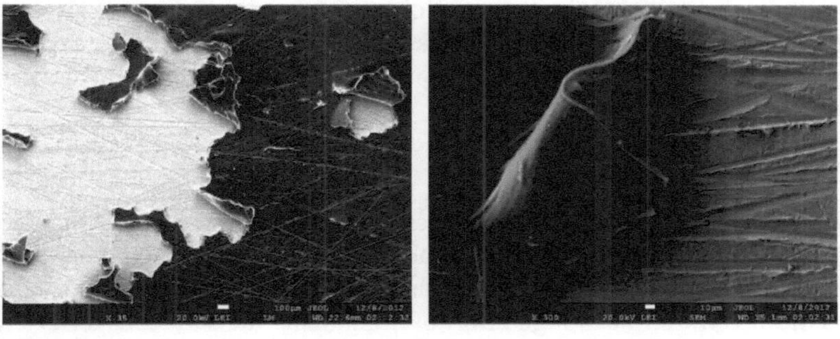

FIGURE 14.7 The fractographic analysis of the fractured adhesive joint of non-nitrooxidized material. (a) Overall view and (b) close-up view.

The results of the fractographic analysis of the non-grinded nitrooxidized steel are documented in Figure 14.8. No oxide layer peeling was observed. The cleavage fracture pattern as well as adhesive fracture morphology was observed. The close-up view on the cleavage fracture is shown in Figure 14.8b.

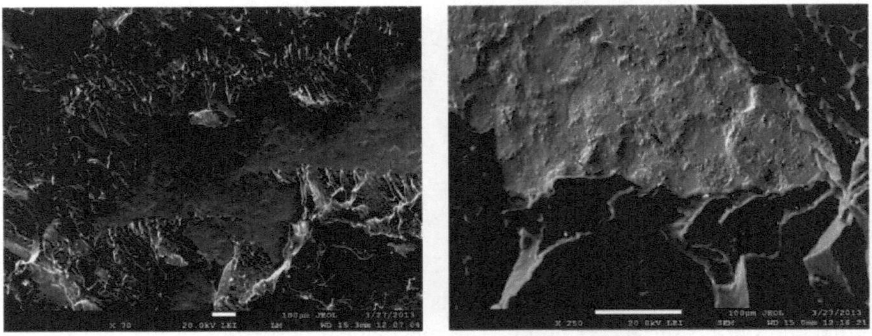

FIGURE 14.8 The fractographic analysis of the fractured adhesive joint of nitrooxidized material. (a) Overall view and (b) close-up view.

Differential scanning calorimetry revealed that three of the adhesives had a very similar glass transition temperature, so the meshing of the adhesives will start in the same way.

The results of mechanical properties evaluation of the joints proved that the material after the nitrooxidation process had a better adhesion to the epoxy adhesives than plain material DC01. Due to this fact, the higher shear strength was achieved. It can be explained by the surface oxide layer porosity, which helped the adhesive to leak in.

On the other hand, increase of mechanical properties of joints prepared from grinded nitrooxidized material can be explained by removing the surface oxide layer, and thus resulting into rapid increase of surface energy.

Only adhesive type of fracture was observed and the fractographic analysis showed that only cleavage type of fractures has been created. It can be stated that the surface energy of the materials was not appropriate for the cohesive fracture pattern.

14.4 CONCLUSION

Joining of steel sheets treated by the process of nitrooxidation represents an interesting technical as well as technological problem. The fusion welding methods with high-energy concentration, for example, laser beam welding are one of the possible options; however, even with high effort of minimizing the surface layer deterioration, it is not possible to completely avoid it.

Adhesive bonding of nitrooxidized steels presents, thus the second alternative, when the surface layer is not damaged and the adhesive joint keeps its properties after it has been cured. Adhesive bonding of metallic substrates often requires removing the surface oxide layer from the areas to be bonded. In case of materials treated by nitrooxidation, this is possible, however, the damage of the formed surface layer will occur. The goal of this paper was to review the effect of surface layer, created by the process of nitrooxidation and or by barrier discharge plasma treatment, on final mechanical properties of the joints, evaluation of fracture morphology, and results comparison for both treated and untreated material.

The acquired results have revealed that the presence of nitrooxidation surface layer caused decrease of free surface energy by 28%. The surface energy and namely its polar component were for barrier plasma-modified steel higher than for unmodified and nitrooxidized steel. On the other hand, this surface layer brings, on the joints, shear strength increase by 10–43% in dependence of the adhesive used. In case of Hysol 9497, the decrease of the joint shear strength by 10% was observed. In the case of barrier plasma-modified steel, the strength of adhesive joint were higher than unmodified and nitrooxidized material. We can presume that the increase of the shear strength was mainly due to porous structure of the surface layer, which enabled the adhesive to leak in.

The adhesive type of fracture morphology was observed during fractographic analysis. Regarding the characteristics of used adhesives, the cleavage fracture morphology of the joints had occurred.

Based on received results, it can be concluded that the epoxy adhesive bonding represents the suitable alternative of creating the high-quality joints of steel sheets treated by nitrooxidation as well as treated by barrier discharge plasma.

ACKNOWLEDGMENTS

This paper was prepared within the support of Slovak Research and Development Agency, grant No. 0057-07 and Scientific Grant Agency VEGA, grant No. 1/0203/11 and 2/0199/14.

This publication was prepared as an output of the project 2013-14547/39694:1-11 "Research and Development of Hi-Tech Integrated Technological and Machinery Systems for Tyre Production—PROTYRE" co-funded by the Ministry of Education, Science, Research and Sport of the Slovak Republic pursuant to Stimuli for Research and Development Act No. 185/2009 Coll.

KEYWORDS

- **steel surface**
- **fracture character**
- **nitrooxidation**
- **adhesive bonding**
- **differential scanning calorimetry**

REFERENCES

1. Michalec, I. CMT Technology Exploitation for Welding of Steel Sheets Treated by Nitrooxidation. Diploma thesis, Trnava, SK, 2010.
2. Konjatić, P.; Kozak, D.; Gubeljak, N. The Influence of the Weld Width on Fracture Behaviour of the Heterogeneous Welded Joint. *Key Eng. Mat.* **2012**, *488–489*, 367–370.
3. Bárta, J. Welding of Special Treated Thin Steel Sheets: Dissertation thesis, Trnava, SK, 2010.
4. Lazar, R.; Marônek, M.; Dománková, M. Low Carbon Steel Sheets Treated by Nitrooxidation Process, *Eng. Extra* 2007, *4*, 86.
5. Palček, P et al. Change of Fatigue Characteristics of Deep-Drawing Sheets by Nitrooxidation. In: *Chemické Listy*; ISSN 0009-2770; Master Journal List, Scopus, 2011; Vol. 105, Iss. 16, pp 539–541.
6. Bárta, J. et al. Joining of Thin Steel Sheets Treated by Nitrooxidation. *Proceeding of Lectures of 15th Seminary of ESAB + MTF-STU in the Scope of Seminars about Welding and Weldability;* Alumni Press: Trnava, SK, 2011; pp 57–67.
7. Marônek, M. et al. *Welding of Steel Sheets Treated by Nitrooxidation, JOM-16,* 16th International Conference On the Joining of Materials & 7th International Conference on Education in Welding ICEW-7, Tisvildeleje, DK, May 10–13, ISBN 87-89582-19-5.
8 Viňáš, J. Quality Evaluation of Laser Welded Sheets for Cars Body. In *Mat/tech automobilového priemyslu: Zborník prác vt-seminára s medzinárodnou účasťou*; Košice, SK, 2005; pp 119–124, ISBN 80-8073-400-3.
9. Marônek, M. et al. *Laser Beam Welding of Steel Sheets Treated by Nitrooxidation*, 61st Annual Assembly and International Conference of the International Institute of Welding, Graz, AT, July 6–11, 2008.
10. Michalec, I. et al. Resistance Welding of Steel Sheets Treated by Nitrooxidation. In *TEAM 2011*, Proceedings of the 3rd International Scientific and Expert Conference with Simultaneously Organised 17th International Scientific Conference CO-MAT-TECH 2011, Oct 19–21, 2011; University of Applied Sciences of Slavonski Brod: Trnava, SK, 2011; pp 47–50, ISBN 978-953-55970-4-9.

CHAPTER 15

SORBTION OF INDUSTRIAL DYES BY INORGANIC ROCKS FROM AQUEOUS SOLUTIONS

F. F. NIYAZI[1] and A. V. SAZONOVA[2]

[1]Fundamental chemistry and chemical technology, South-West State University, 305040 Kursk, street October 50, 94, Russia. E-mail: farukhniyazi@yandex.com

[2]Fundamental chemistry and chemical technology, South-West State University, 305040 Kursk, street October 50, 94, Russia. E-mail: ginger313@mail.ru

CONTENTS

ABSTRACT

The authors of this work have defined chemical composition of inorganic polymeric rocks. They have also studied kinetics of industrial dyes sorption from aqueous solutions by the rocks under investigation. Kinetic curves of the sorption have been stated. This work also considers possibilities to use kinetic equation of pseudo-first and second orders for the description of the sorption process.

Production and use of dyes is connected with the use of huge quantity of water, a great part of which is released in a polluted state.

On an average, about 225 tons of water is spent per 1 ton of dye. During production of each ton of dyes together with industrial sewage, enterprises release dozens and hundreds tons of different mineral and organic compounds in the form of waste.[1]

At present, different types of dyes are widely used for the production of paper, paints, and pigments in textile, shoe-manufacturing, and printing industries. During production processes, 10–40% of dyes being used are released into the sewage, which, in turn, is released into natural pools, thus causing serious violation of biocenose in them.[2]

Once into the water pools, these dyes have a negative effect on the associations of water organisms and also violate their oxygen regime. Most of the organic dyes are highly toxic and have an allergic effect. Besides, as a rule, cationic dyes are more toxic for water ecosystems than anionic ones.[3] Thus, it is imperative to search for new approaches to the problem of sewage treatment containing dyes.

Existing methods of sewage treatment from industrial dyes can be divided into regenerating, destructive, and biological methods. Among them, regenerating methods are better since they allow for the extraction of substances, being constrained in sewage, for their further use. This problem may be solved by using such methods of concentration as extraction, ionic replacement, sorption, and others.

Adsorption on mesoporous activated coals is one of the effective methods of sewage treatment contaminated by dyes.[4] However, such coals are expensive adsorbents and are produced in small quantities. Perspective direction cheaper is utilization, on the one hand, cheap, and, on the other hand, available sorption materials as adsorbents.[5]

Thick chalkbeet stretches through European continent, including the north of France, south part of England, Poland, goes through the Ukraine, Russia, and is displaced into Asia—Syria and Libyan Desert.

Kursk region possesses exceptional variety of natural resources that are able to provide adequate amount of carbonate rocks. That is why inorganic polymeric rocks of Kursk region have been used as sorption material because they are high-dispersion systems.

Carbonate rocks, under investigation, were analyzed for their calcium and magnesium carbonates content—by complexonometric methods such as sulfate ions, ferrous and aluminum oxides; by gravimetric method such as chloride ions; and by methods of turbidimetry. The content of insoluble residue in hydrochloric acid has been discovered, and the electric conduction of carbonate rocks saturated with aqueous solution has been also measured. Electric conduction was defined with the help of conductivity apparatus KSL-101. Received results of the chemical composition of carbonate rocks are given in Table 15.1.

TABLE 15.1 Chemical Composition of Carbonate Rocks.

Deposits of Kursk region	Area of sampling	
	Konyshovka	**Medvenka**
Granulometric composition (%) (0.2–0.06)	73,37–68,49	88,13–79,84
Moisture content (%)	9.6225	6.3125
Content of $CaCO_3$ и $MgCO_3$ in terms of $CaCO_3$ (%)	49.50	41.95
Insoluble residue in HCl	1.075	13.566
Sesquilateral content ferrous and aluminum oxides	0.273	1.530
Water-soluble substances (%)	3.1823	2.15997
Electric conduction (мкСм/см)	79.04	88.26
Salt content in terms of NaCl (mg/L)	37.22	41.59
Sulfate ions (%)	0.00445	0.01403
Chloride ions (мг/л)	0.005	0.0015

The analysis of sorbent shows that calcites form a great share of the systems being studied. Deposits of carbonate rocks in Kursk region differ by low content of insoluble residue and high content of carbonates.

Structural and microscopic characteristic of polymeric rocks were different as determined by the method of homogeneous field with the help of polarizing-interference microscope BIOLAR. The investigation has shown that original rocks consist of dolomite trigonal and rhombic crystals.[6]

The aim of this work is to study the process of industrial dyes of different sorption nature by carbonate rocks.

Dyes (cationic blue, cationic red, acidic brightly green, and antraquinone blue) widely used in industry were used as an adsorptive.

Experiments in sewage treatment from industrial dyes were carried out using model aqueous solutions of industrial dyes and sewage from a dye-finishing shop of knitted fabric incorporation «Seim» (Kursk). Dyes of «pure-for-analysis» qualification without additional cleaning were used for preparation of aqueous solutions. Aqueous solutions of dyes with concentration of 0.01 g/l at 298°C with a solution volume $V = 50$ ml were investigated.

Effects of sorbent mass on dyes sorption and kinetics of the process of sewage treatment from industrial dyes have been studied in order to find optimal parameters of sorption and also to state sorption properties of carbonate rocks. Method of one-step static sorption was used in this work. Sorption was carried out by adding carbonate rocks samples, grinded to granules to a size of 0.06–2.0 mm, to the dyes solution being studied. Then all this was stirred by a magnetic mixer and after definite intervals, samples were taken and the defined residual concentration of dyes by spectrophotometric method was carried out.

Readings of absorption spectra on coordinates: optical density (A) and wave length (λ) on the device СФ-26 have been taken; wavelengths of maximum light absorption for cationic blue (610 nm), cationic red (490 nm), acid brightly green (670 nm), antraquinone blue (590 nm) were chosen. Subordination bound of dyes solution to the main law of light absorption—Beer–Lambert–Bonguer law—has also been found.

Data on carbonate rocks mass affect on the sorption of industrial dyes are given in Table 15.2.

TABLE 15.2 Effect of Carbonate Rocks Mass on the Sorption of Industrial Dyes at Phase's Relation: $t = 30$ min, $V = 50$ ml, and $C = 0.01$ g/L.

Mass (g) Dye	0.05	0.1	0.25	0.5	1.0	2.0
Cationic blue	93.0	99.0	99.5	100	100	100
Cationic red	66.0	84.4	95.98	100	100	100
Antraquinone blue	76.0	88.0	100	100	100	100
Acid brightly green	0	20.0	38.6	40.2	46.6	49.6

Data of the given researches helped us in finding sorption capacitance of carbonate rocks: for cationic dyes—1 mg/g of sorbent, for antraquinone blue dye—2 mg/g of sorbent, for acid brightly green dye—0.1 mg/g of sorbent.

Cleaning efficiency (CO, %) shows the share of absolute substance quantity, which is caught by sorbent, and gives rather a complete notion of the process and its character.[7] This factor is an important criterion while defining optimal condition of the sorption process and is calculated by the following formula:

$$CO = \frac{(C_0 - C)\times 100\%}{C_0},$$

where C_0 is the initial concentration in g/l and C is the residual concentration in g/l.

These data allow choosing optimal mass of sorbent for definition of industrial dyes kinetics by carbonate rocks. Results of these studies are given by kinetic curves in Figure 15.1.

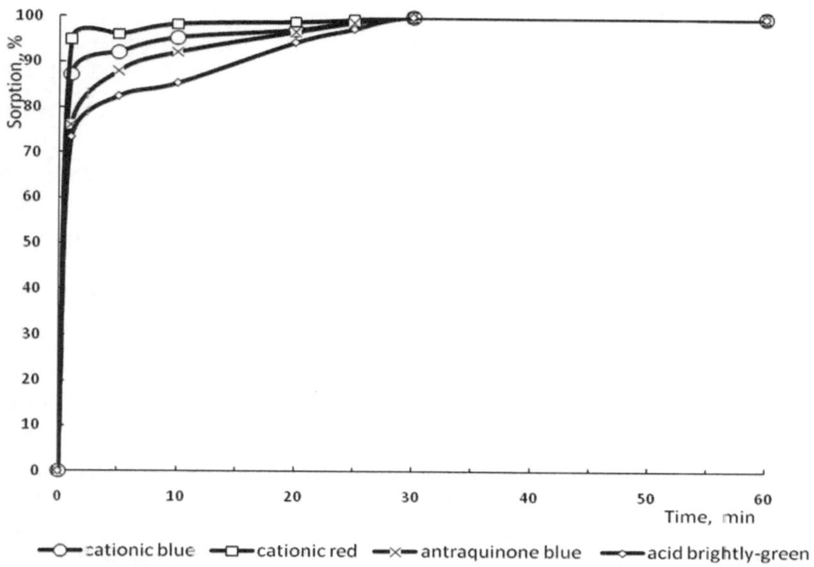

FIGURE 15.1 Kinetic curves of industrial dyes sorption by carbonate rocks.

Sorption lasts for the first 2–3 min after the beginning of contact phases regardless of the dye nature at the first section of the stepped kinetic curve. But this sorption quickly increases at the second section during the following 3–30 min. Complete (100%) sorption of dyes takes place after 30 min of interaction and there is sorption equilibrium. Further increase of contact time is pointless.

Comparison of studied literature data allows us to come to a conclusion that the stepped character of sorption kinetic cures is the result of the fact that adsorption of large ions takes place on microporous carbonate rocks. Mass transfer on the phase's boundary and interaction of dyes with carbonate rock surface play an important role at the first stage of sorption, and at the second stage there is internal diffusion of dye into sorbent pores of accessible size, allowing the dye to be sorbet again on the external surface of sorbent.[8]

It is stated that for all stepped kinetic curves, the first section can be described by the equation of pseudo-first order; but the second section is not described by integral equation of kinetic models neither of the pseudo-first, nor pseudo-second order.[9]

It is known that hydrogen index of environment (pH) has a great influence on the process of sorption and selection of sorbent for sewage treatment. It is connected with the fact that functional groups on the sorbent surface and functional groups of dye molecules may change depending on environmental pH.[10] Change of pH after the process of sorption depending on the time of the contact of sorbents, being studied, and industrial dyes is given in Table 15.3.

TABLE 15.3 Change of pH Depending on the Time of Contact of Carbonate Rocks and Industrial Dyes.

Industrial dye	pH_0	Time of contact (min)					
		1	5	10	20	25	30
Blank test	6.5	7.25	7.67	7.92	8.10	8.54	8.83
Cationic red	4.26	7.29	7.08	7.55	7.17	7.18	7.00
Cationic blue	3.41	7.57	7.60	7.64	7.67	7.44	7.38
Antraquinone blue	5.67	7.45	7.33	7.45	7.38	7.36	7.32
Acid brightly green	5.38	7.32	7.20	7.18	7.05	6.95	6.82

Increase of pH value is observed when carbonate rocks are mixed with water and this is because of hydrolysis of calcium and magnesium carbonates. On the contrary, decrease of pH values is observed during the process of industrial dyes sorption and this may be explained by chemical interaction of carbonate rocks and dyes.

This can be very important when we choose a sorbent for sewage neutralization during the process of acid sewage treatment. Results of the research showed high adsorption ability of carbonate rocks regarding industrial dyes of different classes.

The suggested method of sewage sorption treatment from dyes increases the variety of sorbents being used during treatment, and allows using local carbonate rocks as sorbents.

It is necessary to note that kinetic curves differ from each other to a small extent while comparing sorption ability of carbonate rocks in relation to dyes. However, consumption of carbonate sorbent is decreasing for cationic and antraquinone dyes at the same phase's relation and efficiency of aqueous solution treatment.

As to technological execution, it is necessary to carry out the process of sorption in contact absorbers, equipped by mechanical mixers of batch action. Separation of solid and liquid phases is carried out by decantation and filtration. It is economically pointless to recover waste sorbent that is why it is necessary to utilize it.

One of the ways to utilize this sorbent is to use it as a filling agent. This is connected with the fact that calcium carbonate is a widely used material in the world industry today. Development of the branches of rubber engineering, electric power, glass, paper, polymeric, paint and varnish, and other industries requires an increase in production of high-quality filling agents, and at the first place is chalk. Characteristic feature of this natural material is connected with the fact that it is easier to quarry and process it at rather small expenditures. Its quarrying and processing do not cause serious ecological violations, and its reserves are practically unlimited in many European countries, countries of former UIS, and in Russia.

KEYWORDS

- **inorganic rocks**
- **dye**
- **sorption**
- **kinetic curve**
- **pH**
- **utilization**

REFERENCES

1. Stepanov, B. I. *Introduction in chemistry and technology of organic dyes*; Chemistry: Moscow, RU, 1977; p 488.
2. Ramesh, D. D.; Parande, A. K.; Raghu, S., Prem Kumar, T. *J. Cotton Sci.* **2007,** *11,* 141–153.
3. Hao, O. J.; Chiang, P. C. *Crit. Rev. Environ. Sci. Technol.* **2000,** *30,* 141–153.
4. Soldatkina, L. M.; Sagajdak, E. V.; Menchuk, V. V. Adsorption of Cationic Dyes from Water Solutions on Sunflower. *Chem. Technol. Water* **2009,** *31* (4), 417–426.
5. Litvina, T. M.; Kushnir, I. G. Problem of reset, processing and waste recyclings. Odessa, **2000,** 258.
6. Niyazi. F. F.; Maltsevf, V. S.; Burykina, O. V.; Sazonov, A. V. Kinetics of sorption of ions of copper by cretaceous breeds. News of Kursk State Technical University, (4), Kursk, 2010, 28–33.
7. Gocharuk. V. V; Puzyrnaja, L. N.; Pshinko, G. N; Bogolepov, A. A.; Demchenko, V. J. Removal of Heavy Metals from Water Solutions Montmorillonite, Modified Polyethyleneimine. *Chem. Technol. Water* **2010,** *32* (2), 125–134.
8. Soldatkina, L. M; Sagajdak, E. V. Kinetics of Adsorption of Water-Soluble Dyes on the Active Coals. *Chem. Technol. Water* **2010**, 32 (4), 388–398.
9. Janos, P.; Buchtova, H.; Ryznarova, M. *Water Res.* **2003,** *37* (20), 4938–4944.
10. Bagrovskaja, N. A.; Nikiforova, T. E.; Kozlov, V. A. Influence Acidities of the Environment on Equilibrium Sorption of Ions Zn (II) and Cd (II) Polymers on the Basis of Cellulose. *Mag. Gen. Chem.* 2002, *72* (3), 373–376.

CHAPTER 16

SOME COORDINATION COMPOUNDS OF ARSENIC AND STIBIUM

N. LEKISHVILI[1*], M. RUSIA[1], L. ARABULI[1], KH. BARBAKADZE[1], I. DIDBARIDZE[2], M. SAMKHARADZE[2], G. JIOSHVILI[1], K. GIORGADZE[1], and N. SAGARADZE[1]

[1]Faculty of Exact and Natural Sciences, Institute of Inorganic–Organic Hybrid Compounds and Non-traditional Materials, Ivane Javakhishvili Tbilisi State University, 1 Ilia Chavchavadze Avenue, Tbilisi 0179, Georgia. E-mail: *nodar@lekishvili.info

[2]Kutaisi Akaki Cereteli State University, 59, Tamar Mephe st., 59, Kutaisi 4600, Georgia

CONTENTS

*The paper is dedicated to 75th anniversary of our teacher and senior friend Prof. Dr. Roman Gigauri.

ABSTRACT

Based on arsenic compounds obtained by transformation of arsenic industrial waste and natural resources, [arsenic(III) oxide, arsenic(III) chlorides, alkoxides, stibium(III) oxide], we obtained and studied new coordination compounds, "white arsenic," hyperpure metallic arsenic and various materials with specific properties.

By using obtained nonvolatile inorganic–organic complexes, we created and tested new cheap anti-microbe means and fungicides for protection of archeological items and museum exhibits and biological stuffed. Based on the organic heterochain polymers modified by carbon-functional siliconorganic oligomers, we created antibiocorrosive covers for woodwork, goods from plastics and leather. In the monograph, the possibility of obtaining siloxane–arsenic oligomeric additions for underwater hydrophobic bioactive dye composites is discussed. It showed real perspective to manufacture pharmaceutical preparates, anthelmintes, semiconductors, optical glass fibers, and biomedical nanocomposites based on Georgian region's arsenic industrial waste and natural resources.

16.1 INTRODUCTION

The problem in creation of a system that guarantees protection to human and environment from microorganisms attack for increasing the quality of human life has to be solved by

- proccessing of new effective preparates;
- prevention of materials from biodeterioration and noncontrolled biodegradation;
- inhibition of growth and expansion of microorganisms, which are causal factors of infection—inflammatory sickness of human;
- prophylaxis and treatment of human diseases, provoked by microorganisms and crop's protection against diseases, caused by some microorganisms, and human protection during a contact with them;
- obtaining biologically active compounds and composites with the aim of their inclusion into polymer matrix using different methods of their fixation in the latter;
- study of biological activity of synthesized compounds and screening of biologically active compounds BAC of natural derivation in relation to test cultures of various microorganism genera—biodestructors of polymer materials, etiological factors of the mycosis, plant pathogens: Aspergillus, Penicillium, Cladosporium, Candida, Fusarium, etc.; creation of polymer matrices by structural and chemical modification of industrial polymers.

The perspective sources for obtaining various bioactive compounds are Georgian natural resources.

Three main tendencies are well defined while using production waste of Georgian chemical plants:

1. Regeneration of valuable compounds having noteworthy functions from production waste.[1–5]
2. Recycling of so-called thrown-out and unused accumulated production waste and elaboration of technology of production of various useful cheap compounds from them.[6]
3. Use of production waste for production of various compounds and materials having peculiar properties.[7,8]

Arsenic is among the relatively less widespread elements.[9] Its deposits are located at the Georgian region, in general, such as arsenic industrial waste and realgar–auripigment and arsenic pyrite (As_2S_2–As_2S_3–FeAsS) natural sources. Pyrite, antimonite, nickel, gallium, manganese, titanium, copper etc. contain arsenic in a small amount and are considered as second-rate mineral sources of arsenic.

Realgar (As_4S_4)–auripigment (As_4S_6) ore is world's only one and unique predominant raw material of arsenic. Dominant's content in this ore is especially high and reaches 10–12%.[9] Besides, it is very important that this ore doesn't contain admixtures of other elements and has bright prospects of receipt of not only metallic arsenic and As_2O_3 of high purity, but also of other interesting products (Fig. 16.1).[11] Despite the fact that for extraction of arsenic and its products from realgar (As_4S_4)–auripigment (As_4S_6) ore, first of all it is necessary to burn concentrate in special furnace so that environment pollution with sulfur dioxide and aerosol containing 2–3% of white arsenic cannot be eliminated,[12] in case a revival of corresponding enterprise makes topical manufacturing and putting into operation of catching devices with appropriate design.

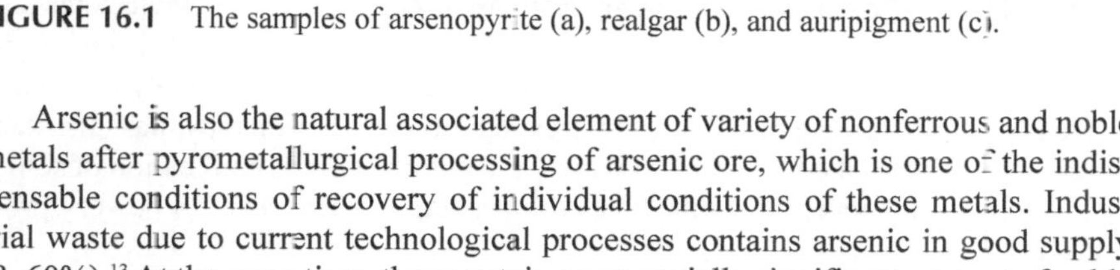

FIGURE 16.1 The samples of arsenopyrite (a), realgar (b), and auripigment (c).

Arsenic is also the natural associated element of variety of nonferrous and noble metals after pyrometallurgical processing of arsenic ore, which is one of the indispensable conditions of recovery of individual conditions of these metals. Industrial waste due to current technological processes contains arsenic in good supply (8–60%).[13] At the same time, they contain commercially significant amount of noble metals. In order to get rid of environmental pollution, the residues are buried in special ground disposals that are associated with considerable material and financial expenses.

During production of chemically pure arsenic sulfide from realgar–auripigment ore using vacuum thermal method[14] along with desired product, coproducts, which contain desired product in significant amount, are originated. Despite the above-mentioned process, neither the technology of extraction of arsenic and its satellite elements (e.g., stibium) from realgar–auripigment ore processing products, nor the application areas for products received from waste are not assimilated to the full extent today and require further development from the viewpoint of creation of cheap raw materials base on their basis.

In the represented article, production waste of former mining and chemical plants of Georgian arsenic production are considered as secondary resources for receipt of important arsenic-containing compounds.

16.2 GEORGIAN REGIONAL NATURAL SECONDARY RESOURCES OF ARSENIC INDUSTRY

There are considerable reserves in Georgia for extraction of arsenic in its various forms (e.g., sulfides, oxides) from arsenic production waste (arsenic pyrite and realgar–auripigment processing waste and waste of pyrometallurgical production) and for production of relatively cheap important compounds and materials with peculiar properties, for example, pharmaceutical preparations, anthelmintics, antimicrobial and parasiticide compounds, conserving agents, as well as industrial use (for technical purposes) and bioactive synthetic compositions and materials.[4,15–17] There is a real prospect of production of arsenic-containing nontraditional advanced materials, antibiocorrosive coatings, nanocomposites and nanomaterials of biomedicine, and microelectronic purposes on the basis of arsenic production waste.[18,19]

We have established that arsenic-containing production waste may be used as starting material for production of gold, barium hydrous arsenate, strontium hydrous arsenate, magnum–ammonium arsenate, sodium monoselenoarsenate, cyclic esters of arsenic acid etc. Among them, barium hydrous arsenate is the best anthelmintic. Strontium hydrous arsenate is used in homeopathy, while it is used as anthelmintic in veterinary. Magnum–ammonium arsenate is used in production of special-purpose glasses, sodium monoselenoarsenate as insecticide, while cyclic esters of arsenic acid are used for production of insecticides, as inhibitors of lubricants' oxidation, as antioxidant for synthetic rubber, for production of some copolymers, etc.

Homeopathy,[20] as one of the most interesting directions of therapy, has attracted a serious attention in the last years, in which, the possibilities of application of some arsenic compounds received on the basis of production waste seem to be very prospective.

Development of above-mentioned directions will promote creation of cheap raw materials base that will make accessible production of well-known and new arsenic-containing compounds having a lot of interesting properties and will simplify their practical implementation (Fig. 16.2).

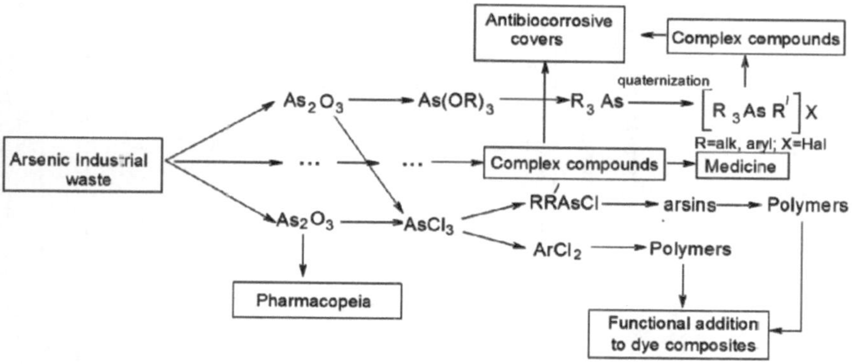

FIGURE 16.2 The scheme of obtaining and use some of important arsenic-and stibium-containing compounds from arsenic industrial waste.

Esters of arsenic acids received from production waste may be used for production of a variety of new arsenic-containing preparations, which have found manifold uses. First of all, the most important is the possibility of gold extraction from pyrometallurgical waste, as well as receipt of barium hydrous arsenate, strontium hydrous arsenate (used in medicine, namely, in homeopathy, while as anthelmintic in veterinary), sodium monoselenarsenate, cyclic esters of arsenic acid, and other substances. Magnum–ammonium arsenate (V) is widely used in production of special-purpose glasses. Demand for this preparation increased day by day in the last years. That's why elaboration of upgraded method of its production[21] is of great practical importance.

The structure of some important inorganic and organic arsenic compounds is given below (Table 16.1).

TABLE 16.1 Basic Physical–Chemical Properties of Coordination Compounds of Some Tertiary Arsines with Mercury (II) Halides ($HgHal_2As(R)Ar_2$).

No	Ar	R	Hlg	Melting point, °C	Molar electric conductivity in dimethylformamide at 25°C, $om^{-1}cm^2 mol^{-1}$	Found, (%)		Brutto formula	Calculated, (%)	
						As	Hlg		As	Hlg
1	m-$CH_3C_6H_4$	m-$CH_3C_6H_4$	Cl	173–174	24.6	11.70	11.52	$C_{21}H_{21}AsHgCl_2$	12.09	11.45
2	m-$CH_3C_6H_4$	C_2H_5	Cl	137–139	25.7	13.15	13.64	$C_{16}H_{19}As\ HgCl_2$	13.44	12.72
3	m-$CH_3C_6H_4$	iso-C_4H_9	Cl	143–144	25.4	12.45	12.02	$C_{18}H_{23}As\ HgCl_2$	12.79	12.12
4	C_6H_5	n-C_4H_9	Cl	122–123	23.1	13.40	12.72	$C_{16}H_{19}As\ HgCl_2$	13.44	12.72
5	C_6H_5	iso-C_3H_7	Cl	150–151	28.4	13.28	13.10	$C_{15}H_{17}As\ HgCl_2$	13.79	13.05
6	C_6H_5	C_6H_5	Br	180–181	22.6	11.17	23.89	$C_{18}H_{15}As\ HgBr_2$	11.24	23.99
7	m-$CH_3C_6H_4$	m-$CH_3C_6H_4$	Br	164–165	25.4	10.20	22.33	$C_{21}H_{21}As_2HgBr_2$	10.57	22.57
8	m-$CH_3C_6H_4$	C_2H_5	BrBr	131–139	27.7	11.49	24.72	$C_{16}H_{19}As\ HgBr_2$	11.59	24.73
9	C_6H_5	n-C_4H_9	Br	147–148	22.4	11.32	24.66	$C_{16}H_{19}As\ HgBr_2$	11.59	24.73
10	m-$CH_3C_6H_4$	iso-C_4H_9	Br	145–148	26.9	11.09	23.42	$C_{18}H_{23}As\ HgBr_2$	11.11	23.70
11	C_6H_5	iso-C_3H_7		158–159	29.5	11.36	24.98	$C_{15}H_{17}As\ HgBr_2$	11.84	25.28

In marketable condition, it can be used as intermediate product. Namely, products of interaction of this compounds and monoatomic spirits of saturated row (ROH, where R = C_5H_{11} or iso-C_5H_{11}) during chemical reactions manifest such behavior as if we have arsenic oxide (III) of high purity. As we already have mentioned earlier, interaction of spirits and white arsenic, including arsenic anhydride contained in mixtures, represents well-defined selective reaction, as a result of which, the ester of arsenic acid of high condition is received. The latter in his turn, as all compound esters easily experience hydrolyzation in water solutions, with formation of white arsenic and corresponding spirit. Arsenic oxide (IV) manifests the same properties. This circumstance can be successfully used not only for production of well known and widely used compounds, but also for synthesis of novel compounds. Use of ester of arsenic acid as source raw material eliminates carrying out of such hazardous and complex processes as the receipt of high-toxicity arsenic compounds (AsH_3, $AsCl_3$), their purification and reduction up to metallic arsenic or transformation into compounds with peculiar properties.[22]

We established that there is no necessity for burying so-called "tails" of arsenic production in appropriate ground disposals, if there is no arsenic in the form of oxide in them. They can be used as intermediate product. Their processing by monoatomic spirits ROH of saturated row, where R = C_4H_9, C_5H_{11}, iso-C_5H_{11}, or by C_6H_{13} using the method of azeotropic dehydration, is possible. This process is based on well-defined selective reaction, since among residues, only arsenic oxides or corresponding acids undergo a reaction with spirits, while coproducts remain on the bottom of reactor in form of slugs. High-purity esters of arsenic acids are received as a result of reaction. They, as all compound esters, are easily hydrolyzed in the water solutions with formation of white arsenic and corresponding spirit. Obtained spirit can be used again as extragent, so this process takes continuous character and at the same time, cyclical pattern. It is clear that there will be a negligible loss of spirit anyway, which is consumed for soaking of spirit. Obtained spirit extracts can be successfully used once more for receipt of both inorganic and organic compounds.

Methods of receipt of very important arsenic compounds—arsenic oxides (e.g., "white arsenic"), arsenic sulfides, metallic arsenic of high purity, and other compounds on the basis of waste of hydro- and pyrometallurgical production are described in the literature.[4,22–26] Mentioned compounds represent basic/starting substances for production of arsenic-containing compounds.[27,28]

Among arsenic products, the receipt of one of the most valuable substances, metallic arsenic of high purity, is of special importance. This production is based on the preliminary treatment of some of the arsenic-containing compound and afterwards on its chemical transformation.[12] Elaboration of intensive technology of metallic arsenic production is one of the most topical problems of arsenic chemistry and technology, all the more if we use production waste transformation products as starting substances.

Use of ester of arsenic acid as source raw material eliminates carrying out of such hazardous and complex processes as are the receipt of high-toxicity compounds (AsH_3, $AsCl_3$), their purification and reduction up to metallic arsenic (Fig. 16.2) or transformation into compounds with peculiar properties.[22]

Thus, in above-mentioned direction, chemists, researchers, and entrepreneurs pay a lot of attention to the development of such progressive and economically substantiated technologies, where:

- amount of operations for receipt of ester of arsenic acid obtained from arsenic-contained waste is reduced to minimum, which considerably simplifies the quality control of obtainable product;
- extraction of highly explosive (e.g., H_2) and toxic coproducts (e.g., HCl, etc.) have to be eliminated.

Use of compounds received in agriculture from arsenic-contained waste (mostly in the form of alkali salts (Na_3AsO_4)) in relation with escalation of fighting with environmental pollution decreasing more and more, while areas of metallic arsenic application are extended. Namely, it is used in up-to-date technologies,[28,29] while Ga_2As_3 and InAs (arsenides) are used for manufacturing of luminescent and laser, microwave and switching diodes, detectors, transistors, solar batteries, integrated circuits, and microelectronic devices.

Ultraclean metallic arsenic is used in semiconductors production, as well as for manufacturing of various alloys.[30]

Arsenic (III) extracted from production waste may be used as intermediate product for receipt of various compounds with reacting capacity. For example, products of interaction of this compounds and saturated monoatomic spirits (R > Bu) manifest such behavior in chemical reactions as arsenic oxides of high quality of purity.[31] We established that interaction of spirits and white arsenic represents well-defined selective reaction, as result of which, the ester of arsenic acid of high condition is received. The latter, in its turn, as all compound esters, easily experience hydrolyzation in water solutions, with formation of white arsenic and corresponding spirit. Obtained spirits can be used again as extragents. This means that process has continuous and at the same time cyclic nature. It is clear that there will be a negligible loss of spirit anyway, which is consumed for soaking of solid phase. Obtained spirit extracts can be successfully used once more for receipt of both inorganic and organic compounds. As it turns out, products of transformation are considerably of high quality than it is accomplished directly on the basis of "white arsenic." Above-mentioned researches have been further developed during last 5–6 years, which was reflected in numerous articles and abstracts of international reports.

Among arsenic-containing compounds, the most interesting are tertiary arsines and their salts.[31–35] As it turns out from the review of corresponding scientific literature, tertiary arsines easily alkylate via "inclusion" into chemical bond of $4s^2$-electrons of central atom, and form arsonium salts of polyfunctional composition. Organic iodine derivatives play the role of electrophile for the most part, since in contradistinction from other halogen derivatives, they easily enter into reaction of nucleophilic addition.

Thus, arsonium salts with general formula $[R_4As]X$, where R = organic radical (alkyl, alkylene or aromatic), while $X = Hal^-$, NO_3^-, CH_3COO^-, etc., should be considered as cationic complexes, since they experience electrolytic dissociation in water or other polar solvents according to the following equation:

$$[R_4As]X \quad [R_4As]^+ + X^-$$

It should be noted that there are no similar radicals among all 4R's. Moreover, the most difficult of receipt, and therefore, of study, is the case when we deal with asymmetric arsonium halogenides. The latter substances are received by alkylation of tertiary arsines:

$$R_3As + R\text{–}Hal \rightarrow [R_4As]Hal$$

Also, noteworthy is the fact that tetraalkyl(aryl)arsonium salts represent the best precipitators for a variety of complex anions.[36]

Receipt of cationic–anionic complexes of tetra-substituted arsenic (arsonium) is possible not only by exchange and addition reactions, but also by substitution reaction.[37]

Proceeding from above-mentioned information and also with the use of arsenic chemical raw materials and other kinds of waste available in Georgia, from the practical viewpoint (biological activity), the coordination compounds on the basis of tertiary alkyl(aryl)arsines are very prospective.[38–40]

Impact of arsenic as toxic element on flora and fauna of Racha and Svaneti (Georgia) is considerable as it makes a recycling of accumulated waste of its production even more topical.

At this moment, in above-mentioned direction, chemists, researchers, and entrepreneurs pay a lot of attention to development of such progressive and economically substantiated technologies, where:

- amount of operations for receipt of ester of arsenic acid obtained from arsenic-contained waste is reduced to minimum, which considerably simplifies the quality control of obtainable product;
- extraction of highly explosive (e.g., H_2) and toxic coproducts (e.g., HCl, etc.) has to be eliminated;
- possibility of formation of toxic intermediate products or waste (which requires neutralization of burying) should be reduced to the minimum in the whole treatment cycle;
- environment should be protected from pollution and poisoning with high-toxic substances.

Processing and transformation of stibium-containing production waste into stable coordination compounds of peculiar properties is important with the purpose of its further use that ranges among topical problems of modern applied and coordination chemistry. Successful solution of this problem not only will create new raw materials base but also will solve important ecological problem, will protect environment from pollution with stibium-containing waste. Certain positive results are obtained by Georgian scientists–chemists in this direction, who carry out a synthesis and study a lot of interesting stibium compounds (Prof. R. Gigauri and colleagues).

There is elaborated by our method of possible separation of arsenic (III) and stibium (III) oxides from each other.[4] Azeotrope effective for this process is selected.

The used method is based on the fact that arsenic oxide quantitatively completely interacts with spirits, for example, with butyl spirit with formation of tributyl arsenite, and as a result of its hydrolysis, white arsenic of high purity is received; arsenic and stibium oxides, which are not included into reaction, turn out to be constituent component of pyrometallurgical waste of many polyelement ores, that's why elaboration of low-temperature methods of their removal and separation from waste is interesting issue for production processes.

Above-mentioned method is rather economically viable compared with other current methods.

Among stibium compounds, stibium (III) oxide and stibium (III) sulfide (Fig. 16.3) are also important from the practical viewpoint. Stibium (III) oxide is used for production of modern fireproof dyes and enamels, brand new functional materials, and composites.[41–47]

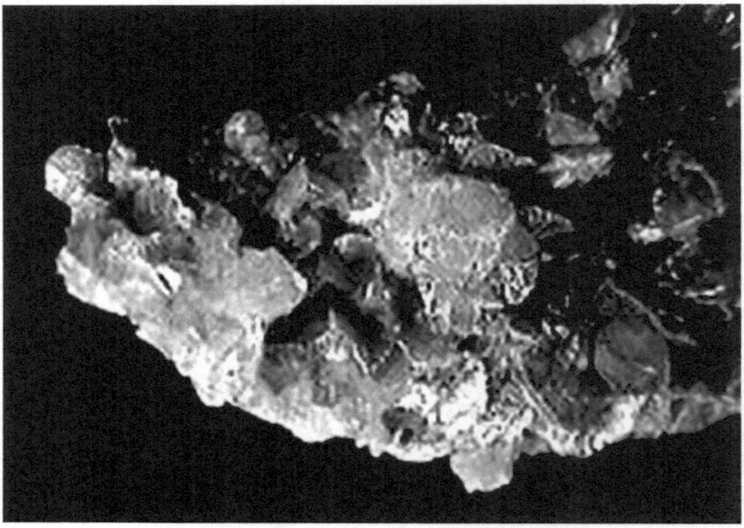

FIGURE 16.3 The crystal of As(III) sulfide.

Stibium (III) sulfide is used in match production, as well as in military industry. Stibium compounds are widely used in pharmaceutical, textile, and other technologies. Stibium is also produced in Georgia in the form of sulfide minerals, which are known under the label of antimonites (Sb_2S_3). Content of basic metal in commercial brands of metal stibium, which are available at world market, reaches 99.95%.

Among stibium (V) compounds, one of the most important is sodium thiostibiate Na_3SbS_4, which is good complex former.[41–43] Due to it, from the economic viewpoint, receipt of stibium compounds (as of arsenic satellite elements) on the basis of arsenic production waste is of special importance, which will make cheaper corresponding final compounds and as a result, coordination compounds derived from them.

Among works published in the area of stibium chemistry, we consider Kanatzidis' work as classical one.[44] Author investigates the possibility of receipt of stibiates (III) with metal triethylenediamines under the action of alkali metals dithioantimonites on coordination compounds of Co (II) and Ni (II). The synthesis of a variety of coordination compounds occurs in a same way, only with the difference that tetrathiostibiates (V) are used as precipitators.[45–47]

In the era of the violent technical progress, wide assortments of synthetic and natural polymeric materials have produced. At the same time, various aggressive microorganisms (more than several hundred) appeared, which can destruct these materials, especially carbon containing. Losses caused by this mean reach enormous amounts and constitute milliards of dollars annually.

Together with above-mentioned information, protection of historical buildings, archeological examples, museum exhibits, and collections is global problem and this topic is actually all over the world. Many Congresses and symposiums are dedicated to this topic. Substantial funds are allocated for solving this problem.

One of the ways to protect from aggressive microorganisms and fungi of wide range of synthetic and natural materials is creation of direct action multifunctional synthetic antibiocorrosive coatings based on various inorganic–organic hybrid composites.

We have developed new antibiocorrosive coatings,[48,49] where bioactive components used are stable, nonvolatile organometallic complexes, safe for human life in conditions of certain concentrations, also highly dispersed bioactive inorganic (possibly nanoparticles) (Fig.-s 1.3-1.5) obtained by transformation of the secondary raw material. Bioactive compounds dropped into the polymeric carriers modified with industrial functional polymers, which are distinguished with good compatibility.

Obtained antibiocorrosive coatings:

- are characterized by a good fixation on various synthetic and natural materials, as well as on museum exhibits and archeological samples;
- do not violate wholeness of samples during hardening of the cover composites;
- are transparent (Fig.-s 1-3) and almost do not change the color during their aging;

- have enough strength, elasticity, and stable mechanical characteristics (they do not scratched easily and do not change relief of the surface) and high hydrophobility (<0.1%), typical for such materials (Fig. 16.3);
- are not dangerous for human; during long time of exploitation, do not produce harmful gases;
- and are relatively cheap and available.

Their thermal stability (in air, isothermal aging at 40 and 60°C), on action of so-called "light weather" (complex action of moisture and of air oxygen, of sun-scattered ultraviolet radiation, and CO_2) showed that during the long period, the initial expression, color, optical transparency, and mechanical properties (homogeneity of the surface without formation of splits), antibiocorrosive coatings did not worsen.

Instead of easily swillable tin- and arsenic-organic low molecular antibacterial additives to dye composition with specific properties, industrial silicon polymer modified by bioactive fragment (Fig. 16.4) were used.

$$(CH_3)_3SiO[((CH_3)_2SiO)_a(\underset{|}{Si}(CH_3)O)_b(Si(CH_3)(C_6H_5)O)_c]_m Si(CH_3)_3$$
$$[(CH_2)_3As(C_6H_5)_2(n\text{-}C_3H_7)]Br$$

FIGURE 16.4 Industrial silicon polymer modified by bioactive fragment.

In case of necessity, it is possible to obtain antibiocorrosive coatings with different color (Figs. 16.5–16.7), which can be adjusted by selecting of appropriate bioactive components.

FIGURE 16.5 Antibiocorrosive coatings containing inorganic–organic bioactive structures obtained by transformation of the arsenic industrial waste (I and II samples: leader, III: wood).

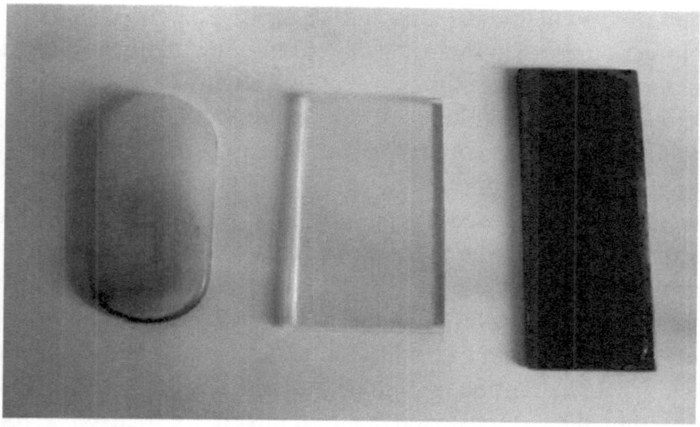

FIGURE 16.6 Antibiocorrosive coatings (first two samples) and due composite contained bioactive silicon polymer: first sample from left—antibiocorrosive coatings for polycarbonate plate based on high dispersed stibium(III) oxide and polyepoxide modified by industrial silicon polymer; second sample from left: optically transparent antibiocorrosive coatings for organic glass plate based on high dispersed stibium(III) oxide and polyepoxide modified by siliconorganic oligomer; third sample from left: dye composite, contained bioactive silicon polymer for protection of underwater part of the ship.

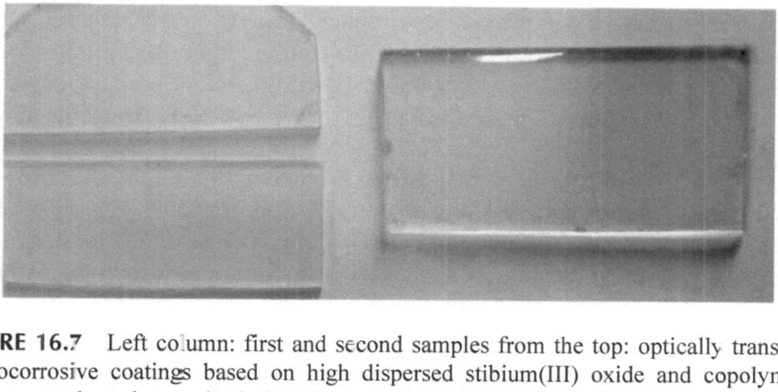

FIGURE 16.7 Left column: first and second samples from the top: optically transparent antibiocorrosive coatings based on high dispersed stibium(III) oxide and copolymer of perfluoro methacrylate and ethyl methacrylate covered on the polycarbonate plate. Right column: optically transparent antibiocorrosive coatings based on high dispersed stibium(III) oxide and copolymer of perfluoro methacrylate and ethyl methacrylate covered on the organic glass plate.

The potential spheres of using of obtained bioactive composites are:

- prevention of materials from biodeterioration, noncontrolled biodegradation, various branches of industry, and human protection during contact with this materials;

- inhibition of growth and expansion of microorganisms, which are causal factors of infection—inflammatory sickness of human and animals; for prophylaxis of above-mentioned diseases provoked by microorganisms;
- protection of museum exhibits and some of cultural heritage;
- for obtaining dye composition for protection of underwater part of the ship toward aggressive microorganisms excreted by some algae.

It must be noted that, in case of need, it will be possible to remove some of the antibiocorrosive covers from the surface without damaging it, which is especially important during the restoration of museum exhibits and archeological samples.

According to the information obtained from scientific literature of last decade, the interest toward stibium and its traditional compounds seems to be reduced relatively worldwide. More attention is switched to arsenic and its compounds.[48,49] This fact can be caused by the circumstance that best efforts are focused on removal from environment and desired use of arsenic as one of the most undesirable pollutant, while stibium is easily removed from almost all satellite elements.

At present, special attention is paid to synthesis and research of hybrid arsenic compounds having biocide properties, since some of them may be successfully used for receipt of bioactive additives, biostable polymers, dyes etc., and part of them is used against phytopathogenic bacteria.[49] Based on this position, it is clear that synthesis of arsenic-containing compounds and study of their biocide properties carries out intensively.[50–54]

In the recent years, the ministry of environment protection and natural resources of Georgia is vitally interested in solution of this problem. For this purpose, a special 6-year program named "Recycling and decontamination of production waste of arsenic mining and chemical plants of Georgia" has been elaborated by the ministry last year. This program is in full compliance with project represented by us. In case of winning of this project, the focuses of environmental pollution with the waste obtained as a result of arsenic ore processing will be deactivated (Fig. 16.6). Their harmful impact on human's health will be maximally decreased, and tourism development and production of ecologically clean agriculture product in the Georgian region (Svaneti and Racha) will be promoted.

16.3 NOVEL BIOACTIVE HYBRID COORDINATION COMPOUNDS OF SOME TERTIARY ARSINES WITH MERCURY (II) HALIDES

We have synthesized[39] two types of coordination compounds: $Ar_2AsR{\bullet}HgCl_2$ and $Ar_3As{\bullet}HgX_2$ (X = Cl, Br) (Table 16.1 Asymmetry tertiary arsines complexes type of $Ar_2AsR{\bullet}HgCl_2$ (where Ar is aryl and R is alkyl) were formed according to the following reaction (Scheme 16.1):

$$Ar_2AsR + HgCl_2 \rightarrow Ar_2(As)R \bullet HgCl_2$$

SCHEME 16.1

It was shown that symmetric trialkyl arsines could not form crystalline compounds with mercury (II) chlorides:

$$R_3As + HgCl_2 \not\rightarrow R_3As \cdot HgCl_2,$$

where R = n-C_3H_7, iso-C_5H_{11}, or C_7H_{25}.

Symmetric triaryl arsines have formed with mercury (II) halides crystalline products:

$Ar_3As + HgX_2 \rightarrow Ar_3As \bullet HgX_2$, where Ar = C_6H_5 or m-CH_3–C_6H_4, X = Cl or Br.

It was established that the change of the ratio of initial reagents (in wide limits) did not influence on the chemical composition of the products, it was always equal to 1:1. The assumption that the obtained products are identical was confirmed by the fact that their melting point did not change. Thus, based on numerous experiments, optimal conditions for this process were elaborated. All synthesized products are white crystalline, stable compound. It was established that the yield in case of triarylarsines is far less than in case of diaryl(alkyl)arsines. The rate of formation of adducts of triphenylarsine with mercury (II) chloride and bromide prevails over the rate of formation of the product of addition of tri-*m*-tolylarsines. This fact can be explained by the spatial factor and by electronegativity of β-substituent. It must be noted that (*m*-CH_3–$C_6H_4)_3As \bullet HgCl_2$ is more coarse crystalline than other synthesized compounds. In aqueous solution, (all the more in organic solvents) mercury (II) halides are weakly ionized and the concentration of Hg^{2+} ions, for example, in extracted solution of sublimate is not more than 10^{-8}. It means that it is non-electrolyte ($\alpha_{HgCl2} < \alpha_{H2O}$) and when $HgCl_2$ undergoes dissociation, it will stop at the first stage: $HgCl_2 \leftrightarrow HgCl^+ + Cl^-$

The addition of triaryl- and diarylalkyle arsines to mercury (II) halides can be considered as nucleophilic process:

$$Ar_3As + HgHal_2 \not\rightarrow [Ar_3As \rightarrow HgHal]Hal$$

The molecular electroconductivity of the obtained compounds was determined in dimethylformamide. The data of the molecular electroconductivity (less than 30 cm^{-1} cm^2) confirmed that these compounds are non-electrolytes. IR spectra of the obtained compounds showed that the absorption band in the region 560–580 (corresponding to As–$C_{alk.}$), disappears, thus the absorption band at 620 cm^{-1} appears, which is characteristic to As–$C_{alk.}$, when As is in sp^3 hybridization state. Based on the data of molar electroconductivity and IR spectra, we can propose that the compound is dimmer bridge complex:

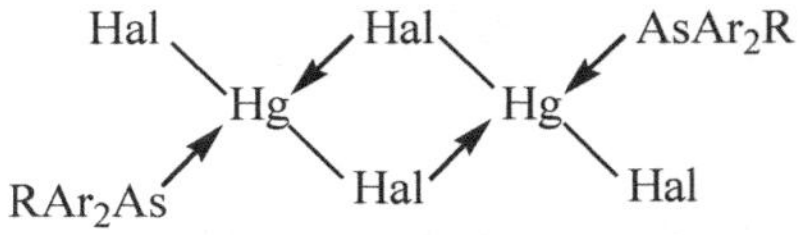

, (Hal = Cl or Br)

We carried out the thermographic analysis of the synthesized compounds. The thermolysis of each synthesized compounds is similar. The process of thermal decomposition can be represented according to the Scheme 16.2:

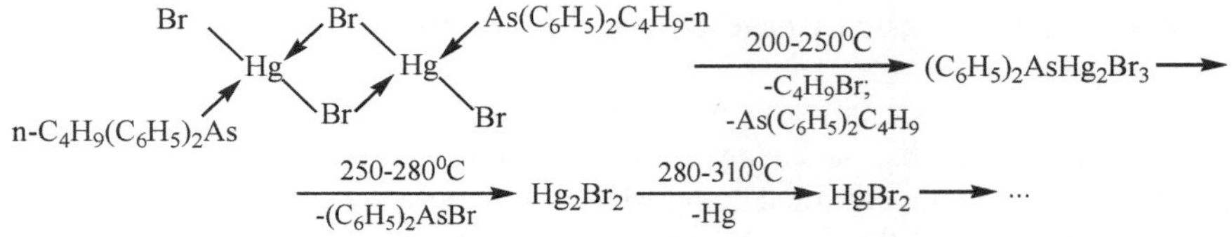

SCHEME 16.2

We have elaborated the method of the algebraic–chemical characterization of typical structures (main characterizing structure) of synthesized compounds R_2AsR, where R = m-$CH_3C_6H_5$; R′ = m-$CH_3C_6H_5$; iso-C_4H_9; C_2H_5. For creation of the bank of data for these chemical structures, we have used the fragmentation matrix (F-matrix) method. We constructed the F-matrix for our systems and use lg(Δ*F*) (*F* called as fragmentation matrix)—effective topologic index for construction and investigation of fragmentation "structure–properties" type correlation equations. As the synthesized compounds are crystalline, we have constructed the correlation equation and it has the form:

$$T_{melt} = 78 \cdot \lg(\Delta F) - 123 \text{ (A)}; \quad T_{melt.} = 72 \cdot \lg(\Delta F) - 110 \text{ (B)},$$

where A: for the complex of $HgCl_2$, B: for the complex of $HgBr_2$.

The values of the $T_{melt.}$ and lg(Δ*F*) for these compounds are presented in the Table 16.2

TABLE 16.2 The Values of the T_{melt}. and lg(Δ*F*) for Tested Compounds.

No	Coordination compounds*	lg(Δ*F*)	$T_{melt.}$, °C
I	$(CH_3C_6H_4)_2AsC_2H_5 \cdot HgCl_2$	3.35	137
II	$(CH_3C_6H_4)_2AsC_4H_9 \cdot HgCl_2$	3.62	143
III	$(CH_3C_6H_4)_3As\ HgCl_2$	3.81	173
IV	$(CH_3C_6H_4)_2AsC_2H_5 \cdot HgBr_2$	3.35	131
V	$(CH_3C_6H_4)_2AsC_4H_9 \cdot HgBr_2$	3.62	145
VI	$(CH_3C_6H_4)_3As\ HgBr_2$	3.81	164

*Correlation coefficient $r = 0.921519$; $k = 72.4423$; $b = -111.476$.

We have carried out the preliminary virtual (theoretical) bioscreening of obtained structures (using the above-mentioned characterizing of arsenic-containing fragments) by using Internet-system program PASS C&T.[50] The estimation of probability of activity of compounds was carried out via determination of parameters P_a (active) and P_i (inactive). Based on virtual bioscreening, the synthesized compounds (Table 16.2, I–VI), with experimentally high probability (P_a = 0.55–0.80), showed virtually antibacterial, antifungal, anthelmintic, antiviral, and anticarcinogenic activity. Preliminary microbiologically study of the investigated compounds confirmed the virtual concepts (anticarcinogenic activity was not tested).

By using synthesized bioactive compounds (1–3–5%) and selected polymer matrix (polyurethanes and polyester urethanes non-modified and modified by siliconorganic oligomers), we prepared novel inorganic–organic antibiocorrosion covers.[51] We studied tribological properties[52] of polymer matrices and antibiocorrosion covers based on them. It was established that the friction of polyurethane and polyester urethane matrices modified by siliconorganic oligomers and antibiocorrosion covers, is higher for non-modified polyurethanes. So, the modification can be used for improvement of tribological properties of antibiocorrosive covers. By using gravimetric method, the water absorption ability of obtained antibiocorrosive covers was determined. It was established that during 720 h, their water absorption ability was 0.01–0.03%. Their testing on photochemical and thermal stability (in air, isothermal aging at 40 and 60°C), on action (on definite time) of "light weather" (complex action of moisture and of air oxygen, of ultraviolet eradiation of sun or of "scattered" sunlight, CO_2) showed that during 1200 h, the initial expression, color, optical transparency, and mechanical properties (homogeneity of the surface without formation of splits) of antibiocorrosive covers has not deteriorated.

16.4 COORDINATION COMPOUNDS OF MERCURY (II) NITRATE WITH TRIARYL-AND DIARYL-ALKYLARSINEOXIDES

One of the most important properties of oxides of tertiary arsines is a formation of coordination compounds with salts of transition metals (d elements). It should be emphasized that this kind of compounds is well studied in case of phosphorus-containing ligands. As to arsenic-containing compounds, like arenes, they are almost unstudied, if we didn't take into account several researches, which have fragmentary character and are not carried out on regular basis at all, in order to make some substantial conclusion. It is noteworthy too that complex compounds, which contain arsenal group As_O from the very beginning, have been widely used in practice as the best extragents of different metals. It seems we have to seek the cause in good capabilities of d-metals to attach oxides of tertiary arsines with formation of quite stable compounds. In all cases of extragents $R_3As = O$, where R = organic radical, complex formation occurs according to electron donor–acceptor interaction mechanism,[249] at that donor function is performed by oxygen atom which is entered into arsinal group, while acceptor is represented by complex former (metal) along with its vacant (unoccupied) orbitals. As it is supposed,[250] properties of mentioned ligands should be caused by the quality of As_O bond order, which, in its turn, is caused by the possibility of transition of two nonbonding electrons of negatively charged substituent (O) to the vacant 4d-orbital of arsenic.

Taking above-mentioned information into account, at this time, we have tried to study the complex formation ability of mercury (II) salts with oxides of triaryl- and diaryl-alkylarsines. If test will prove its value, of course we also study synthesized coordination compounds by means of accessible physical and chemical methods.

The well-known technique of triaryl(alkyl)arsine oxidation in the medium of hydrogen peroxide acetone was used in order to receive oxides of resulting tertiary arsines.[251] Reaction runs according to the following equation:

$$Ar_3As + H_2O_2 \rightarrow Ar_3As{=}O + H_2O$$

Azeotrope-former—benzol is removed from water of reaction mixture using the method of azeotropic dehydration.

The tertiary arsines are received using the action of Grignard reagent with esters of arsenic acid[57]:

$$(RO)_3As + 3ArMgBr \rightarrow Ar_3As + 3ROMgBr,$$

where R = n-C_5H_{11} or iso-C_5H_{11}, while diarylalkylarsines have been received by following method:

$$Ar_2AsCl + RMgBr \rightarrow Ar_2AsR + MgClBr$$

On the basis of many experiments, it was established that possibility of mercury (II) ion complex formation from oxides of tertiary oxides is considerably depended on the nature of ions entering into salt composition. Namely, most parts of corresponding halogenides are either insoluble in water and spirit or slightly soluble. Among soluble salts, mercury (II) acetate also have not came up to our expectations—complex compounds are formed, but goal will be achieved only after gradual drying of reaction mass that eliminates the possibility of receipt of basic product in chemically pure condition.

Thus, all our attempts to separate in individual condition, the products of mercury (II) salts interaction with arsinal compounds has ended without result.

Positive results have been obtained in the experiment only in case of mercury (II) nitrate's use as original substance. Regardless of ratio between reacting substances, right from the beginning of mixtures interaction, the addition products are precipitated almost in one and the same quantitative content.

It is known that oxides of tertiary arsines are monodentate ligands,[53] while coordination number 4 is characteristic for mercury (II) ions. That's why, as far as we assured ourselves of possibility of desired product's receipt due to action of tertiary arsine oxide on mercury (II) nitrate, we have carried out basic experiments by taking into account following molar ratio: $Hg(NO_3)_2{:}Ar_3As{=}O = 1{:}4$

Reaction has been carried out in non-aqueous solutions (spirit solutions). It turns out that in all cases, coordination number is equal to 4. Formation of desired products has to be explained according to the Scheme 16.3:

$$Ar_2AsCl + RMgBr \rightarrow Ar_2AsR + MgClBr$$

SCHEME 16.3

Composition and structure of synthesized compounds was determined by methods of both chemical and physical–chemical analysis. Results of element analysis are given in Table 16.3. Formulas of obtained coordination compounds were determined via study of molar electrical conductivity of samples dissolved in dimethylformamide. As it turns out from table 2.9, μ is changed within the limits of 112–123 $Ohm^{-1}\ cm^2\ mole^{-1}$, which is characteristic for three-ionic electrolytes.[58] This fact unequivocally points at the circumstance that investigated coordination compounds experience this kind of electrolytic dissociation in dimethylformamide:

$$[Hg(O\text{–}AsAr_3)_4](NO_3)_2 \rightarrow [Hg(O\_AsAr_3)_4]_2^{+} + 2NO_3^{-}$$

As to coordination rule of ligands, with the purpose of solution of this problem, we have studied the IR spectra of synthesized compounds. Their analysis shows that, as it was expected, a ligand is coordinated with mercury (II) ion via oxygen atom of arsinal group. For instance, intensive absorption band in 880–900 cm^{-1} area, which is referred to valence vibrations ($V_{As=O}$) of As–O bond is manifested on all spectrograms of initial oxides of tertiary arsines. [254,55] Presence of phenyl groups in the composition of synthesized compounds is confirmed by existence of absorption bands in 1590, 1490, 1440, 1170, 1156, 1080, 1030, 1000, 745, and 690 cm^{-1} areas. As–C_{alk} bond is confirmed by existence of absorption bands in 475 cm^{-1} area. It should be noted that all above-mentioned bands, except of arsinal group, are basically kept in spectra of interaction of oxides of tertiary arsines and mercury (II) nitrate. It is not surprising, since, namely, by means of oxygen atom of arsinal group, a ligand bounds with complex former. In particular, frequency of valence vibrations of As–O bonds in resulting complexes' spectra is decreased by 20–50 cm^{-1}. This fact, in its turn, points at the circumstance that ligands are monodentate.

TABLE 16.3 Some Physical Constants and Data of Element Analysis of $[Hg(OAs(R)Ar_2)_4](NO_3)_2$.

No	Ar	R	Melting point, °C	Molar electro-conductivity in di-methylformamide at 25°C, $omi^{-1}sm^2\,mol^{-1}$	Found, (%)		Brutto formula	Calculated, (%)	
					As	Hlg		As	Hlg
1	m-$CH_3C_6H_4$	n-C_4H_9	95–96	112.5	18.05	12.20	$C_{72}H_{92}As_4HgN_2O_{10}$	18.23	12.21
2	m-$CH_3C_6H_4$	iso-C_3H_7	121–122	114.7	18.66	12.65	$C_{68}H_{84}As_4HgN_2O_{10}$	18.87	12.64
3	p-$CH_3C_6H_4$	n-C_4H_9	87–88	118.5	18.11	12.18	$C_{72}H_{92}As_4HgN_2O_{10}$	18.23	12.21
4	p-$CH_3C_6H_4$	iso-C_4H_9	118–119	116.9	18.14	12.18	$C_{72}H_{92}As_4HgN_2O_{10}$	18.23	12.21
5	C_6H_5	C_6H_5	128–129	120.4	18.41	12.44	$C_{72}H_{60}As_4HgN_2O_{10}$	18.59	12.46
6	m-$CH_3C_6H_4$	C_2H_5	108–109	122.6	19.23	13.14	$C_{64}H_{71}As_4HgN_2O_{10}$	19.56	13.11

While considering absorption IR spectra, we make sure that in $[Hg(O{=}AsAr_3)_4](NO_3)_2$ and $[Hg(O{=}As(ROAr_2)_4](NO_3)_2$ compounds, a nitrate-ion $(NO_3)^-$ is located in second sphere of complex (see Fig. 2.14), which is determined by means of IR spectral data and points at the splitting of absorption band in 1300–1500 cm^{-1} area, while $V_{(NO3)}$ absorption band in 840 cm^{-1} is singlet. The fact that bond between complex former and ligand is strengthened by means of oxygen atom of arsinal of arsenic-containing compounds is confirmed by availability of absorption band in ~930 cm^{-1} area of IR spectra.

All above-mentioned information gives us an opportunity to make an univocal conclusion that synthesized coordination compounds represent three-ionic complexes that, along with molar electric conductivity, are confirmed by spectral analysis of researched substances.

Behavior of synthesized compounds during heating was also studied. Tetra (di-m-tolyl-buthylarsinoxymercury (II) nitrate thermograph is given by us as a sample (Fig. 2.15). As it turned out from this figure, thermolysis occurs quite difficult and contains no more than five exothermal effects and four endothermic effects. Using detailed analysis, we make sure that exothermal effect indicated at the DTA, as a rule, is not in compliance with weight gain. Therefore, the mentioned effect cannot be caused by addition reaction, for example, by involvement of air oxygen in chemical reaction. We guess that, in this case, we deal with even more complicated transformations that, maybe, should be caused by intramolecular regrouping into thermodynamically advantageous condition. This fact in our opinion can be explained by retro–Arbuzov regrouping[53] in ligand molecule subsequent to which takes place complex melting during compound's heating up to high temperature.

It is not impossible that diphenyl ester of arsenic acid would be received by further oxidation of Ar_2AsOR-type compound, but in this case, the role of oxidizing agent may be played not by air oxygen but nitrate-ion oxygen according to the general pattern:

$$Ar_3As + H_2O_2 \rightarrow Ar_3As{=}O + H_2O$$

As it was mentioned earlier, we have an opportunity to make such conclusion due to the fact that weight gain doesn't occur, though there is a pronounced exo-effect at DTA curve. Besides, it is well known from literature[59] that such transformations are quite allowable and are caused by system's transition to so-called energetically "advantageous condition."

Taking into account this consideration, we can represent the thermolysis of, for example, tetra(di-m-tolyl-n-buthylarsinoxide) mercury (II) nitrate as follows: sample decomposition begins at 90°C and gradually continues up to 600°C. Weight loss in the range of 90–445°C is 21.66% that corresponds with breakaway (theoretically 23.70%) of N_2O_3 and tertiary arsines $(m\text{-}CH_3C_6H_4)_2AsC_4H_9$ from the sample. In the same range, especially from 220°C, "intramolecular transformations" takes place, which, as we mentioned, have to be explained by retro–Arbuzov regrouping.

Separation of organic fragment takes place under 465–600°C conditions. At that, mercury (II) oxide also experiences final decomposition that is succeeded by complete "vaporization" of residue. Thus, at the last stage, not only ligand's removal occur, but also the decomposition and vaporization of inorganic residue itself: HgO → Hg + $1/2O_2$, which is one of the basic distinctive features of mercury (II) complex compounds in contradistinction from similar compounds of all other d-elements.

Hence, taking into account the above-mentioned information, the most likely pattern of thermolysis of considered complex compound can be represented as follows:

$$[Hg(O{=}As(C_6H_4{-}CH_3\text{-m})_2C_4H_9\text{-n})_4](NO_3)_2 \rightarrow -N_2O_3;\ (\text{m-}CH_3C_6H_4)_2AsC_4H_9\text{-n}$$

465–6000°C

$$HgO(\text{m-}CH_3C_6H_4As\text{-}OC_4H_9CH_3\text{-m})_3 \rightarrow Hg + 3(\text{m-}CH_3C_6H_4O)_2AsOC_4H_9\text{-n}$$

Thus, as a result of our research, it was established that products of interaction of mercury (II) nitrate and oxides of tertiary arsines are corresponding complex compounds.

Composition and structure of synthesized substances were determined using methods of IR spectroscopy, element analysis, and molar electric conductivity (Table 16.3). It is shown that bond between complex former and ligand is strengthened by means of oxygen atoms of arsinal group. Nitrate-ion is located at the second sphere of complex.

16.5 COORDINATION COMPOUNDS OF SOME D-METAL TETRATHIOARSENATES (V) WITH PYRIDINE

As we have mentioned in the introduction to the article, searching and creation of new cheap raw materials base becomes more and more topical at the modern stage of chemical science's development. Under the conditions of deficit, the use of products obtained as a result of transformation of arsenic production waste, for receipt of coordination compounds having peculiar properties, will promote not only an elaboration of commercially interesting forms of appropriate compounds, but also will solve important problem, will protect environment from pollution with arsenic-containing residues. One of the most interesting ways of solution of this problem is the receipt of new bioactive coordination compounds from arsenic-containing production waste.

As is known from literature, polyelement coordination compounds manifest a lot of interesting properties that gives us an opportunity of their wide practical application. In recent years, among this type of compounds, the complex formers generate special interest, which contain tetrathioarsenate groups along with nitrogen and d-metals.[56–58]

Goal of our research is a receipt and study of coordination compounds of zinc, mercury (II), copper (II), and nickel (II) with pyridine—one of the most prospective and accessible nitrogen-containing ligand received on the basis of products obtained via transformation of above-mentioned production waste. These compounds may manifest high bioactivity.

As the initial compounds, we have used zinc acetate, mercury (II) nitrate, copper (II) sulfate, and nickel (II) chloride, that is, salts soluble in water with above-mentioned metals. Among arsenic-containing compounds, we have used sodium tetrathioarsenate $Na_3AsS_4 \cdot 8H_2O$, which we received using processing of arsenic production waste[59–60]: $AsCl_3 \uparrow + (AsCl_3, H_2S) + (NaOH, H_2O)$ (Industrial waste) $As_2O_3 \rightarrow (RO)_3As \rightarrow As_2S_3 \rightarrow Na_3AsS_4 \cdot 8H_2O$

IR spectra have been obtained on the spectrometer "SPECORD IR-75" in Vaseline oil. X-ray studies were carried out on diffractometer "DRON-3M", thermographic studies were carried out on derivatograph "Q-1500" ("F. Paulik, J. Paulik, L. Erdey", Hungary), by heating of test specimen in the air up to 1000°C with the 10°C/min rate. Reference substance is corundum.

Synthesis of d-metal pyridinates occurs according to exchange reaction via action of equivalent quantity of sodium tetrathioarsenate saturated solution on the pyridinates of salts of transition metals:

a) $MX_2 + nC_5H_5N \rightarrow [M(C_5H_5N)_n]X_2$

b) $3[M(C_5H_5N)_n]X_2 + 2Na_3AsS_4 \cdot 8H_2O \rightarrow [M(C_5H_5N)_n]_3(AsS_4)_2\downarrow + 6NaX + 16H_2O$

or summarily:

$3MX_2 + 3nC_5H_5N + 2Na_3AsS_4 \cdot 8H_2O \rightarrow [M(C_5H_5N)_n]_3(AsS_4)_2\downarrow + 6NaX + 16H_2O$,

Obtained complex compounds represent solid, colored, finely crystalline substances, which are dissolved in dimethylformamide, are not dissolved in water, alkali, ethyl spirits and liquid saturated hydrocarbons, and are disintegrated at high temperatures up to melting. During acid treatment, they experience transformation with formation of arsenic (V) sulfide, for example (Table 16.4):

$[Zn(Py)_4]_3(AsS_4)_2 + 6HCl \rightarrow 3[Zn(Py)_4]Cl_2 + 2H_3AsS_4$

$2H_3AsS_4 \rightarrow As_2S_5 + 3H_2S$

Or summarily:

$[Zn(Py)_4]_3(AsS_4)_2 + 6HCl \rightarrow 3[Zn(Py)_4]Cl_2 + As_2S_5 + 3H_2S$

Composition and structure of obtained compounds was established by means of data of element analysis (Table 16.5), infrared spectroscopy, and X-ray phase studies.

TABLE 16.4 Element Analysis Data and Output of Synthesized Complex Compounds of $[M(Py)_n]_3(AsS_4)_2$-Type.

#	Formula complex compounds	Color	Date of elemental analysis calculated/found, (%)				The yield, (%)
			M	As	N	S	
I	$[Cu(C_5H_5N)_4]_3(AsS_4)_2$	Grey	12.33/ 12.11	9.71/ 9.94	10.87/ 10.64	16.57/ 16.71	93.4
II	$[Zn(C_5H_5N)_4]_3(AsS_4)_2$	Light yellow	12.64/ 12.77	9.67/ 9.49	10.83/ 10.58	16.51/ 16.69	96.5
III	$[Cd(C_5H_5N)_4]_3(AsS_4)_2$	Carroty	19.22/ 20.02	8.87/ 8.81	9.93/ 9.84	15.13/ 15.26	95.3
IV	$[Hg(C_5H_5N)_4]_3(AsS_4)_2$	Grey	30.77/ 29.89	7.66/ 7.85	8.58/ 8.23	13.08/ 13.02	95.2
V	$[Ni(C_5H_5N)_6]_3(AsS_4)_2$	Black	8.82/ 9.19	7.48/ 7.71	12.56/ 12.32	12.76/ 12.51	92.8

As is seen from IR spectra of synthesized compounds, valence vibration absorption bands, which are characteristic for AsS_4^{3-} group, are manifested in them at 420 cm^{-1}, while absorption bands of deformation vibration at 470 cm^{-1} area.[62] Besides, as is known,[64] value of absorption band during free ligand's coordination with central atom experiences shift by ~8–30 s m^{-1}. Absorption band of uncoordinated pyridine υ(C=N) is located in 1580 cm^{-1} area,[64] while the same absorption band in spectra of compounds synthesized by us is manifested in 1600–1610 cm^{-1} area that unequivocally confirms the existence of pyridine coordinated with metal ions in these compounds.

Very close agreement of tetrathioarsenate-ions absorption bands with the absorption bands of sodium tetrathioarsenate-ions indicates the fact that mentioned anion creates second sphere of coordination compound. Data of X-ray phase analysis of synthesized substances (Table 16.5), based on classification given in the literature,[65] testifies that obtained compounds according to their structural nature belong to the subgroup of sulfosalts. Cations' impact on ordering of crystal structure deserves special attention. For example, copper (II) and mercury (II) promotes formation of finely crystalline phase, which approaches to X-ray amorphous state.[66]

TABLE 16.5 Data of Roentgen phase Analysis Tetrathioarsenates (V) of Pyridine Complex Some *d*-Metal (Table16.5a and Table 16.5b).

Table 16.5a

$[Cu(Py)_4]_3(AsS_4)_2$	
I/I_0	*d a/n*
20	11.33
100	8.19
20	7.23
20	6.56
10	5.71
10	5.33
10	5.07
20	4.38
20	4.04
15	3.87
5	3.24
5	3.16
5	3.00
5	2.95
5	2.75
8	2.53
5	2.44
10	2.11
10	1.97
10	1.92

Table 16.5b

$[Cd (Py)_4]_3(AsS_4)_2$	
I/I_0	*d a/n*
20	13.0
15	9.07
10	7.0
25	6.41
15	5.96
10	4.,33
35	3.90
25	3.27
15	3.17
15	3.05
20	2.98
20	2.94
15	2.43
30	2.23
30	2.19
20	2.15
10	2.08
15	1.97
15	1.94
40	1.83
100	1.77

$Zn(Py)_4]_3(AsS_4)_2$		$[Hg(Py)_4]_3(AsS_4)_2$	
I/I_0	I/I_0	I/I_0	*d α/n*
20	20	4	11.6
15	15	2	7.69
10	10	2	7.26
25	25	5	6.10
15	15	5	5.62
10	10	6	5.00
35	35	8	4.73
25	25	10	4,62
15	15	7	4.33
15	15	9	3.92
20	20	3	3.75
20	20	3	2.82
15	15	3	2.46
30	30	2	2.18
30	30	2	2.04
20	20	3	1.93
10	10	3	1.86
15	15		
15	15		
40	40		
100	100		

At the initial stage, breakaway of ligand from complex compound's molecule takes place, and afterwards, stage-by-stage thermal decomposition of tetrathioarsenate (V) continues.[56]

As is seen from results of thermographical study, processes of thermal decomposition (thermolysis) of complex compounds of copper and zinc are almost similar (Fig. 16.8).

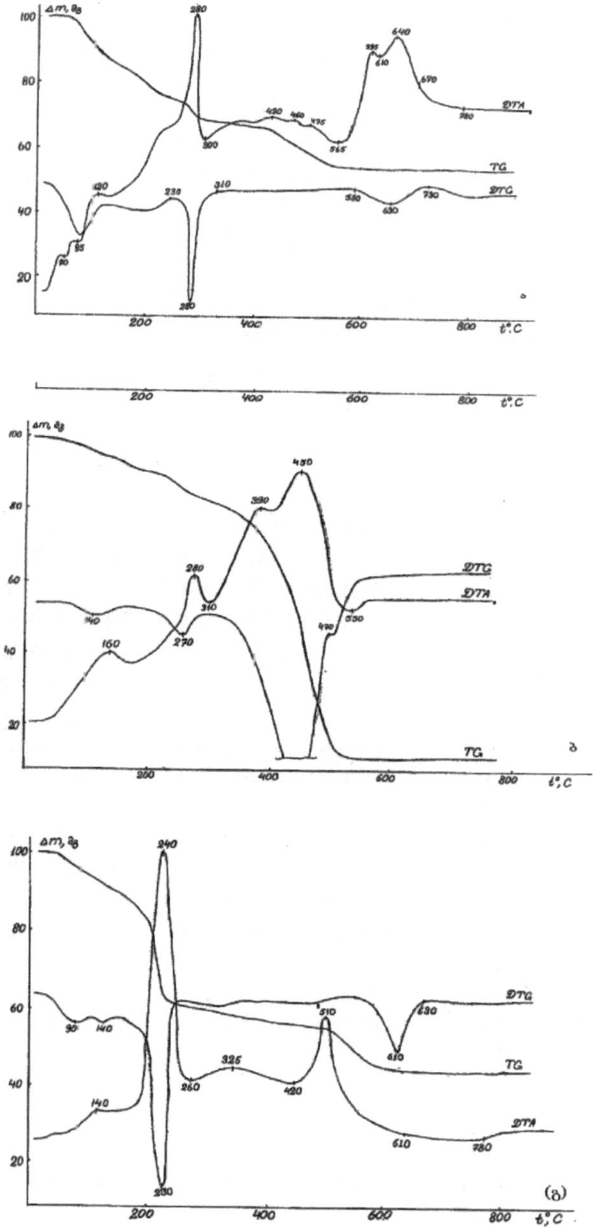

FIGURE 16.8 Thermogravimetric graphs of synthesized compounds: (a) $[Zn(Py)_4]_3(AsS_4)_2$; (b) $[Hg(Py)_4]_3(AsS_4)_2$; and (c) $[Ni(Py)_6]_3(AsS_4)_2$

There is a possibility to differ from each other, two types of complex compounds' decomposition are obtained: thermolysis of nickel tetrathioarsenate (V) pyridine complex (Scheme 16.4) and of mercury (II) tetrathioarsenate (V) pyridine complex:

$$[Ni(Py)_6]_3(AsS_4)_2 \xrightarrow[-6Py]{100-200^0C} [Ni(Py)_4]_3(AsS_4)_2 \xrightarrow[-12Py]{200-240^0C} Ni_3(AsS_4)_2 \xrightarrow[-3S]{250-420^0C}$$

$$\xrightarrow[-3S]{250-420^0C} Ni_3AsS_5\ (3MS{\cdot}As_2S_2) \xrightarrow[-AS_2S_2]{420-920^0C} 3MS$$

$$[Hg(Py)_4]_3(AsS_4)_2 \xrightarrow[-6Py]{100-200^0C} [Hg(Py)_2]_3(AsS_4)_2 \xrightarrow[-6Py]{200-240^0C} Hg_3(AsS_4)_2 \xrightarrow[-2S]{240-330^0C}$$

$$Hg_3As_2S_6 \xrightarrow[-2Hg]{330-430^0C} HgAs_2S_6 \xrightarrow[-HgS,-As_2S_5]{>430^0C} \text{sruli daSla}$$

SCHEME 16.4

Thermolysis of $[Ni(Py)_6]_3(AsS_4)_2$ sample (Fig. 16.8) begins from 100°C. Gravimetric analysis showed that first 6 moles of ligand breakaway at 200° that corresponds with 23.0% of weight loss (theoretically 23.6%). Very important endothermal effect is observed in the range of 200–250°C, with minimum at 235°C. The rest 12 moles of ligand breakaway in this interval. Therefore, sample's weight loss is 46% (theoretically 47.28%). Since then, thermal decomposition of the sample continues in a way as that in case of corresponding neutral salt.

As is seen from comparison of Diagram 4.3 and Diagram 4.4, thermal decomposition of pyridine complex of g(II) tetrathioarsenate (V) $\{[Hg(Py)_4]_3(AsS_4)_2\}$ must occur in different way. Ligand breakaway can also be represented step-by-step here: at the first step, in the range of 100–200°C, complex loses 6 moles of pyridine, while loses other 6 moles, in the range of 200–240°C. Weight loss at both the stages is equal to 31% (theoretically 30.77%). After ligand's breakaway, the process of thermal decomposition continues in a way as that in case of corresponding neutral salt.[56]

Along with thermogravimetric analysis, data obtained results are confirmed by the processes of thermal decomposition of intermediate products, which is described in literature, as well as by analysis of composition of residues formed at the various stages.[56,61–65]

Using the method of density functional theory, total energy of tetrahedral complex ions $[Zn(C_5H_5N)_4]^{2+}$, $[Cd(C_5H_5N)_4]^{2+}$, $[Hg(C_5H_5N)_4]^{2+}$, $[Cu(C_5H_5N)_4]^{2+}$, and rows of metal–nitrogen bonds in them was calculated (Table 16.6). For calculation, a basis with pseudopotential (only valent electrons) that implies relativistic corrections was used.

TABLE 16.6 Metal–Nitrogen Bonds.

Ion	Full energy, kJ/mol	Bond order			
		N_1	N_2	N_3	N_4
Zn^{2+}	−1026272.49	0.47	0.47	0.47	0.47
Cd^{2+}	−870624.67	0.35	0.35	0.35	0.35
Hg^{2+}	−836100.01	0.36	0.36	0.36	0.36
Cu^{2+}	−948390.59	0.54	0.54	0.54	0.54

Theoretical estimation of bioactivity of metal–pyridine fragments of obtained coordination compounds has been carried out using computer program PASS C&T,[66] which is performed by $P_{a\,(act.)}$ and $P_{i\,(inact.)}$ parameters. Prior to that, Na_3AsS_4 was calculated, which manifested $P_a = 0.507–0.898$ antiprotozoal (amoeba) and antihelmintic (nematodes, Fasciola) activity. Calculations show that metal–pyridine fragment of synthesized coordination compound, presumably, also manifests high bioactivity. Selected structures with high probability ($P_a = 0.570–0.900$) (Table 16.7) may manifest the following kinds of biological activity: atherosclerosis treatment, antineoplastic, antiseborrheic, antiviral picornavirus, cytoprotectant, etc. Conjugation of above-mentioned fragments apparently makes prospective creation of wide range of coordination compounds with antimicrobial and antihelmintic properties.

16.6 SYNTHESIS AND STUDY OF $[(I\text{-}PR)_2(ET)_4AS_2(PH)][CO(NCS)_4]$ USING X-RAY DIFFRACTION AND IR SPECTROSCOPY ANALYSES

16.6.1 AIM OBJECTIVE AND BACKGROUND:

Many inorganic/organic arsenic compounds are used as therapeutic agents against variable diseases, especially as antitumor drugs,[67] biological active substances,[68] in material science,[69,70] as auxiliaries in asymmetric synthesis,[71–75] as catalysts,[76] etc.

Compounds containing lone pairs can be considered electron-rich and trialkyl(aryl)arsines (AsR_3) act as nucleophilic toward haloalkanes to produce tetraalkyl(aryl)arsonium salts (AsR_4^+), which contain As(V).[77]

Chemically active (with high toxicity) three-coordinated organoarsenic compounds (e.g., arsines) can be stabilized by forming fourth bond with electrophilic substituents and four-coordinated arsenic compounds are chemical stable and less toxic. Arsonium yields with four-coordinated arsenic atom are successfully used in synthetic inorganic chemistry as bulky cations to stabilize bulky anions.[78]

In compounds, arsenic atoms have the following electronic configurations: p^3, sp^3, sp^3d, and sp^3d^2. Tetracoordinate arsenic derivatives—sulfides, oxides, and arsonium compounds have a tetrahedral configuration and it has been suggested that

p_π-d_π conjugation is characteristic of the corresponding compounds.[79] Nevertheless, the aryl- and ($E(CH_3)_3$), alkylarsines such as trimethylarsane, are encountered as ligands in d-metal complexes.

The order of affinity of these soft Lewis bases for a d-metal ion generally follows the order: $PR_3 > AsR_3 > SbR_3 > BiR_3$.[78] Because of soft-donor nature, many aryl and alkylarsane complexes of the soft species Rh(II), Ir(I), Pd(II), and Pt(II) have been prepared and studied.[78] The series of tri- and four-coordinated arsenic compounds were studied,[80,81] for example, in trimethylarsine $(CH_3)_3As$, As–C bond lengths are equal 1.98 Å and C–As–C angles around 96°, the geometry is pyramidal, respectively. Among the trivalent arsenicorganic compounds, the bisarsines (RAs(R)AsR) are important as useful bidentate ligands, especially o-phenylenebis(dimethylarsine) or 1,2-(arsino)benzene ($C_6H_4(As(CH_3)_2)_2$), known as diars is very often used in complexes.[82–85]

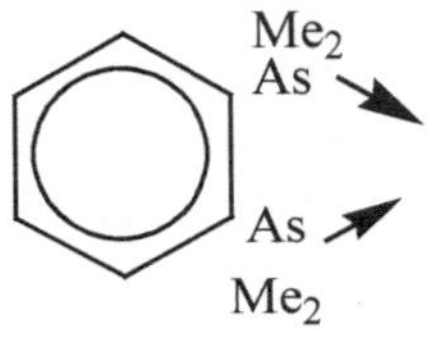

1,2-Bis(dimethylarsino)benzene or o-phenylenebis[(dimethyl)arsine]

Numerous complexes are studied with o-phenylenebisarsines with different substituents.[86–89]

It must be noted that (p-phenylene) derivatives of arsenicorganic compound do not show complexing ability, but at that time they could be converted easily into +5 species such as R_3AsO and $[R_4As]^+$ due to their basicity and nucleophilicity. In addition, most arsenic compounds have four substituents located in tetrahedron apexes around the central atom. In this case, the fourth bond is formed by coordination of nonbonding $4s^2$ electrons of the trivalent arsenic atom with a Lewis acid. Nevertheless, electrophilic addition of alkyl halides to tertiary arsines still remains one of the main methods for preparing tetraalkyl- and alkylarylarsonium salts $[R_4As]^+X^-$ (X=halg).[90] The effect of the nature of the halogen in alkyl halides on their addition to trialkyl(aryl)arsines is poorly studied, but it is established that alkyl iodides are the most active then alkyl chlorides.[90] As a result of alkylation of arsines or bisarsines cationic complexes- arsonium yields form which are stable than trivalent arsines. Due to this fact, first we obtained cationic bisarsonium diiodides and then converted into cationic–anionic complex, containing Co(II) and NCS-pseudohalide group.[90]

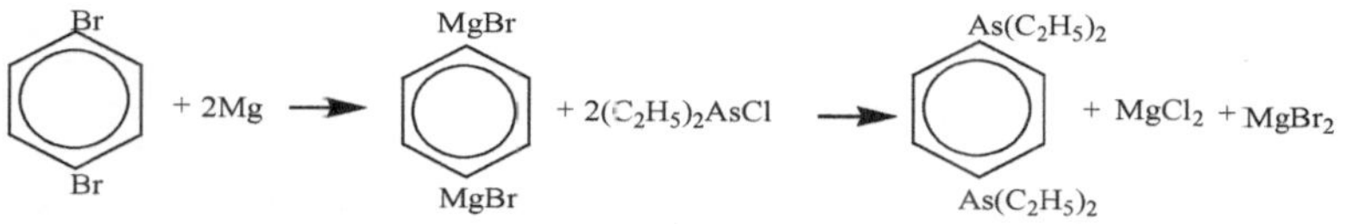

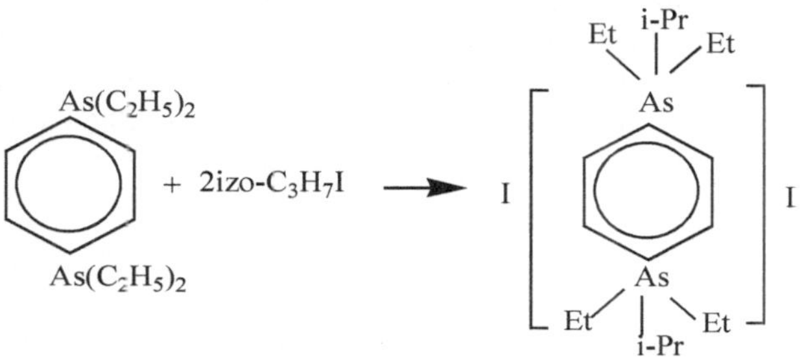

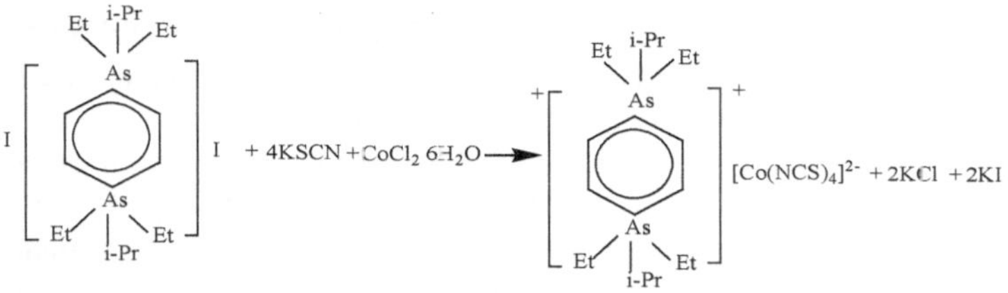

Bis(diethyl)arsine was synthesized by common Grignard reaction,[91] in diethyl ether environment and under cooling, the product was purified by distillation (liquid) under a reduced pressure. Diethylchlorarsine was synthesized according to literary,[92] through the following consecutive reactions:

$$R_3As + Cl_2 \rightarrow R_3AsCl_2 \xrightarrow{t^{o}C} R_2AsCl + RCl$$

Chlorine was obtained and dried according to the methodic,[93] in synthesized compounds, arsenic was determined quantitatively by Evins' method.[94] nitrogen by Duma's method,[95] cobalt by complexonometry, [96] iodine by mercurymetry,[97] and sulfur by gravimetric method.[98] Required solvents, ethanol and diethylether, were purified and dried according to the procedure described in Ref 99.

Dark blue single crystals of the complexes selected from the bulk before filtration were used for data collection. The cell determination and data collection were carried out on a Nonius Kappa CCD diffractometer using graphite monochromated MoK_α radiation ($\lambda = 0.71070$ Å). The phase problem was solved by SIR-97[100] and the structure refinement was carried with full-matrix least-squares on F^2 using the SHELXL-97[101] program. All non-hydrogen atoms were refined anisotropically.

16.6.2 GENERAL RESULTS:

IR SPECTROSCOPIC STUDY

In the IR spectrum of $[(Ph_3AsCH_2I]I_3$, the aromatic C–H stretching bands appear in the 3047 cm^{-1} and the aliphatic C–H, in the 2877 (asym.) and 2946 (sym.) cm^{-1} regions. Skeletal vibrations, representing aromatic C=C absorb in the 1581–1434 cm^{-1} range. The C–H_{ar} bending bands appear in the regions 1241–1025 cm^{-1} (in plane bending) and 833–686 cm^{-1}(out-of plane bending).[102] The week and medium stretching bands appear at the 462–481 cm^{-1} (As–$C_{ar.}$) and 685 cm^{-1} (As–$C_{aliph.}$), characterized for As–C_4 bonds in tetrahedral position.

X-RAY STUDY:

The structures of cation and anion of the obtained complexes are shown in Figures 16.9 and 16.10.

Compound crystallizes in the monoclinic, space group $P2_1/n$ (No. 14) with a = 10.197 (1) Å, b = 13.152 (1) Å, c = 16.882 (1) Å, β = 93.01 (1)°, and four formula units per unit cell. The crystal structure was solved via the Patterson method. For refinement, full-matrix least-squares methods were applied.

Despite the identification of various arsonium cations, few examples have been comprehensively studied and rarely mentioned in the literature.[103–112] The well-known and studied cation is tetraphenylarsonium, in which four independent arsenic atoms with As–C distances in the range 1.910–1.921 and angles at arsenic between 106.1 and 110.7 assigned to the tetrahedral geometry. Tetrahedral configuration is also observed in arsonium cations carrying benzyl, methyl, 4-methyl-phenyl, and naphthalen-1-yl groups with As–C bond lengths 1.889–1.949 and C–As–C angles 106.9–112.5. A variety of complexes has been isolated with structures based on a hydrogen bonded $(Ph_3AsO)_2H]^+$ cation with I_3, BF_4, and $AlCl_4$ counterions. The cation–anion contacts (3.363 and 3.58 Å) in 1,3-dimethyldiaza-2-arsenanium tetrachlorogallate are within the sum of the Van der Waal's radii for As and Cl (As, 2.0 Å; Cl, 1.7 Å). Recently, we have isolated $[R_3AsCH_2I][I_3]$ (a) and $[R_2(R')AsCH_2I][I_3]$ (b) (R = Pr, iPr, Bu, iBu, Ph, am; R' = Bu, Ph) arsonium compounds in As–C interatomic distances range in the 1.906–1.924 Å and C–As–C angles of the cation: 105.3–109.8°(a) as well as the As–C interatomic distances range in the 1.921–1.953 Å and C–As–C angles of the cation: 106.8– 114.7° (b). The As–C bond lengths of Ph-substituted compound are shorter than izo-Bu-substituted compound. The anion (in contrast of comp. a) has symmetric linear structure: $[I\text{-}_{2.91}\text{–}I\text{-}_{2.91}\text{–}I]^-$, as it is in $[As(C_6H_5)_4]^+$ $[I\text{-}_{2.90}\text{–}I\text{-}_{2.90}\text{–}I]$.

For some transition complexes containing ortho-phenylenebis(dimethylarsine) and its phenyl analogues, as chelates, the octahedral arrangement dominates, $[Rh(diars)_2Cl(CO_2)]$ has an octahedral geometry, containing a molecule of carbon dioxide coordinated via carbon atom, as well as cobalt complex $[Co(diars)_3][BF_4]_3$ $2H_2O$, with As–Co distances ranging between 2.365 and 2.395 Å and the platinum complex $[Pt(diars)_2I_2][BF_4]_2$, which has trans-octahedral geometry (As–Pt 2.45 Å).

The arsenic–carbon bond lengths and angles of the synthesized bisarsonium tetra-izo-thiocyanatocobaltate (II) complex are corresponding to the tetrahedral configuration, therefore, the As–C bond lengths range in the 1.911–1.959 Å and angles 107.4–110.2°; as well as, in the counterion, Co–N bond lengths are equal 1.964 Å and angles 105.4°, respectively. The crystal packing is due to the cation–anion contacts and week intra-or intermolecular interaction forces.

The structures of $^{+}[(i\text{-}C_3H_7)(C_2H_5)_2As(C_6H_5)As(C_2H_5)_2(i\text{-}C_3H_7)]^{+}$ cation and $[Co(NCS)_4]^{2-}$anion and crystal packing of $[(i\text{-}Pr)(Et)_2As(Ph)As(Et)_2(i\text{-}Pr)]$ $[Co(NCS)_4]$ are given below (Fig. 16.9, Fig. 16.10)

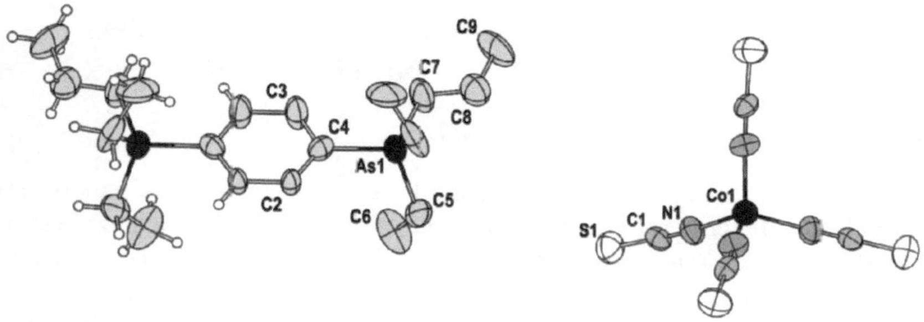

FIGURE 16.9 Structures of $^{+}[(i\text{-}C_3H_7)(C_2H_5)_2As(C_6H_5)As(C_2H_5)_2(i\text{-}C_3H_7)]^{+}$ cation and $[Co(NCS)_4]^{2-}$ anion.

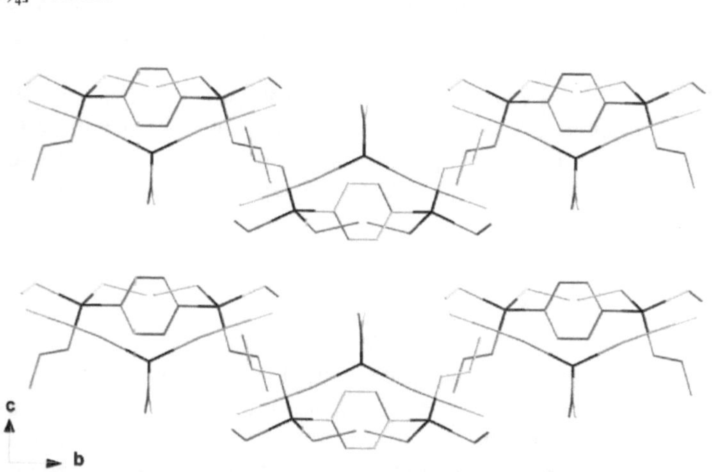

FIGURE 16.10 Crystal packing of $[(i\text{-}Pr)(Et)_2As(Ph)As(Et)_2(i\text{-}Pr)][Co(NCS)_4]$.

The preliminary researches showed that the synthesized compounds have bactericide properties. They may be used against some phytopathogenic microorganisms. Synthesized arsenic–manganese complexes, dropped into the several functional polymer matrixes, may serve as a basis to prepare new bioactive materials with regular duration of the action: antibiocorrosion coatings for multifunctional purpose foamed polyurethane materials and various dipping compounds.

The preliminary researches showed that the synthesized compounds have bactericide properties. They may be used against some phytopathogenic microorganisms. Synthesized arsenic–manganese complexes, dropped into the several functional polymer matrixes, may serve as a basis to prepare new bioactive materials with regular duration of the action: antibiocorrosion coatings for multifunctional purpose foamed polyurethane materials and various dipping compounds.

16.7 STIBUM-CONTAINING BIOACTIVE COMPLEX COMPOUNDS BASED ON *D*-METALS AND SOME NITROGEN-CONTAINING LIGANDS: SYNTHESIS, STRUCTURE, AND PROPERTIES

16.7.1 AIMS AND BACKGROUND:

Transformation of stibium containing industrial waste into stable coordination compound with specific properties with the purpose of their further application belongs to a variety of topical issues of applied and coordination chemistry. Successful solution of this problem will not only create new raw materials resource base, but also will solve important ecological problem, will protect environment from pollution by stibium containing waste. The presented work is devoted to this problem, in which, reaction–responsive stibium compounds extracted from arsenic production waste are considered, on the basis of which new nitrogen-containing bioactive coordination compounds are received.

2,2'-dipyridyl as a ligand, for a great while is studied during the synthesis of coordination compounds.[113] The reason of this study is nonuniform and contains both pure chemical and applied aspects. The ample opportunities of 2,2'-dipyridyl for creation of coordination compounds of different types and behavior should be sought in its structure:

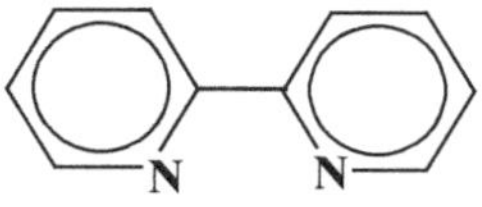

Because of the fact that nitrogen atoms hold such a position in molecule that optimum conditions are created for formation of five-member cycle, 2,2'-dipyridyl from the very beginning became the focus of scientists interest, and due to this fact, coordination compounds with metal halogenides, nitrates, sulfates, and almost every other soluble salt containing it are studied in details.

Sodium tetrathioarsenates (V) water solutions action with the products of interaction of *d*-metals soluble salts with bidentate ligand 2,2'-dipyridyl leads us to the formation of appropriate coordination compound[2]. Having used the same method, we implemented the synthesis of coordination compounds tetrathiostibiates (V) with 2,2'-dipyridyl.

Salts dissolved in *d*-metals water are used as basic substances: Ag(I) and Hg(II) nitrates, Zn, Fe(II) and Cu(II) sulfates, Mn(II), Cd(II), Ni(II), and Co(II) chlorides; sodium tetrathiostibiate is used as precipitator, while 2,2-dipyridyl $(C_5H_4N)_2$ (shortly dipy) is used as nitrogen-containing ligand. We also used ethylenediamine (H_2N–CH_2–CH_2–NH_2), one of the best bidentate ligand, for obtaining of complex compounds of *d*-metals (II).

16.7.2 RESULTS AND DISCUSSION:

For preliminary estimation of relative complex-forming ability of the selected organic ligands and study of their electronic structure, we have carried out their quantum-chemical investigation. Quantum-chemical calculations were performed on PC with an AMD processor with the built-in coprocessor by using Mopac 2000 and CS Chem3D Ultra, v8.[114] We gave the following key words to guide each computation: EF GNORM = 0.100 MMOK GEO-OK AM1 MULLIK LET DDMIN = 0.0 GNORM = 0.1 GEO-OK.

We have calculated energetical and geometrical parameters, effective charges on atoms, and electron occupation of atomic orbitals (electronic density) in 2,2'-dipyridyl and ethylenediamine molecules.

In the molecule of 2,2'-dipyridyl (Fig. 16.11), the bond lengths and valence bond angles of nitrogen atoms N_1 and N_{12} with neighbor atoms ($\angle C_2$–N_1–C_6 = 118.1° and $\angle C_7$–N_{12}–C_{11} = 118.1°) indicate mainly their sp^2 hybridized position.

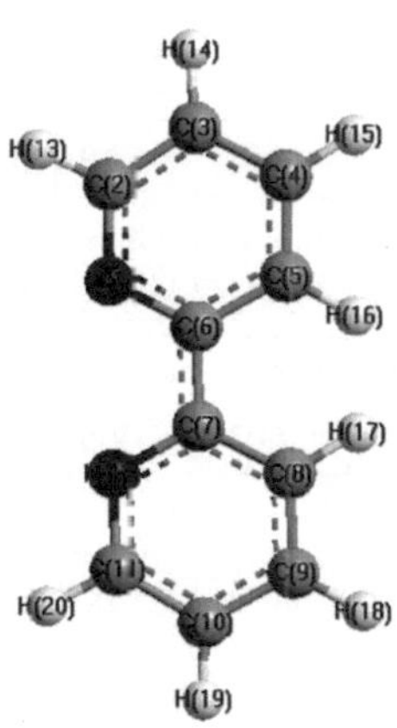

FIGURE 16.11 3-D model of 2,2'-dipyridyl.

Analysis of values of effective charges on atoms (Table 16.8) shows that potentially electron donor atoms are N_1 ($q_1 = -0.105052$), C_3 ($q_3 = -0.177231$), C_5 ($q_5 = -0.167740$), C_8 ($q_8 = -0.167764$), C_{10} ($q_{10} = -0.177228$), and N_{12} ($q_{12} = -0.104952$). But electron occupation of atomic orbitals (Table 16.9) shows that in spite of comparatively high negative relative charge of carbon atoms (C_3, C_5, C_8, C_{10}), in comparison with nitrogen atoms, are unable to form σ-bond with metal atom by donor-acceptor mechanism as they don't have an electron pair on the second energetical level. The electron pairs of nitrogen atoms is located on 2*s* orbital (electron occupation 1.71732 (N_1) and 1.71738 (N_{12})) and only they have the capacity to form σ-bond with metal atom by donor–acceptor mechanism.

TABLE 16.7 The Results of Calculated Virtual Bioactivity of Metal-Pyridine Fragments of Tetrathioarsenates (V) Some of *d*-Metals.

Compound	Atherosclerosis treatment	Antineoplastic	Anemia, sideroblastic	Antiseborrheic	Antiviral (picornavirus)	Cytoprotectant	Antineurotic	Insulin promoter
	P_a/P_i							
$[Cu(C_5H_5N)_4]_3(AsS_4)_2$	–	–	0.874/ 0.002	0.870/ 0.007	0.666/ 0.009	0.655/ 0.013	0.679/ 0.046	0.625/ 0.012
$[Zn(C_5H_5N)_4]_3(AsS_4)_2$	0.991/ 0.002	0.892/ 0.005	0.859/ 0.003	0.852/ 0.010	0.639/ 0.012	0.637/ 0.018	0.624/ 0.063	0.570/ 0.018
$[Cd(C_5H_5N)_4]_3(AsS_4)_2$	0.987/ 0.002	0.953/ 0.004	0.822/ 0.004	0.852/ 0.010	0.583/ 0.023	0.598/ 0.031	0.624/ 0.063	0.570/ 0.018
$[Hg(C_5H_5N)_4]_3(AsS_4)_2$	0.987/ 0.002	0.953/ 0.004	0.852/ 0.010	0.852/ 0.010	0.563/ 0.023	0.578/ 0.031	0.624/ 0.063	0.563/ 0.018
$[Ni(C_5H_5N)_6]_3(AsS_4)_2$	–	0.945/ 0.004	0.867/ 0.003	0.839/ 0.012	0.574/ 0.025	0.602/ 0.030	0.601/ 0.071	0.579/ 0.025
$[Co(C_5H_5N)_6]_3(AsS_4)_2$	–	–	0.859/ 0.003	0.886/ 0.005	0.645/ 0.011	0.640/ 0.017	0.725/ 0.034	0.673/ 0.008

TABLE 16.8 Relative Charges and Electronic Density on the Dipyridyl Atoms.

Atom (i)	Relative charges on atoms (q_i)	Electronic density (q_i (d))	Atom (i)	Relative charges on atoms (q_i)	Electronic density (q_i (d))
N_1	−0.105052	5.1051	C_{11}	−0.071521	4.0715
C_2	−0.071475	4.0715	N_{12}	−0.104952	5.1050
C_3	−0.177231	4.1772	H_{13}	0.161749	0.8383
C_4	−0.095118	4.0951	H_{14}	0.144233	0.8558
C_5	−0.167740	4.1677	H_{15}	0.139053	0.8609
C_6	0.025459	3.9745	H_{16}	0.146123	0.8539
C_7	0.025459	3.9745	H_{17}	0.146118	0.8539
C_8	−0.167764	4.1678	H_{18}	0.139062	0.8609
C_9	−0.095095	4.0951	H_{19}	0.144225	0.8558
C_{10}	−0.177228	4.1772	H_{20}	0.161696	0.8383

Thus, the molecule contains two potential electron donor atom, N_1 and N_{12}, because of that it represent as a bidentate ligand and is capable to form coordination compounds with *d*-metals in the form of five-member cycle:

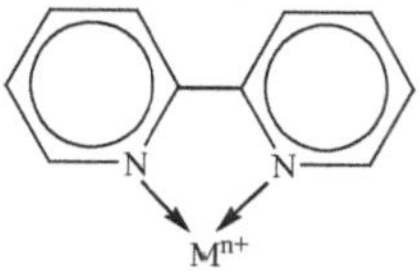

In the molecule of ethylenediamine (Fig. 16.12), the bond lengths and valence bond angles of nitrogen atoms N_3 and N_4 with neighbor atoms ($\angle C_1–N_3–H_9$ = 109.5°, $\angle C_1–N_3–H_{10}$ = 109.5°, $\angle H_9–N_3–H_{10}$ = 110.6°, $\angle C_2–N_4–H_{11}$ = 110.3°, $\angle C_2–N_4–H_{12}$ = 111.3°, and $\angle H_{11}–N_4–H_{12}$ = 109.3°) indicate mainly theirs sp^3 hybridized position.

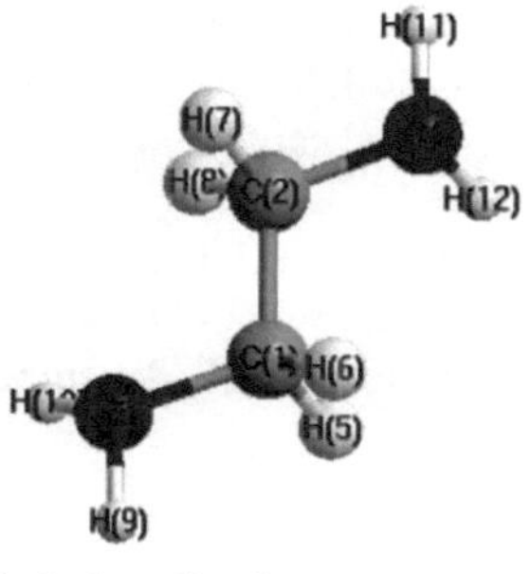

FIGURE 16.12 3-D model of ethylenediamine.

Analysis of values of effective charges on atoms (Table 16.10) shows that potentially electron-donor atoms are N_3 ($q_3 = -0.349295$) and N_4 ($q_4 = -0.349288$).

TABLE 16.9 Electron Occupation of Atomic Orbitals.

Atom	2*s*	2p_x	2p_y	2p_z
N_1	1.71732	1.06728	1.22193	1.09853
C_3	1.22757	1.00124	0.92718	1.02123
C_5	1.22712	0.94049	0.97761	1.02252
C_8	1.22715	0.94466	1.00769	0.98827
C_{10}	1.22752	1.00113	0.99355	0.95502
N_{12}	1.71738	1.05008	1.13945	1.19804

TABLE 16.10 Relative Charges and Electronic Density on the Ethylenediamine Atoms.

Atom (i)	Relative charges on atoms (q_i)	Electronic density (q_i (d))	Atom (i)	Relative charges on atoms (q_i)	Electronic density (q_i (d))
C_1	−0.080091	4.0801	H_7	0.105135	0.8949
C_2	−0.080043	4.0800	H_8	0.035121	0.9649
N_3	−0.349295	5.3493	H_9	0.149137	0.8509
N_4	−0.349288	5.3493	H_{10}	0.140076	0.8599
H_5	0.035061	0.9649	H_{11}	0.149103	0.8509
H_6	0.105054	0.8949	H_{12}	0.140031	0.8600

Electron occupation of atomic orbital (Table 16.11) shows that the electron pairs of nitrogen atoms are located on 2*s* and 2p_z orbitals (electron occupation for N_3 atom is 1.58211 (2*s*) and 1.49979 (2p_z), but for N_4 atom is 1.58220 (2*s*) and 1.52119 (2p_z). Nitrogen atoms, N_3 and N_4, have capacity to form σ-bond with metal atom by donor–acceptor mechanism.

Thus, the molecule contains two potential electron-donor atoms, N_3 and N_4, because of that it represents as a bidentate ligand and is capable to form coordination compounds with *d*-metal ion in the form of five-member cycle:

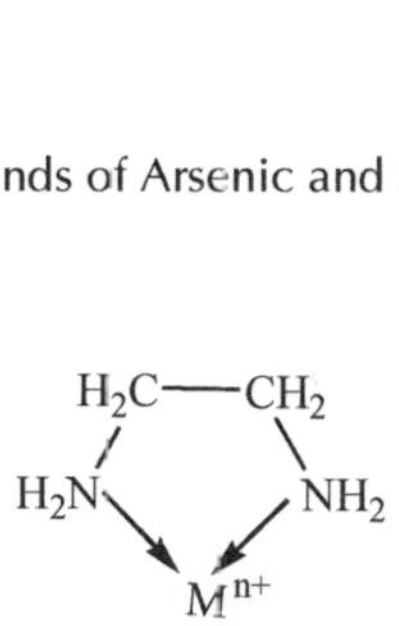

For comparison of the complex formation ability for selected ligands, we calculated their complexes with Cu^{2+}. The N_i–Cu^{2+} bond orders for dipyridyl $P_{Cu_1\text{-}N_2}$ = 0.333011 and $P_{Cu_1\text{-}N_{13}}$ = 0.335304, the correspond charges: q_1 = 0.633404, q_2 = −0.184639, and q_3 = −0.184586. For ethylenediamine, $P_{Cu_1\text{-}N_2}$ = 0.566363-oβ and $P_{Cu_1\text{-}N_3}$ = 0.566054, the charges: q_1 = 0.775783, q_2 = −0.191420, and q_3 = −0.190861. The above-mentioned information allowed us to conclude that the complex of Cu^{+}-ion with ethylenediamine is more stable than with dipyridyl.

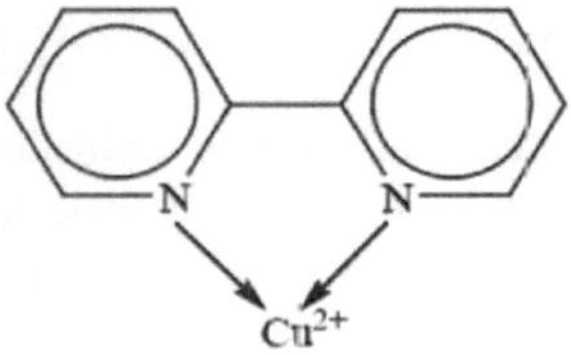

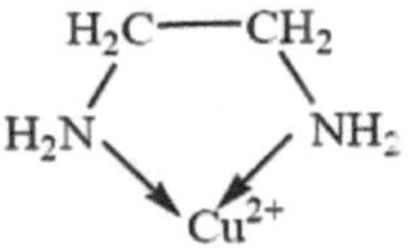

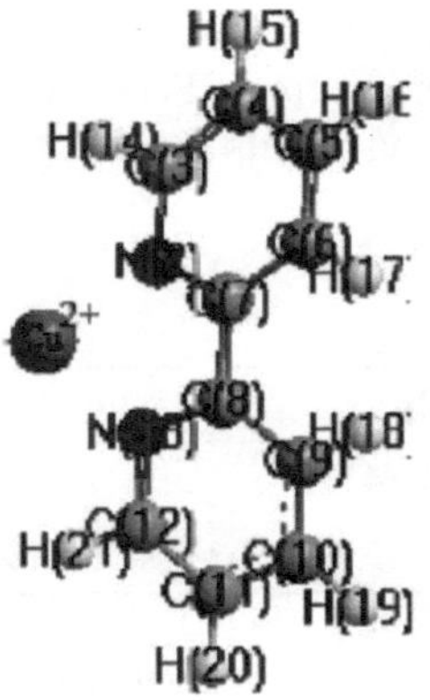

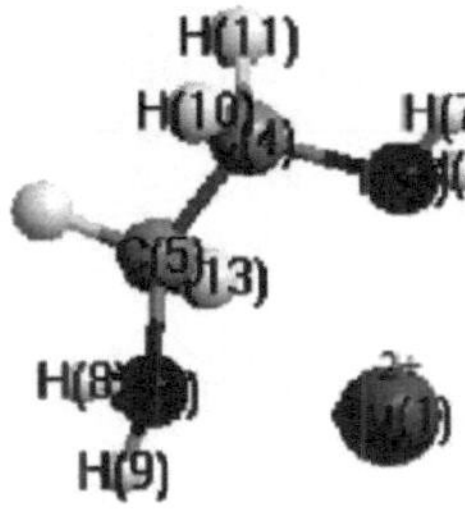

Synthesis of *d*-metals tetrathiostibiate coordination compound with 2,2'-dipyridyl was carried out by exchange reaction, as result of which tetrathiostibiate complexes of corresponding *d*-metals are precipitated, formation of which can be explained by the unity of the following consecutive reactions:

$$AgNO_3 + dipy \rightarrow \left[Ag(dipy)\right]NO_3 \quad (a)$$

$$3\left[Ag(dipy)\right]NO_3 + Na_3SbS_4 \cdot 9H_2O \rightarrow \left[Ag(dipy)\right]_3 SbS_4 \downarrow + 3NaNO_3 + 9H_2O \quad (b)$$

or in total:

$$3AgNO_3 + 3dipy + Na_3SbS_4 \cdot 9H_2O \rightarrow \left[Ag(dipy)\right]_3 SbS_4 \downarrow + 3NaNO_3 + 9H_2O$$

As to other metals (II) dipyridylates, in particular when M = Fe, Co, Ni, Zn, Cd, Hg, Cu, and Mn:

$$MX_2 + ndipy \rightarrow \left[M(dipy)_n\right]X_2 \quad (a)$$

$$3\left[M(dipy)_n\right]X_2 + 2Na_3SbS_4 9H_2O \rightarrow 6NaX + \left[M(dipy)_n\right]_3 (SbS_4)_2 \downarrow + 18H_2O \quad (b)$$

or in total:

$$3MX_2 + 3ndipy + 2Na_3SbS_4 \cdot 9H_2O \rightarrow \left[M(dipy)_n\right]_3 (SbS_4)_2 \downarrow + 6NaX + 18H_2O$$

The essence of this method comprises in the fact that nitrates, chlorides, and sulfates of the mentioned metals form stable, but water-soluble coordination compounds with 2,2'-dipyridyl. That is why their extraction from mother solution in chemically pure form for thio stibiates (V) of some *d*-metals

The obtained coordination compounds represent finely dispersed substances, insoluble in differently colored water and ethanol (Tables 16.12 and 16.13). All of them are extracted without crystallization water, except of Fe(II) dipyridylates, which adds to three molecules of water. They have no certain melting temperature, since they resolve before melting.

TABLE 16.11 Electron Occupation of Atomic Orbitals.

Atom	2s	$2p_x$	$2p_y$	$2p_z$
N_3	1.58211	1.12209	1.14530	1.49979
N_4	1.58220	1.12078	1.12511	1.52119

TABLE 16.12 Some Basic Characteristics of the Synthesized Ethylenediamine Complexes of Tetrathiostibiates (V) of Some *d*-Metals.

No	Compound	Yield(%)	Color	Elemental analysis data (%)							
				Found				Calc.			
				M	Sb	N	S	M	Sb	N	S
1	$[Ag(en)]_3SbS_4$	94.1	Black	42.94	16.16	11.15	17.02	43.07	16.21	10.90	16.81
2	$[Zn(en)_2]_3(SbS_4)_2$	88.7	Yellow	18.57	23.06	15.91	24.29	18.68	22.99	15.79	24.19

TABLE 16.11 *(Continued)*

3	$[Cd(en)_2]_3(SbS_4)_2$	93.9	Orange	28.17	20.34	14.03	21.42	28.09	20.40	14.15	21.25
4	$[Hg(en)_2]_3(SbS_4)_2$	95.2	Brown	41.17	16.66	11.49	17.55	41.21	16.54	11.70	17.34
5	$[Cu(en)_2]_3(SbS_4)_2$	94.6	Grey	18.14	23.18	15.99	24.41	18.07	23.27	15.82	24.69
6	$[Co(en)_3]_3(SbS_4)_2$	89.7	Black	14.53	20.01	20.71	21.08	14.67	20.16	20.48	21.45
7	$[Ni(en)_3]_3(SbS_4)_2$	91.5	Black	14.51	20.02	20.72	21.08	14.59	19.96	20.69	20.84

The composition of the synthesized complexes was established by elemental analysis: stibium was determined by the Evins method, sulfur by gravimetric method, metal by volumetric method, and nitrogen by the Duma' micromethod.

The composition and structure of the synthesized complexes, except of elemental analyses, was determined by physicochemical research methods. Study of IR spectra of these complexes adsorption shows that SbS_4^{3-} group in the studied substances represents exteriorly spherical tetrathiostibiate (V) ion. In the long wave, 380 and 384 cm^{-1} spectral region absorption band is observed, which correspond to v_3 oscillation of SbS_4^{3-} ion.

The obtained dipyridyl coordination compounds are finely divided into various color compounds dissolved in water, ethanol, and other usual organic solvents. All these compounds, besides dipyridylates of Fe(II), separated without crystallization water. They have distinct melting temperature because they are dismissed until melting.

Comparison of free (incoordinate) ligand: 2,2'-dipyridyl spectrum with *d*-metals (Fe, Zn, Mn) tetrathiostibiate (V) dipyridylates spectra shows displacement of high-frequency absorption bands (Fig. 6.3). Free ligands absorption band is at 1584 cm^{-1}, while in the complexes it is displaced to 1600–1610 cm^{-1}, which is related to heterocyclic nitrogen coordination with metals atoms 5–7.[115]

By taking into account the above-mentioned information, we come to a conclusion that the formulas of the synthesized coordination compounds can be presented as follows:

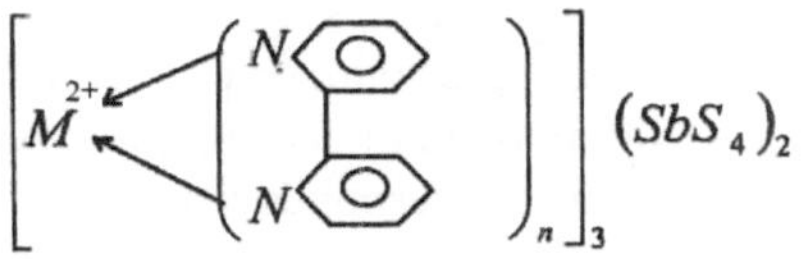

For Fe(II) complex, as for crystalline hydrate, absorption bands are observed in 1630 cm^{-1} region (Fig. 16.13) that points to the presence of crystallization water in the compounds.

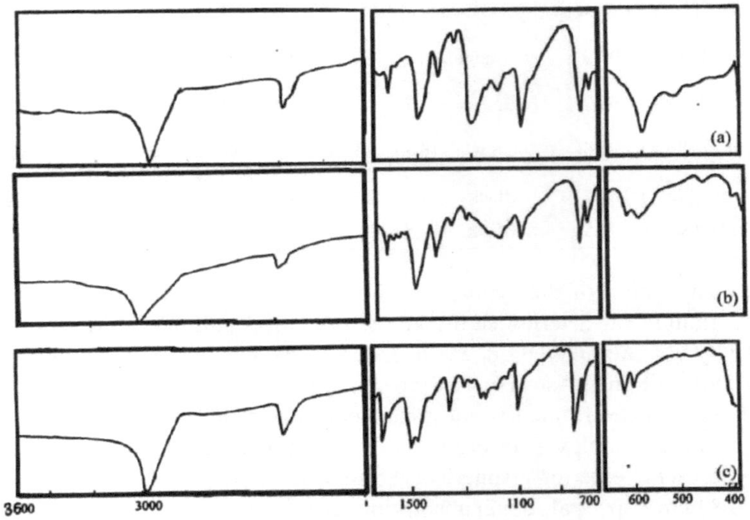

FIGURE 16.13 IR spectra of the synthesized compounds. (a) $[Fe(dipy)_3]_3(SbS_4)_2 \cdot 3H_2O$; (b) $[Zn(dipy)_2]_3(SbS_4)_2$; and (c) $[Mn(dipy)_3]_3(SbS_4)_2$.

Individuality of investigated substance is testified by the results of X-ray phase studies (Table 6.7).

Thermal studies of the synthesized compounds assured us that dipyridyl complexes do not contain crystallization water, except of dipyridylate of Fe(II) tetrathiostibiates, which is testified by the IR spectroscopic data (Fig. 16.13), too.

The thermal behavior of these compounds is almost similar. Thermolysis of Fe(II) tetrathiostibiate(V) dipyridylates $[Fe(dipy)_3]_3(SbS_4)_2 \cdot 3H_2O$ was considered as an example (Fig. 16.14a). The destructive process began at 80°C; in the temperature interval 80–150°C of the corresponding DTA curve, one can observe the exothermic peak with maximum The mass loss equal to 3.57% hat corresponds to 3 moles crystallization water (theoretically: 2.54%). The next stage of the thermolysis process in the temperature interval 150–380°C proceeds with difficultly. On the DTA, one can observe three exothermic peaks with maximum at 210, 250, and 360°C. The mass loss at this moment is about 45.00% that corresponds to the removal of 6 mole crystallization water (theoretically: 44.03%). In the temperature interval 380–530°C, one can observe an exothermal effect with maximum at 480°C with mass loss of 27.57%. Based on the theoretical evaluation (25.03%), the corresponding peak may belong to the loss of 3 moles ligands and 2 moles sulfur. After this process, the mass loss of the investigated sample comprises 1%, which is equal to the removal of 1 mole sulfur (theoretically 1.5%).

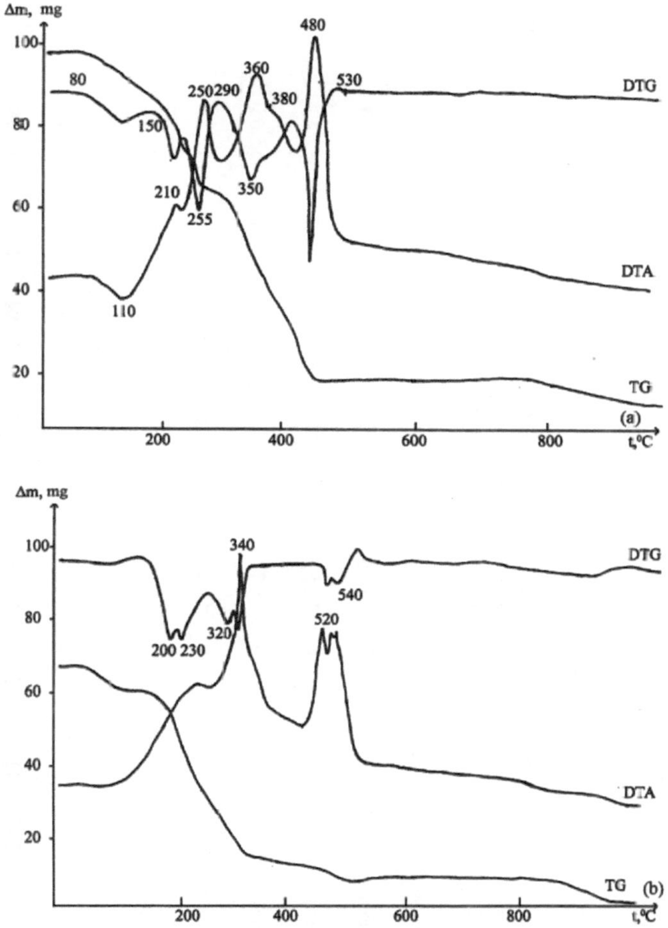

FIGURE 16.14 DTA and TGA curves of the synthesized coordination compounds. (a) $\left[Fe(dipy)_3\right]_3(SbS_4)_2\cdot 3H_2O$; (b) $\left[Zn(dipy)_2\right]_3(SbS_4)_2$.

Foreseen from the above-mentioned information, the thermolysis of dipyridylate of Fe(II) tetrathiostibiate may be represented by the following approximate(Table 16.14) Scheme 16.5:

$$\left[Fe(dipy)_3\right]_3(SbS_4)_2\cdot 3H_2O \underset{-3H_2O}{\overset{80-150^0C}{\rightarrow}} \left[Fe(dipy)_3\right]_3(SbS_4)_2 \underset{-6dipy}{\overset{150-380^0C}{\rightarrow}}$$

$$\rightarrow \left[Fe(dipy)\right]_3(SbS_4)_2 \underset{-3dipy;-2S}{\overset{380-530^0C}{\rightarrow}} Fe_3Sb_2S_6 \underset{-S}{\overset{>530^0C}{\rightarrow}} \ldots$$

SCHEME 16.5

TABLE 16.13 Some Basic Characteristics of the Synthesized Dipyridyl Complexes of Tetrathiostibiates (V) of Some *d*-Metals.

No	Compound	Yield (%)	Color	Elemental analysis data (%)									
				Found					Calc.				
				M	Sb	N	S	H_2O	M	Sb	N	S	H_2O
1	$[Ag(dipy)]_3SbS_4$	88.7	Black	31.07	11.69	8.06	12.31	–	30.83	11.92	8.03	12.46	–
2	$[Zn(dipy)_2]_3(SbS_4)_2$	93.3	Yellow	12.02	14.92	10.29	15.71	–	11.77	15.05	10.36	15.87	–
3	$[Cd(dipy)_2]_3(SbS_4)_2$	86.3	Orange	19.02	13.73	9.47	14.46	–	18.81	13.49	9.27	14.30	–
4	$[Hg(dipy)_2]_3(SbS_4)_2$	96.6	Brown	29.53	11.95	8.24	12.59	–	29.71	12.07	8.02	12.44	–
5	$[Cu(dipy)_2]_3(SbS_4)_2$	95.8	Black	11.72	14.97	10.33	15.77	–	11.85	15.01	10.11	15.65	–
6	$[Mn(dipy)_3]_3(SbS_4)_2$	95.6	Brown	7.97	11.77	12.18	12.40	–	8.02	11.98	12.01	12.27	–
7	$[Fe(dipy)_3]_3(SbS_4)_2 \cdot 3H_2O$	89.8	Red	7.88	11.46	11.86	12.07	2.54	7.66	11.37	11.73	11.91	2.59
8	$[Co(dipy)]_3SbS_4$	90.2	Black	8.50	11.70	12.11	12.33	–	8.43	11.53	11.86	12.14	–
9	$[Ni(dipy)_3]_3(SbS_4)_2$	85.7	Black	8.47	11.71	12.12	12.33	–	8.32	11.84	12.32	12.18	–

TABLE 16.14 Results of X-ray Analysis of the Synthesized Dipyridyl Complexes of Tetrathiostibiates (V) of Some *d*-Metals (F, Ni, Hg, and Ag).

$[Fe(dipy)_3]_3(SbS_4)_2{\cdot}3H_2O$		$[Ni(dipy)_3]_3(SbS_4)_2$		$[Ag(dipy)]_3SbS_4$		$[Hg(dipy)_2]_3(SbS_4)_2$	
I/Io	dα/n	I/Io	dα/n	I/Io	dα/n	I/Io	dα/n
100	11.05	10	11.0	5	9.15	3	9.8
5	9.31	2	7.08	15	6.18	33	6.33
5	7.69	2	6.44	32	3.63	6	3.83
4	4.79	2	5.55	25	3.56	43	3.36
5	4.28	1	5.07	100	3.34	100	3.186
5	4.18	1	4.28	35	3.13	10	2.915
5	3.708	2	3.95	5	2.978	8	2.765
10	3.562	2	3.86	15	2.90	10	2.529
4	3.36	3	3.74	15	2.80	5	2.127
6	3.10	2	3.63	5	2.675	16	2.057
4	2.54	1	2.69	28	2.564	5	1.953
5	2.227	1	2.51	10	2.368		
4	2.156	1	2.44	8	2.22		
				40	2.054		
				10	1.898		
				30	1.758		
				30	1.744		

Only exception is presented by Hg(II) complex, thermolysis begins on relatively low (100°C) temperature and completes by total decay (Figs. 16.15a and 16.16a). This fact is caused by instability of Hg(II)-containing compounds at high temperatures.

Thus, the obtained results allow to make a conclusion that molecules of 2,2'-dipyridyl are coordinated with *d*-metals atoms by means of nitrogen atoms, while SbS_4^{3-} group is located in the external (second) sphere of the complex.

Since ethylenediamine ($H_2N–CH_2–CH_2–NH_2$) is one of the best bidentate ligand, we have set, as a goal, the reception of complex compounds of *d*-metals tetrathiostibiates in the system $M^{+2} - SbS_4^{-3} - en - H_2O$. Sodium tetrathiostibiate $Na_3SbS_4 \cdot 9H_2O$, ethylenediamine (50% water solution) and salts of water-soluble *d*-metals have been used by us as a mother (basic) substances.

Coordination compounds of *d*-metal tetrathiostibiates with ethylenediamine have been received by means of exchange reaction by the action of sodium tetrathiostibiate with the products of interaction of 50% ethylenediamine and *d*-metals salts, without extraction of the latter in individual status. Ag(I), Cd, Zn, Hg(II), Cu, Co, and Ni(II) complexes with ethylenediamine have been synthesized according to the following reactions:

$$AgNO_3 + en \rightarrow \left[Ag(en)\right]NO_{3\,h}$$

$$3[Ag(en)]NO_3 + Na_3SbSu \cdot 9H_2O \rightarrow [Ag(en)]_3 SbS_4 \downarrow + 3NaNO_3 + 9H_2O \quad (a)$$

$$MX_2 + nen \rightarrow [M(en)_n]X_2$$

$$3[M(en)_n]X_2 + 2Na_3SbS_4 \cdot 9H_2O \rightarrow [M(en)_n]_3 (SbS_4)_4 \downarrow + 6NaX + 18H_2O \quad (b)$$

where $M = Zn$, Cd, Hg, Cu, Ag, Co, Ni; $X = Cl^-$, $1/2SO_4^{2-}$ or NO_3^-.

Since the Fe(II) ethylenediamine complex is extracted from water solution in the form of precipitate, while received precipitate is insoluble in the sodium tetrathiostibiates, we were not able to receive Fe(II) tetrathiostibiate complex by means of exchange reaction with ethylenediamine, as it was achieved during synthesis of other *d*-metals aminates.

The synthesized complexes are finely crystalline compounds of various coloring; they are insoluble in the water, spirit, and other ordinary organic solvents.

Synthesized compounds, except of elemental analysis, have been studied by IR spectroscopy, X-ray graphical studies, and thermogravimetric analysis.

Study of the IR spectra of the synthesized compounds (Fig. 16.17) shows that the NH_2 group absorption bands are significantly shifted in comparison with uncoordinated ligand. For free, uncoordinated ethylenediamine absorption bands in 1595 and 3510 cm^{-1} regions are characteristic, while in coordination compound under investigation, the absorption bands of this group are located in 1620 and 3370 cm^{-1} region[113] that is characteristic for valence vibration of $H_2N \rightarrow M^+$ bond. And since silver(I) coordination number is equal to 2, while in case of other *d*-metals, proceeding from the quality of their oxidation, this number increases up to 4, we can draw a conclusion that ethylenediamine plays the role of cyclic bidentate ligand, and the synthesized complexes have the following structure:

Individuality of obtained products has been checked by means of X-ray phase analysis. Cu(II), Ni(II), and Ag(I) compounds show sufficiently defined X-ray pattern, while as to Hg(II), Zn(II), and Co(II) tetrathiostibiates of ethylenediamine (Table 16.8), they turn out to be X-ray amorphous.[116]

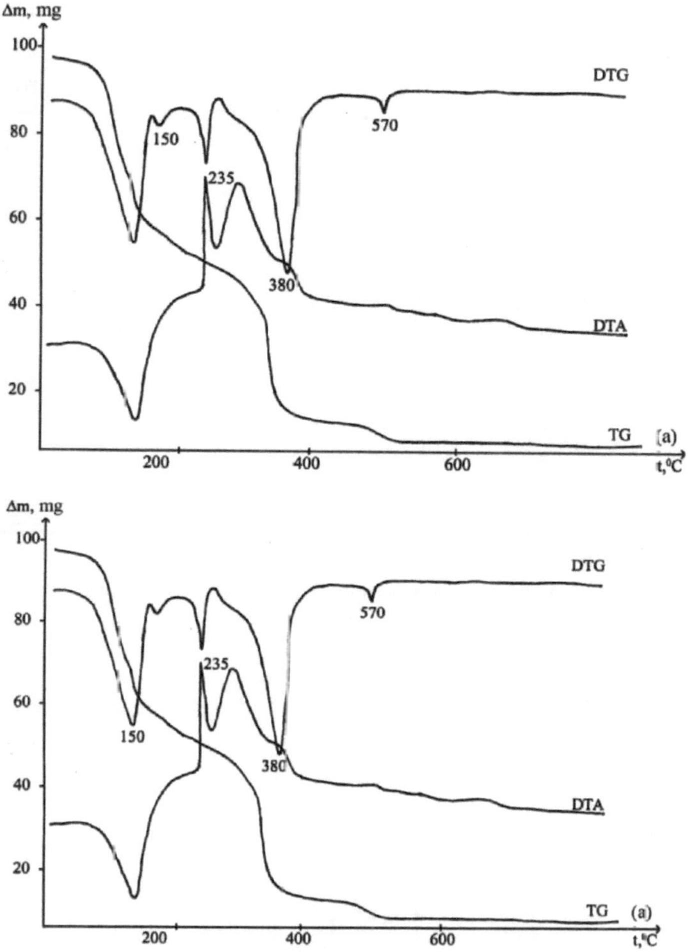

FIGURE 16.15 DTA and TGA curves of the synthesized coordination compounds. (a) $[Hg(dipy)_2]_3(SbS_4)_2$; (b) $[Ni(dipy)_3]\ (SbS\ )_2$.

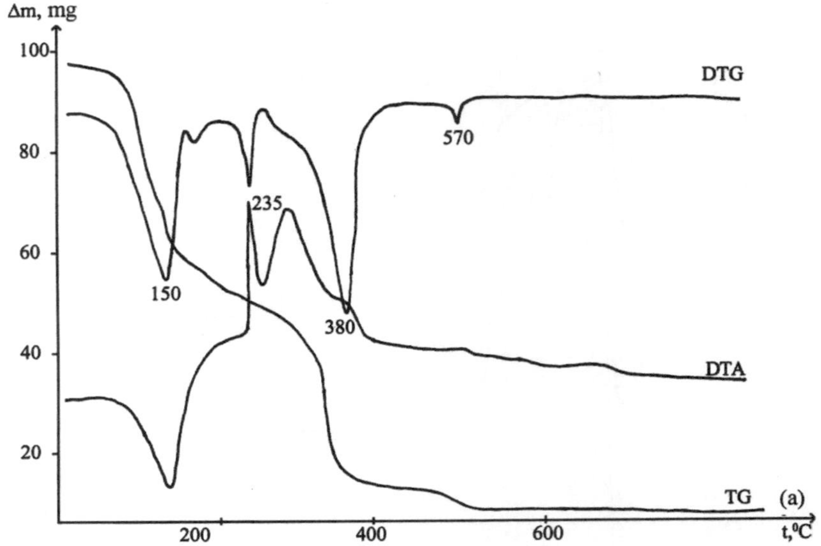

FIGURE 16.16 DTA and TGA curves of the synthesized coordination compounds. (a) $\left[Hg(dipy)_2\right]_3(SbS_4)_2$; (b) $\left[Ni(dipy)_3\right]_3(SbS_4)_2$.

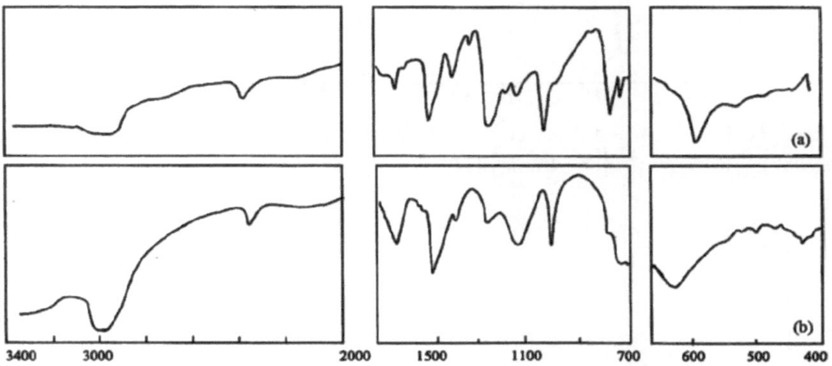

FIGURE 16.17 IR spectra of the synthesized compounds: (a) $\left[Zn(en)_2\right]_3(SbS_4)_2$; (b) $\left[Co(en)_3\right]_3(SbS_4)_2$.

X-ray diagram of the American card-catalogue ASTM 20-1692 ($C_3H_8N_2$·2HCl pure ligand) has been used with the purpose of study of received X-ray diagrams. It turns out that, in our case, correlation takes place, but the certain amount of X-ray reflections is not deciphered (decoded), and for this purpose, the method of homology has been used. More perfect X-ray diagram existing in the American card-catalogue ASTM 24-1670 partially filled the gap and gave us the picture that is almost similar to Ni(II) and Cu(II) tetrathiostibiate complexes with ethylenediamine. It may be noted that the samples are virtually similar to the compared references and represent ethylenediamine complexes.

The thermolysis of $\left[Co(en)_3\right]_3(SbS_4)_2$ as a sample (Fig. 16.18a) is also considered below. Removal of 9 mol of ligand occurs at 100–280°C temperature range in three stages: at 100–130, 130–180, and 180–280°C. At the first stage, the mass decreases by 15.00% (theoretically 14.79%); at the second, by 16.42%; and at the third, by 15.01%.

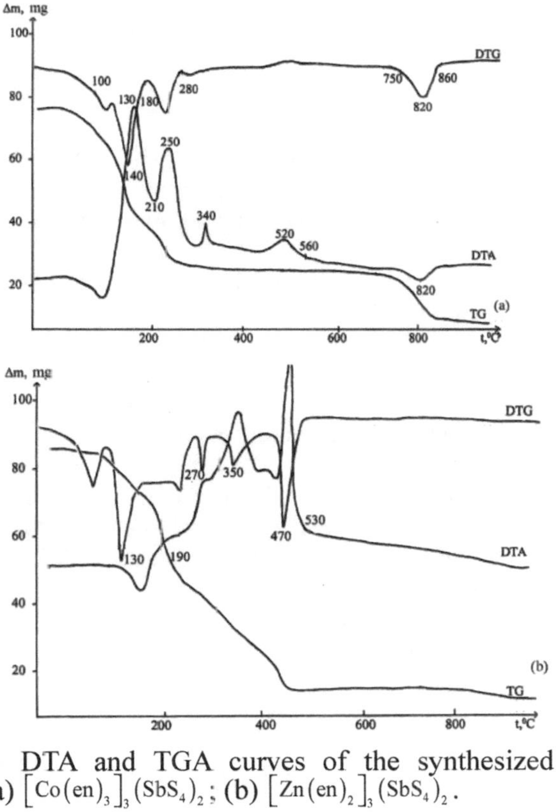

FIGURE 16.18 DTA and TGA curves of the synthesized coordination compounds. **(a)** $\left[Co(en)_3\right]_3(SbS_4)_2$; **(b)** $\left[Zn(en)_2\right]_3(SbS_4)_2$.

It may be noted that the removal of 3 moles of ligand corresponds with each of these stages. After 280°C, as it was mentioned, the process of thermolysis proceeds as well as in case of normal (neutral) salts. In 280–560°C temperature range, the mass decreases by 7.14% that corresponds to the removal of 3 moles of sulfur (theoretically 7.89%). At DTA curve, two exothermic effect are observed within this range, with peaks at 340 and 520°C.

Further decay of the samples continues above 520°C, endothermic effect is observed at DTA curve with peak at 820°C, the mass loss comprises of 24.28% that should be caused by removal of stibium sulfide form (theoretically 25.27%).

On the assumption of above-mentioned information, the probable Scheme 16.6 of thermolysis of Co(II) tetrathiostibiates (V) of ethylenediamine can be presented as follows:

$$\left[Co(en)_3\right]_3(SbS_4)_2 \underset{-3en}{\overset{100-130^0C}{\rightarrow}} \left[Co(en)_2\right]_3(SbS_4)_2 \underset{-3en}{\overset{180-280^0C}{\rightarrow}}$$

$$Co_3(SbS_4)_2 \underset{-3S}{\overset{280-560^0C}{\rightarrow}} 3CoS\cdot Sb_2S_2 \underset{-Sb_2S_2}{\overset{560-860^0C}{\rightarrow}} 3CoS$$

SCHEME 16.6

As well as in other cases, here, the process of $\left[Hg(en)_2\right]_3(SbS_4)_2$ thermolysis (Fig. 16.19a) differs from the corresponding processes of other ethylenediamine complexes that during heating experiences total decay without any residue.

Decay of $\left[Hg(en)_2\right]_3(SbS_4)_2$ sample begins from 120°C and becomes especially intense in the 210–270°C temperature range. At that time, the mass loss is equal to 9.33% (theoretically 8.21%) that corresponds to the removal of 2 moles of ethylenediamine.

In the 270–450°C temperature range, the exothermic peak has several maxima at DTA curve that points out that complex process runs on. During this process, total loss of remained mass takes place.

Study of the synthesized complexes [Cu(II), Cd, and Ag(I) (ethylenediaminate tetrathiostibiates] thermographs shows that removal of ethylenediamine in all compounds occurs in two or three stages. For these complexes with ethylenediamine (as well as in case of normal salts) mass increase is observed in the process of thermolysis (Table 16.15).

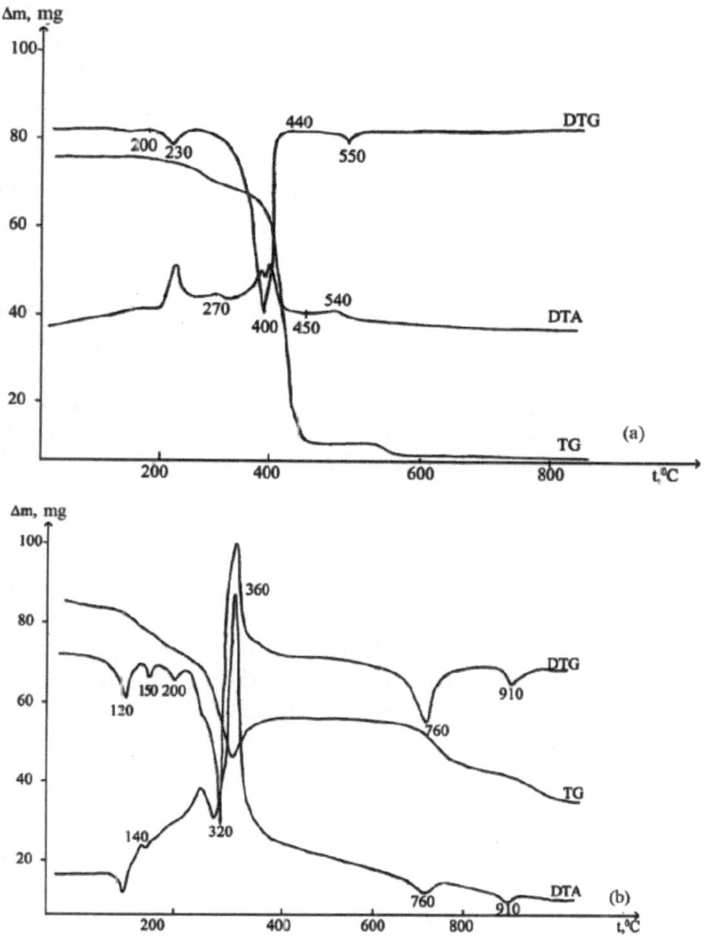

FIGURE 16.19 DTA and TGA curves of the synthesized coordination compounds. (a) $[Hg(en)_2]_3(SbS_4)_2$; (b) $[Cu(en)_2]_3(SbS_4)_2$.

Thus, the investigations show that in case of *d*-metal ethylenediamine complexes treatment by sodium tetrathiostibiates water solution corresponding compounds $[Ag(en)]_3SbS_4$ and $[M(en)_n]_3(SbS_4)_2$ are received in the form of precipitate (where $n = 2, 3$).

We have carried out preliminary virtual (theoretical) bioscreening of obtained structures by using Internet-system program PASS C&T.[50] The estimation of probability of activity of compounds is carried out via parameters P_a (active) and P_i (inactive); when $P_a > 0.5–0.7$, the compound also will show activity experimentally and probably will be analog of known pharmaceutical agents too. Evaluated relative bioactivities of some synthesized compounds are given in Table 16.16. Microbiological study of the investigated compounds confirmed the evaluated virtual concepts.[117]

16.8 OBTAINING AND STUDY OF ARSENIC-CONTAINING BORATES BASED ON TRANSFORMATION PRODUCTS OF INDUSTRIAL WASTE

In the chemistry of boron, great place takes of important representatives of boron–organic compounds: alkyl(aryl)borates $(RO)_2BR'$ and $ROB(OR')_2$ (R = alk., R' = aryl, $R = R'$, $R \neq R'$). Synthesis and study of properties of these compounds have begun from XIX century and did not lose actuality up to day. They have not only preparative, but also theoretical and practical significance. They are used in many organic synthesis as borating agents and Lewis acids—for obtaining the boronhydrids of potassium and sodium trimethoksiborokcol—and also as gaseous flux at the welding of metals.[118–121]

By standpoint of bioactivity, special interest provokes borates containing other elements, privately arsenic.[122,123] We synthesized arsenic-containing borates based on ethylene glycol, pyrocatechin, 2,3-dihydroxynaphthalene, p-oxyphenylarsonic acid, and salicylic acid. The synthesis carried out in the presence of toluene azeotrop generated azeotrop with water, for removing of isolated water at the boiling of reaction mixture, and for carrying out process in the set direction.

The reaction may be expressed by the following general Scheme 16.7:

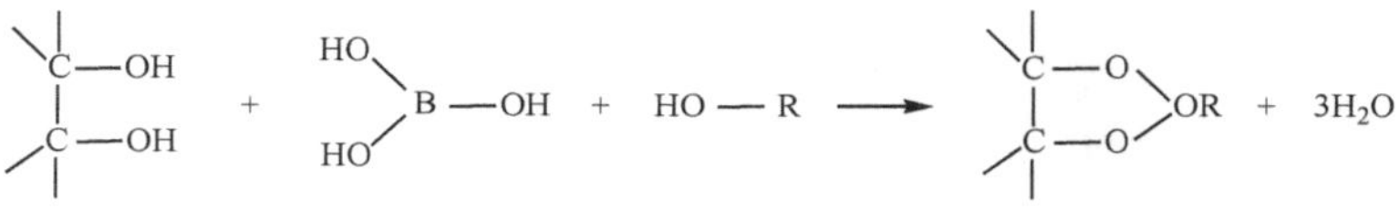

I II

SCHEME 16.7

All obtained compounds are solid amorphous substance (Table 16.17), insoluble in nonpolar and low polar solvents (benzene, toluene), well dissoluble in dimethyl formamide and dimethyl sulfoxide, moderately in ethanol. They are hydrolyzed in water by heating and products of hydrolysis supposedly must be a mixture of the initial products.

TABLE 16.15 Results of X-ray Analysis of the Synthesized Ethylenediamine Tetrathiostibiates (V) of Some *d*-Metals (Cu, Ni, and Ag).

$[Cu(en)_2]_3(SbS_4)_2$		$[Ni(en)_3]_3(SbS_4)_2$		$[Ag(en)]_3SbS_4$	
I/Io	dα/n	I/Io	dα/n	I/Io	dα/n
3	5,67	5	8.29	2	9.5
4	5,41	5	7.69	1	8.23
1	5,10	4	5.69	2	3.18
2	4,796	4	5.38	6	2.82
10	4,70	10	5.10	10	2.58
1	4,23	5	4.74	3	2.42
5	3,95	6	4.04	2	2.37
5	3,70	9	3.91	1	2.20
1	3,59	9	3.75	2	2.07
1	3,70	6	3.53		
3	3,528	10	3.36		
2	3,30	5	3.186		
4	3,028	4	2.529		
4	2,80	3	2.449		
2	2,734				
1	2,64				
2	2,584				
1	2,413				
2	2,127				
2	2,02				
	1,824				

TABLE 16.16 Virtual Relative Bioactivity of Some Synthesized Coordination Compounds.

Compound	Antiseborrheic	Urethanase inhibitor	Antihelmintic (nematodes)	Antiviral (arbovirus)	Neurotoxin	Antibacterial activity enhancer	Antineurotoxic	Cytoprotectant	Antineoplastic (brain cancer)
	P_a/P_i								
$[Fe(dipy)_3]_3(SbS_4)_2 \cdot 3H_2O$	0.865/ 0.021	0.820/ 0.005	0.586/ 0.053	0.687/ 0.046	0.719/ 0.029	0.695/ 0.010	0.672/ 0.041	0.661/ 0.035	0.637/ 0.047

TABLE 16.16 *(Continued)*

$[Ag(dipy)]_3(SbS_4)_2$	0.824/ 0.041	0.766/ 0.011	0.496/ 0.105	0.636/ 0.089	0.670/ 0.065	0.657/ 0.025	0.581/ 0.071	0.615/ 0.063	0.522/ 0.084
$[Mn(dipy)_3]_3(SbS_4)_2$	0.865/ 0.021	0.820/ 0.005	0.586/ 0.053	0.687/ 0.046	0.719/ 0.029	0.695/ 0.010	0.672/ 0.041	0.661/ 0.035	0.637/ 0.047
$[Ag(en)]_3(SbS_4)_2$	0.768/ 0.066	0.651/ 0.034	0.543/ 0.075	0.681/ 0.049	0.663/ 0.070	0.640/ 0.034	0.636/ 0.052	0.650/ 0.041	–

TABLE 16.17 Mass Spectral Data of Arsenic-Containing Borates.

Compound	Mass spectral data, m/z ($I_{rel.}$, %)
I	288.0 [M^+] (4.71), 243.9 (17.36), 227.9 (51.42), 167.9 (100.0), 133.8 (67.49)
II	384.0 [M^+] (3.63), 275.9 (100.0), 182.9 (24.53), 167.9 (25.48), 139.9 (7.36), 135.9 (3.11)

We show one of the possible scheme of obtaining boron and arsenic-containing compounds from industrial waste below (Scheme 16.8):

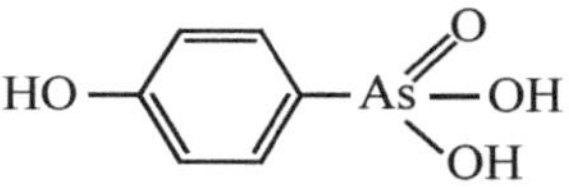

ROH H_2O HNO_3 C_6H_5OH

(industrial waste -As_2O_3) → 2As$(OR)_3$ → As_2O_3 → $2H_3AsO_4$ →

The composition and structure of the synthesized compounds that we have established by chemical (B_{found}/B_{calc}: I-3.61/3.81; II-3.23/3.27; III-2.73/2.84; As_{found}/As_{calc}: I-25.75/26.04; II-21.18/19.43), infrared (IR), NMR, and Mass spectral analyses (Table 16.17).

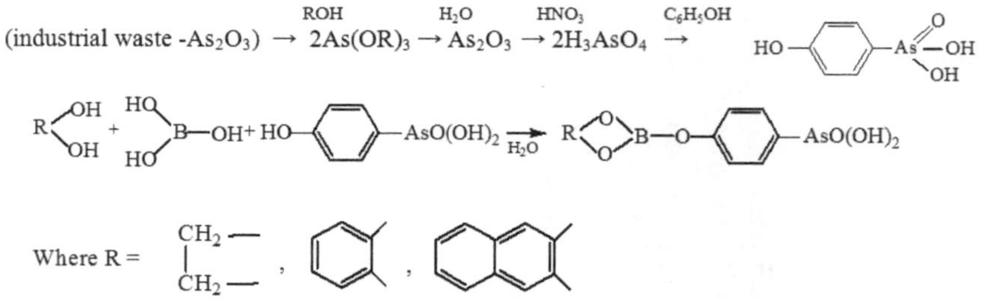

SCHEME 16.8

In the IR spectra of ethylene glycol(p-oxyphenylarsonic acid) borate, we have observed intensive absorption band 1344–1345 cm^{-1} related to boron (III) valence wave. The absorption band 1520 cm^{-1} characterized for 1.4-substituted benzene ring. In the IR spectra, we also observed the characteristic absorption bands for the CH-groups (848 cm^{-1}); middle intensive absorption band 832 cm^{-1}, which must be related to frequency wave of oxoarsenate ions. The absorption bands 416–420 cm^{-1} are also related to oxoarsenate ions.

Mutual comparison of the IR spectra of the borates shows some likeness and difference among them, for example, in the IR spectra of the compound I (Table 16.17), we have observed absorption band 2880–2890 cm^{-1} related to CH_2-groups valency wave, which are also characterized with deformation vibration (1460 cm^{-1}). Mentioned absorption bands are provoked by existence of ethylene glycol fragment and was not observed in other spectra of borates.

In the ^{1}H NMR spectra of ethylene glycol(p-oxyphenylarsonic acid) borate, one can observe signal with chemical shift 3.98–4.04 ppm for the protons in the CH_2 groups of the ethylene glycol fragment; signals with chemical shifts at 6.86–7.67 ppm related to the protons of CH groups of benzene, the resonance signal in the range chemical shift 2.46–2.49 ppm for the protons of the As–OH. In the ^{13}C NMR spectra, one can observe a signal with four chemical shifts within the range 59.22–62.56 ppm typical for the carbon atoms of CH_2 of the ethylene glycol fragment. In the ^{13}C NMR spectra, we also observe chemical shifts within the range 161.50–161.83 and 123.84 ppm related to carbon atoms and 115.20–116.53 and 130.51–131.18 ppm related to CH groups of benzene ring.

In the ^{13}H NMR spectra of pyrocatechin(p-oxyphenylarsonic acid) borate, one can observe a signal with chemical shifts within the range 6.45–6.53 and 6.72–6.99 ppm typical for the carbon atoms of twice substituent benzene ring; signals with chemical shifts at 6.88–6.79 and 7.01–7.38 ppm related to the protons of CH groups of benzene, the resonance signal in the chemical shift within the range 2.39–2.42 ppm for the protons of the As–OH. In the ^{13}C NMR spectra, we also observe chemical shifts within the range 118.25–118.92, 121.72, and 144.45–145.15 ppm related to carbon atom of twice substituent benzene group, the chemical shifts within the range 162.34–162.34, 123.89, 115.54–116.92, and 130.04–132.37 ppm related to carbon atoms of benzene ring.

In the ^{13}H NMR spectra of 2.3-dihydroxynaphthalene(p-oxyphenylarsonic acid) borate, one can observe a signal with chemical shifts within the range 6.81–6.96 and 7.45–7.48 ppm, typical for the carbon atoms of naphthalene group; signals with chemical shifts at 6.68–6.74 and 7.01–7.03 ppm related to the protons of CH groups of benzene, the resonance signal in the range chemical shift 2.40–2.54 ppm for the protons of the As–OH. In the ^{13}C NMR spectra, we also observe chemical shifts within the range 110.85–111.15, 124.52–126.27, 129.25–130.08, and 147.86–148.89 ppm related to carbon atom of naphthalene group, the chemical shifts within the range 160.94–161.73, 124.09, 116.07–116.35, and 131.48–131.86 ppm related to carbon atoms of benzene ring.

The important data have given mass spectral analysis method of synthesized compounds too. Privately, the mass-spectrogram data of synthesized compounds (I–IV) show that the mass of molecular (M^+) and fragmental ions correspond with obtained structures of above-mentioned compounds and clearly represents succession of splitting of different atoms or groups of atoms (Table 16.17).

Additional information about structure and composition of arsenic-containing borates gives us quantum-chemical semiempirical AM1 method. By means of this method, we have calculated energetical (formation enthalpy ΔH_f, ionized potential) and geometrical parameters (angles between the atoms, valence, and dihedral angles) effective charges on atoms (q), dipole moment (M), and bond number between the atoms. Quantum-chemical calculations were performed by CS MOPAC (Chem3D Ultra-version 8.03).[115] In benzene ring of ethylene glycol(p-oxyphenylarsonic acid) borate, distances between C–C atoms (1.39–1.40), valence bond angles of C–C–C (119–121°C) and bond number (1.30–1.40) corresponds to the sp^3 hybridized position of carbon atoms. The quantum-chemical calculations confirmed partial deformation of five-member boron cycles, which is caused by inhomogeneity of distances between the atoms. Distances between B(12)-0 (13) and B (12).0 (16) 1, 38 A, also 0 (13) C (14), and 0 (16) C (15), correspondingly are 1.49 A and 1.46 A. Analysis of electronic charges shows that biggest deficiency of electrons are characterized carbon atoms C(3) and C(6), which is explained by their connection with the strong electronegative atoms. Calculated E = 7.94.107 kJ/mole and $D_{\text{(dip. moment)}}$ = 7.48 Debai.

The study of thermal properties of obtained compounds by thermogravimetric (TGA) and differential-thermal analysis (DTA) method shows that they, except compound I (Table 16.1), do not have distinctly expressed melting points and before melt were decomposed, which is easy to detect visually too. The character of thermal processes current in temperature intervals 20–600°C are almost identical and consists of three stages: soften (not melt), partial reduce to coal, and final destruction. By gas–liquid chromatography analysis, isolation of CO_2 and H_2O (350°C) were established. Comparatively high temperature of beginning of intensive destruction (450–570°C), how it seems, is conditioned by content of cyclic structure of borates in the molecule structure. In the 550–650°C temperature range with the maximum 600°C (exothermic effect) must be caused by oxidation process of residual carbon (with organic component). By this time, the mass decreases are 55%. The remaining black welded residue, as expected, is a mixture of boron and arsenic oxides, 43.8%. A similar picture, except in the case of borates of aromatic dioles, with the difference that the endothermic peak respective to their melting points, also the peak respective to initial and intensive destruction is shifted toward higher temperatures (580–670°C).

16.8.1 VIRTUAL (THEORETICAL) BIOSCREENING OF ARSENIC-CONTAINING BORATES

We have carried out the preliminary virtual (theoretical) bioscreening of synthesized compounds by using Internet-system program PASS C&T.[50]

The estimation of probable bioactivity of chosen compounds was carried out via parameters P_a (active) and P_i (inactive); when $P_a > 0.7$, the compound also could be shown bioactivity experimentally.[124–130] Following from above-mentioned virtual bioscreening, based on analysis of obtained results, the synthesized compounds (I–IV) with experimentally high probability (Table 20) (P_a = 0.70–0.98) possibly will show the following bioactivity: *antibacterial (antispirochetal, antitreponemal, subtilisin inhibitor, Salmonella), antiparasitic (antiprotozoal—amoeba, Histomonas, Trypanosoma, Trichomonas), growth stimulant etc.*

The synthesized compounds were tested as inhibitors of growth of some phytopathogenic bacteria. As test objects, the following microorganisms were used: *Agrobacterium tumefaciens* (causing a grapevine disease), *Xanthomonas campestric* and *Pectobacterium aroideae* (striking some melons and gourds), and *Streptomyces spp.* (destroys some water dissolved polymers).

Biocide properties of compounds were determined by an alveolar standard method. Test results are presented in the Tables 16.18–16.19. Results of experiences showed that the synthesized compounds of I and II in various degree oppress growth of studied test organisms and can be used against these phytopathogenic microorganisms.

TABLE 16.18 Relative Bioactivity of Synthesized Compounds (I–III).

Compounds	Antibacterial				Antiparasitic					
	Antispirochetal	Antitreponemal	Subtilisininhibitor	Non mutagenic, Salmonella	Antiprotozoal	Antiprotozoal (amoeba)	Antiprotozoal (Histomonas)	Antiprotozoal (Trypanosoma)	Antiprotozoal (Trichomonas)	Growth stimulant
	P_a / P_i									
I	0.840 0.000	0.900 0.000	0.837 0.004	0.880 0.010	0.984 0.001	0.983 0.001	0.907 0.000	0.985 0.001	0.769 0.001	0.853 0.001
II	0.818 0.001	0.880 0.000	0.809 0.004	0.739 0.025	0.978 0.001	0.975 0.001	0.878 0.000	0.977 0.001	0.702 0.001	0.802 0.002
III	0.837 0.000	0.899 0.000	0.825 0.004	0.717 0.029	0.982 0.001	0.981 0.001	0.904 0.000	0.984 0.001	0.769 0.001	0.835 0.001

TABLE 16.19 Influence of the Synthesized Compounds on the Growth of Various Microorganisms.

No	Microorganisms	Compounds					
		I			II		
		Concentration of compound, g/l					
		0.1	0.01	0.001	0.1	0.01	0.001
		Zone of suppression of test microorganisms, mm					
1	Xanthomonas campestric	6.0	5.0	3.0	7.0	6.0	3.0
2	Pectobacterium aroideae	5.0	4.5	3.0	8.0	6.5	4.0
3	Agrobacterium tumefaciens	7.0	6.0	4.0	8.0	7.0	4.0
4	Streptomyces spp.	4.0	4.0	3.0	6.0	4.0	2.0

KEYWORDS

- **arsenic**
- **stibium**
- **chemical compounds**
- **coordination**
- **inorganic–organic complexes**
- **mechanism of reactions**
- **biomedical nano**

REFERENCES

1. Gigauri, R.; Chachava, G.; Gverdtsiteli, M.; Laperashvili, I. Extract of the Arsenic from Industrial Waste. *Proc. Acad. Sci. Georgia. Sieria Chem.* 2007, *33* (4), 395.
2. Tskhakaia, N. Sh. Prospects on Production of Arsenic and its Compounds. *Proceedings of the All-Union Conference "The Prospects of Production of Arsenic and its Compounds including the Ultra-Pure Ones*", Tbilisi, Georgia; 1978, pp 3–18.
3. Piotrovsky, K. B.; Tarasova, Z. N. In *Aging and Stabilization of Synthetic Rubbers and of Vulcanizates*. Khimiya: Moscow, RU, 1980.
4. Brostow, W.; Gakhutishvili, M.; Gigauri, R.; Lebland, H. E. Y.; Japaridze, S.; Lekishvili, N. A New Possibility of Separation of Natural Oxide forms of Arsenic and Antimony. *Chem. Eng. J.* (USA) **2010,** *159*, 24–26.
5. Jokai, J.; Hegoszki, P. F. Stability and Optimization of Exstractrion of Four Arsenic Species. *Microchem. J.* **1998,** *59*, 1.
6. Gakhutishvili, M. Arsenic Utilizing Production Waste: A Foundation for Development of Science and Society, *Intelecti* **2000,** *17*, 23.

7. Hazziza-Laskar, J.; Helary, G.; Sauvet, G. "Biocidal Polymers Active by Contact. IV. Polyurethanes Based on Polysiloxanes with Pendant Primary Alcohols and Quaternary Ammonium Groups". *J. Appl. Polym. Sci.* **1995,** *1* (58), 77–84.
8. Lekishvili, N.; Brostow, W.; Gigauri, Rus.; Rusia, M.; Lobzhanidze, T.; Arabuli, L.; Gakhutishvili, M.; Kezherashvili, M.; Barbakadze Kh.; and Gigauri, N. Rational and Effective Use of Georgian Region Arsenic Industrial Waste for Obtaining Compounds and Materials with Specific Properties. *Proceedings of the First Online International Conference on "The Real Environmental Crisis—Effects in Tourism Development, Conflicts and Sustainability"*, Craiova, RO, Nov 30, 2011; pp 25–59.
9. Morin, G.; Calas, G. Arsenic in Soils, Mine Tailings, and Former Industrial Sites. *Elements* **2006**, *2,* 97–101.
10. Zakharchenko, P. I.; Yashunskaya, F. I.; Evstratov, V. F.; Orlovsky, P. N. In *Spravochnik Rezinshchika;* Chemistry: Moscow, RU, 1971; 342–395.
11. Gigauri, N.; Pareshishvili, N.; Gigauri, R. Chemical Study of the Arsenic-Containing Production Waste of Racha Mining Plant. *Georgian Chem. J.* **2006,** *6* (6), 599–604.
12. Gigauri, R. Arsenic and the Environment. Tbilisi University Publishers: GE, 2004; pp 15–22.
13. Gakhutishvili, M.; Gigauri, R. Vachnadze, E.; Chikovani, O. Arsenic Eduction from Realgar-Auripigment Ores of Racha (Georgia), *Georgian Eng. News* **2000**, *2,* 174.
14. Grund, S. C.; Kanusch, K.; Wolf, H. U. Arsenic and Arsenic Compounds. In: *Ull-mens Encyclopedia of Industrial Chemistry*. VCH-Wiley: Weinheim, 2006.
15. Lekishvili, N.; Rusia, M.; Pichkhaia, B.; Gakhutishvili, M.; Arabuli L.; Barbakadze, Kh. Creation of Bioactive Arsenic and Antimony Compounds from Arsenic Processing Industrial Waste and Materials with Predefined Properties, International Conference on Materials and their Recycling, Sept 17–19, 2008, Tbilisi.
16. Barbakadze, Kh.; Brostow, W.; Gakhutishvili, M.; Gigauri, R.; Gventsadze, D.; Lekishvili N.; Lobzhanidze, T. Inorganic-Organic Hybrid Materials from Low-Cost Industrial Waste and Secondary Resources, to be Presented at POLYCHAR 17 World Forum on Advanced Materials, April 20–24, 2009, Rouen.
17. Lekishvili, N.; Arabuli, L.; Beruashvili, T.; Lobzhanidze, T.; Barbakadze, Kh.; Kezherashvili, M. Antibiocorrosive Covers Based on Some Arsenic Compound Derivatives. 1st International Caucasian Symposium on Polymers and Advanced Materials. Sept 11–14, Tbilisi, Georgia, pp 74–75.
18. Kasradze, K.; Bezarashvili, G.; Lekishvili, N.; Rusia M.; Pichkhaia, B. Synthesis, Use and Study of Kinetics of Triisoamyl-Arsenate Obtained by Etherification Orthoarsenic Acid with Isoamyl Alcohol. Georgian International Journal in Science and Technology. Nova Sci. Publ., Inc: New York, 2008; 1, 1, pp 63–69.
19. Kundu, S.; Ghosh S. K.; Mandal M.; Pal, T. Micelle Bound Redox Dye Marker for Nanogram Level Arsenic Detection Promoted by Nanoparticles. *New J. Chem.* **2002,** *26*, 1081–1084.
20. Gigauri, R. D.; Pareshishvili, N. G.; Gigauri, R. J. New Possibilities of Processing Pyrometalurgical Wastes of Production of Non-Ferous and Noble Metals: Production of Arsenic(III) Oxide From the Waste. *Georgian Eng. News* **2006,** *1*, 232–237.
21. Isakov, V.; Nesterov, V. *Color Metalurgy;*. Macne: Kazaxstan, 1962; 5, pp 72–103.
22. Sakhvadze, L.; Adamia, T.; Buachidze, Z.; Gvakharia, V.; Chirakadze, A. Bacterial Leaching of Arsenic from Arsenic Ore Waste and Industrial Production of Arsenic Compounds. *Proceedings of Microbiology Institute of the Azerbaijan National Academy of Sciences*, Bacu. 2009.
23. Nasibulina, S. S. Razvitie Issledovanii v Oblasti Chimii Mishiakoorganicheshix Soedinenii. *Dis. Kand. Chim. M.* **1978**, *45*, 190.

24. Gigauri, R.; Gigauri, Rus. et. al. Investigation of Arsenic-contaned Industrial Waste of Mining Factory in Racha (Georgia). *Georgian Chem. J.* **2006,** *6* (6), 599–604.
25. Gigauri, Rus.; Gigauri, N.; Lekishvili, N. Obtaining of the "White Arsenic" from Arsenic Industrial Waste. International Conference "Compounds and Materials with Specific Properties Based on Industrial Waste, Secondary and Natural Recourses". Book of Abstracts. Tbilisi, July 15–16, 2010; p 15. Technological Regulations No.I2: White Arsenic Production (Georgia, Racha Mining-Chemical Plant), 1982; p13.
26. Gigauri, R. Synthesis and Transformation of the Arsenic Organic Compounds Based on As_4O_6. Diss. Doctor of Thechnical Sci. Тбилиси, 1987.
27. Gigauri, R.; Gakhutishvili, M.; Giorgadze, K.; Machaidze, A. A Method of Sodium Dithioarsenate(V) Production from Hydrometaiiurgical Industrial Wastes of Non-ferrous and Noble Metals, *Proc. Georgian Acad. Sci.* **2004**, *30*, 205.
28. Akhmetov, N.S. *General and Inorganic Chemistry (rus.)*; HS: Moscow, RU, 2004.
29. Khidasheli, A. Oxidation of arsenic row sulfide forms. PhD diss. Tbilisi, 1996, p 153.
30. Chelidze, I. Study of the Reactions of the Obtaining of Arsenic Sulfides. PhD Dissertation, Moscow, RU, 1975.
31. Tomas, H. Patent USA. 2976382.
32. Gigauri, Rus.; Chelidze, I.; Gakhutishvili, M.; Chirakadze, A. Bacterical-Chemical Extraction of the Arsenic from Arsenic Industrial Waste and the Technology of the Obtaining of Superclean Arsenic. *Nanochemistry and Nanotechnology;* Publ. House "Universali": Tbilisi, GE, 2011; pp 183–187.
33. Gakhutishvili, M.; Gigauri, R.; Machaidze, Z.; Koranashvili, G.; Kokhreidze, M. Synthesis and Properties of Amine Complexes of Dithioarsenates. *Bull. Georgian Acad. Sci.* **2001**, *164*.
33. Gakhutishvili, M; Gigauri, R. Synthesis and Thermal Investigation of Complexes of d-Metals(III) of Dithioarsenates. *Georgian Chem. J.* **2007**, *7*, 21.
34. Baláž, P.; Sedlák, J. Arsenic in Cancer Treatment: Challenges for Application of Realgar Nanoparticles (A Minireview). *Toxins* **2010**, *2*, 1568–1581.
35. Chowdhury, S. R.; Yanful, E. K. Arsenic and Chromium Removal by *Mixed* Magnetite-Maghemite Nanoparticles and the Effect of Phosphate on Removal. *J. Environ. Manag.* **2010**, *91*, 2238–2247.
36. Lobzhanidze, T.; Gigauri, R. Synthesis and Some Properties of Tetrabromocadmiates (II) of Iodmethylentrialkyl(aryl)arsenates. *Bull. Acad. Sci. Georgia*. **2003**, *2* (168), 265–268.
37. Chachava, G.; Lomtatidze, Z.; Gigauri, R.; Arabuli, L. Antibacterial Properties Some Tetrasubtitueted Arsonium Nitrate. *Georgia Chem. J.* **2004**, *3* (4), 213–214.
38. Freidlina, F.Kh. Synthetic Methods in the Field of Metalorganic Arsenic Compounds. *Moscow-Leningrad: Acad. Sci. USSR* **1945**, 164.
39. Lekishvili, N.; Rusia, M.; Barbakadze, Kh; Lekishvili, G.; Gverdtsiteli, M. Novel Bioactive Hybrid Composites Based on Coordination Compounds of Some Tertiary Arsines with Mercury(II) Halides. *Asian J. Chem*. **2012**, *24* (9), 4235–4237.
40. Lobzhanidze, T.; Gigauri, R. Synthesis and Physical and Chemical Propertiesof Tetraiodzincatov of Iodmethylenetrialky(Aryl)Arsoniev. *Chem. J. Georgia*. **2002**, *2*, 112–116.
41. Samkharadze, M.; Rusia, M.; Gigauri, Rus.; Meterevili, J.; Gigauri, R. Synthesis and Research of Tetrathioantimonate. *Proc. Acad. Sci. Georgia. Chem. Serie*. **1998**, *24* (4), 34–38.
42. Samkharadze, M.; Didbaridze, I.; Koranashvili, G.; Rusia, M.; Gigauri, Rus. Coordination Compounds Tetrathioanthimonates(V) of d-metals with Ethylenediamine. Synthesis and Research. *Georgia Chem. J.* **2004**, *4* (2),101.

43. Gigauri, R. D.; Kamai, G.; Ugulava, M. M. The Question to the Ethers of Arsenic Acid. *Bull. Georgian Acad. Sci.* **1970**, *60* (3), 585.
44. Kanatzidis, M. G. High Thermopower and Low Thermal Conductivity in Semiconducting Ternary K-Bi-Sb Compounds. *Chem. Mater.***1997**, *9* (12), 3050–3071.
45. Gigauri, R. D.; Tigishvili, Z. L.; Injia, M. A.; Gurgenidze, N. I.; Chernokolski, B. D. The Synthesis and Properties of Some Iodides of Methylhalogenylalkyl(Aryl)Arsonious. *JOCh* **1980**, *50* (11), 2514–2516.
46. Samkharadze, M.; Lekishvili, N.; Rusia, M.; Barbakadze, Kh.; Kakhidze, N.; Pachulia, Z.; Gigauri, R. Novel Bioactive Coordination Compounds Some of d-Metals and Nitrogen-Containing Ligands. *Oxid. Commun.* **2012**, *35* (3), 633–650.
47. Baláž, P.; Sedlák, J. Arsenic in Cancer Treatment: Challenges for Application of Realgar Nanoparticles (A Minireview). *Toxins* **2010**, *2*, 1568–1581.
48. Gigauri, R. D.; Tigishvili, Z. L.; Injia, M. A.; Gurgenidze, N. I.; Chernokolski, B. D. The Synthesis and Properties of Some Iodides of Methylhalogenylalkyl(Aryl)Arsonious. *JOCh* **1980**, *50* (11) 2514–2516.
49. Kamai, G.; Gigauri, R.; Chernokalskii, B. D. New Method of Synthesis of Trialkyl(Aryl) Arsins. *Russ. J. Gen. Chem.* **1971,** 39, 94–103.
50. Sadim, A. B.; La, A. A.; Filiminov, A. A.; Poroikov, V. V. Internet-System of Prognose of the Spectrum of Bioactivity of Chem. Comp. *Chem. Farm. J.* **2002**, *36*, 21–26.
51. 41st IUPAC World Chem. Congress. Chem. Protection Health, Natural Environment and Cultural Heritage. Programme and Abstracts. Turin (Italy), August 5–11, 2007.
52. Lekishvili, N.; Lobzhanidze, T.; Barbakadze, Kh. Synthesis, Study, and Use of New Type of Biologically Active Arsenicorganic Complex Compounds. World Forum on Advanced Materials (POLYCHAR 20). *Book of Abstracts*; Dubrovnik: Croatia, March 26–30, 2012; p 65.
53. Lekishvili, N.; Jioshvili, G.; Barbakadze, Kh.; Turiashvili, L.; Giorgadze, K.; Lomtatidze, Z. Arsenic-Containing Borats Based on Industrial Waste Transformation Products. *The 6th China–Korea International Conference on Multi-functional Materials and Application,* Daejeon, KR, Nov 22–24, 2012.
54. Arabuli, L.; Lekishvili, N.; Rusia, M. Synthesis, Structure and Properties of Quaternary Arsonium Triiodides for Antibiocorrosive Covers and Conservers. In *Advanced Biologically Active Polyfunctional Compounds and Composites for Health, Cultural Heritage and Environmental Protection;* Nova Science Publishers, Inc.: New York, 2010; ch. 2. Web site: www. novapublishers.com.
55. Antibiocorrosive Covers and Conservators Based on some Adamsite Derivatives (L. Arabuli, Nodar Lekishvili, M. Kadagidze, N. Kebuladze, R. Gigauri). 41st IUPAC World Chemistry Congress. August, 2007. Turin, Italy. Programme and Abstracts, p. 62.
56. Didbaridze, I.; Khelashvili, G.; Chubinidze, A.; Gigauri, R. Synthesis and Study of Tetrathioarsenates of d^{10}-Metals *Bull. Georgian Acad. Sci.* **1998**, *157* (1), 56–59.
57. Samkharadze, M.; Rusia, M.; Lekishvili, N.; Barbakadze, Kh.; Kakhidze, N.; Pachulia, Z.; Gigauri, R. Bioactive Complex Compounds Based on Transition Metals and some Nitrogen-containing Ligands: Synthesis, Structure and Properties. *Oxid. Commun. (Intern. J.).* **2012**, *35* (3), 633–650.
58. Didbaridze, I.; Khelashvili, G.; Rusia, M.; Endeladze, N.; Gigauri, R. Sodium Tetrathioarsenate as a Precipitate of Ommoniate Ions of Transitional Metals. *Bull. Georgian Acad. Sci.* **1998,** *157* (2), 238–240.
59. Brauer, G. M. Руководство on the Inorganic Chemistry. *MIR* **1985**, *2*, 126–127.
60. Didbaridze, I.; Samkharadze, M.; Samkharadze, M.; Kakhidze, N.; Gigauri, R. Coordination Compounds of Tetrathioarsenites of Zn, Cd and Hg with Ortho-phenylenediamin. *Proc. Acad. Sci. Georgia. Chem. Serie.* **2008**, *34* (1), 9–12.

61. Nakamoto, K. Infrared Spectra of Inorganic and Coordination Compounds. *Moscow, MIR* **1966**, 411.
62. Shagidulin, P. P.; Izosimova, I. (As=S) in IR and KR Spectra. Izvestia of the Academy of Sciences of USSR. *Chem. Serie*. **1976**, *5* (I), 863.
63. Belami, L. Infrared Spectra of Difficult Molecules. *Leningrad, Inostr. Liter*. **1963**, 591.
64. Lipson, G.; Stipl, G. Interpretation of Pouder Roentgenogrames. *Moscow: MIR*. **1972**, *2*, 384.
65. Sadim, A.; Lagunin, A.; Filiminov, D.; Poroikov, V. *Chem.-Farm. J.* **2002**, *36* (10), 21–26.
66. Michael, J. S.; Dewar, E. G.; Zoebisch, E. F. H.; and James J. P. S. *J. Am. Chem. Soc.* **1985**, *107*, 3902–3909.
67. Waxman, S.; Anderson, K. C. *Oncologist* **2001**, *6* (2), 3–10.
68. Gregus, Z.; Cyurasics, A.; Csanaky, I. *Toxicol. Sci.* **2000**, *57*, 22–31.
69. Valiulina, V. A.; Gavrilov, V. I. *Vestnik Kazan Technol. Univ.* **1988**, *1*, 28–38.
70. Caeter, M.; Baker, N.; Bunford, R. *J. Appl. Polym. Sci.* **2003**, *58* (11), 2039–2046.
71. Jain, V. K. *Bull. Mater. Sci.*, **2005**, *28* (4), 313–316.
72. Kojina, A.; Boden, Ch. D.; Shibasaki, M. *Tetrahedron Let.* **1997**, *38* (19), 3459–3460.
73. Aggarwal, V. K.; Patel, M.; Studly, J. *Chem. Comm.* **2002**, 1514–1515.
75. Butler, I. S.; Harrod, J. F. *Inorganic Chemistry. Principles and Applications;* The Benhamin/Cummings Publishing Company, Inc.: California, 1989; p 650.
76. Yambushev, F. D.; Savin, V. I. *Russ. Chem. Rev.* **1979**, *48* (*6*), 123-134.
77. Hill N. J.; William, L.; Gallian, R. *Dalton Trans.* **2002**, *6*, 1188–1192.
78. Sandhu, S. S.; Baweja, S.; Parmar, S. S. *Trans. Met. Chem.* **1980,** *5*, 299–302.
79. Jewiss, H. C.; Levason, W.; Webster, M. *Inorg. Chem.* **1986**, *25*, 1997–2001.
80. Furlani, C. *Coord. Chem. Rev.* **1968,** *3*, 14.
81. Basolo, F. et al. *J. Am. Chem. Soc.* **1963,** *85*, 1700.
82. Wiederhold, M.; Behrens, U. *J. Organomet. Chem.* **1990**, *384*, 48.
83. Parmar, S. S.; Kaur, H. *Transit. Met. Chem.* **1982**, *7*, 167–169.
84. Burmesiter, J. L.; Basolo, F. *Inorg. Chem.* **1964,** *3*, 1587.
85. Maiset, P. B. H. et al. *Chem Ind (London)* **1963**, *78*, 1204.
86. Kepert, D. L.; Keith, R. Trigwell. *Aust. J. Chem.* **1975**, *28*, 1359–1361.
87. Kirby, A.; Yorren, S. Organic Chemistry of Phosporus. *Moscow: MIR* **1971**, 38.
88. Gigauri, R. D.; Injia, M. A.; Chernokalski, B. D.; Ugulava, M. M. *Russ. J. Gen. Chem.* **1975**, *45*, 2179.
89. Gigauri, R. D.; Chachava, G. N.; Chernokalski, B. D.; Ugulava, M. M. *Russ. J. Gen. Chem.* **1974**, *42*, 1537.
90. Arabuli, L.; Matoga, D.; Lekishvili, N. Synthesis and Study of the Structure of Novel Bis-arsenic Complex Compounds. *Asian J. Chem.* (India) **2011**, *23* (7), 2999–3002.
91. Freydlina, R. Kh. Synthetic Methods in Region of Arsenic Metallorganic Compounds. M.-L.: Izd AN USSR, 1945, p 164.
92. Guben-Veil. *Methods in Organic Chemistry*; Khimia: Moscow, RU, 1967; p 180.
93. Umland, F.; Iansen, A. Complexes in Analytical Chemistry. *Moscow: Mir* **1975**, 240.
94. Gigauri, R. D.; Arabuli, L. G.; Rusia, M. Sh.; Kikalishvili, M. A. *Georgia Chem. J.* **2002**, *29* (3), 195.
95. Vaisberger, L.; Proskauer, E.; Riddyk, J.; Tups, E. Organic Solvents. *M.: IIL* **1958**, 518.
96. Altomare, A.; Burla, M. C.; Camalli, M.; Cascarano, G.; Giacovazzo, C.; Guagliardi, A.; Polidori, G. *J. Appl. Crystallogr.* **1994**, *27*, 435.
97. Sheldrick, G. M. *SHELXL-97, Program for the Refinement of Crystal Structures*; University of Göttingen: Germany, 1997.
98. Muller, U.; Diorner, H.-D. *Z. Naturforsch., Teil B* **1982**, *37*, 198.

99. Bogaard, M. P.; Pterson, J.; Rae, A. D. *Acta Crystallogr., Sect. B* **1981**, *37*, 1357.
100. Allen, D. G.; Raston, C. L.; Skelton, B. W.; White, A. H.; Wild, S. B. *Aust. J. Chem.* **1984**, *37*, 1171.
101. Kostick, A.; Secco, A. S.; Billinghurst, M.; Abrams, D.; Cantor, S. *Acta Crystallogr.* **1989**, *45*, 1306.
102. Nakamoto, K. IR and Raman Spectra of Inorganic and Coordination Compounds. *Moscow, Mir* **1995**, 128.
103. Glidewell, C.; Harris, G. S.; Holden, H. D.; Liles, D. C.; McKechnie, J. S. *J. Fluorine Chem.* **1981**, *18*, 143.
104. Jones, P. G.; Olbrich, A.; Schelbach, R.; Schwartzmann, E. *Acta Crystallogr., Sect. C*, **1988**, *44*, 2201.
105. Burford, N.; Macdonald, Ch. L. B.; Parks, T. M.; Wu, G.; Borecka, B.; Kwiatkowski, W.; Cameron, T. S. *Canad. J. Chem.* **1996**, *74*, 2209–2216.
106. Arabuli, L.; Lekishvili, N.; Rusia, M.; Pichkhaia, B. In *Chemistry of Advance Compounds and Materials;* Nova Science Publishers, Inc.: New York, 2010; in press (www.novapublishers.com).
107. Drozdov, A.; Zlomanov, V. *Chemistry of Main Group Elements of Periodic System. Halogens;* MGU: Moscow, RU,1998.
108. Calabrese, J. C.; Herskovitz, T.; Kinney, J. B. *J. Am. Chem. Soc.* **1983**, *105*, 5914.
109. Hanton, L. R.; Lewason, W.; Powell, N. A. *Inorg. Chim. Acta* **1989**, *160*, 205.
110. Rolf, M.; Hirsch, C.; Barends, T. *Eur. J. Inorg. Chem.* **1999**, 2249–2254.
111. Cullen, W. R.; Deacon, G. B.; Green, J. H. S. *Canad. J. Chem.* **1965**, *43*, 3193–3200.
112. Didbaridze, I.; Khelashvili, G.; Rusia, M.; Injia, M.; Gigauri, G. Coordination Compounds of Tetrathioarsenates of *d*-Metals(II) with 2,2'-Dipyridyl. *Georgian Eng. News* **1997**, *4*, 97.
113. Nakamoto, K. IR and Combinatio n Scattering Spectrum of Inorganic and Coordination Compounds. *MIR, Moscow*, RU, **1991,** 13, 169–181.
114. Dewar, M. J. S.; Zoebish, R. E. G.; Healy, E. F.; Stewart, J. J. P. Development and use of quantum mechanical molecular models. 76. AM1: A New General Purpose Quantum Mechanical Molecular Model. *J. Am. Chem. Soc.* **1985**, *107*, 3902.
115. Iremadze, A.; Chilogidze, I.; Kharitonov, Iu. Oscillatory Spectra of Complexes of Nickel with Oxymethylnikatinamides. *Coord. Chem.* **1978**, *8* (4), 1239.
116. Schwartz, A. G.; Dinzburg, B. N. In *The Combination of Rubbers with Plastics;* Khimiya: Moscow, RU, 1972; p 224.
117. Tsintsadze, M.; Machaladze, T.; Kereselidze, M.; Skhirtladze, L.; Kurtanidze, R.; Varazashvili, V.; Palavandishvili, T.; Tsarakov, M. Coordination Compounds of Zinc Sulphates with *Ortho*-Amino-4 and 5-Methylpirydines. *Proc. Acad. Sci. Georgia* **1999**, *25* (1–2), 33.
118. Oscillatory Spectra of Cyanates Complexes of Metals with Hydrazides of Isonicotinic Acid. *Coord. Chem.* **1975**, *1* (4), 525.
119. Datta, A.; Giri, A. K.; Chakravorty, D. AC Conduction Mechanism in Borate Glasses Containing Antimony and Arsenic Ions. *Jpn. J. Appl. Phys.* **1995**, *34*, 1431–1435.
120. Gerard, V. *Chemistry of Organic Compounds of Boron*; Khimia: Moscow (Translated from English), 1966, p 26.
121. Thomas, L. H. *I. Chem. Soc.* **1976,** *65*, 820–823.
122. Arabuli, L. G.; Jioshvili, B. D.; Gigauri, R. I.; Gigauri, R. D. *Georgian Eng. News.* **2002**, *4*, 203.
123. Gross, J. H. *Mass Spectrometry;* Springer, 2004; p 536.

124. Kiani, F. A.; Hofmann, M. Periodic Trends and Easy Estimation of Relative Stabilities in11-Vertex Nido-p-Block-Heteroboranes and -Borates. *J. Mol. Model.* **2006**, *12*, 597–609.
125. Mihajilov-krstev, T.; Zlatkovic, B.; Stankov-jovanovic, B.; Ilic, M.; Mitic, V.; Tojanovic, G. S. Antioxidant and Antimicrobial Activities of Almond-leafed Pear (*Pyrus spinosa* F o r s s k.) Fruits. *Oxid. Commun.* **2013**, *4*, 1079.
126. Chuchkov, S. K. Hydroxycinnamides of Some Amino Acids and Their Antioxidant Activity. *Oxid. Commun.* **2008**, *4*, 798–803.
127. Srivastava, A. K.; Jaiswal, M.; Archana, M.; Pathak, V. K. Molecular Modelling of Anti-HIV-Activity of Bifunctional Betulinic Acid Derivatives with Physicochemical Parameters. *Oxid. Commun.* **2009**, *2*, 455–462.
127. Joshi, S. H.; Manikpuri, A. D.; Khare, D.; Khadikar, P. V. Synthesis and Structural Characterisation of the Mannich Bases of 5-Uriedohydantoin as Potential Antibacterial Agents. *Oxid. Commun.* **2009**, *3*, 714–723.
128. Chari, M. A.; Shobha, D.; Prakash, K. M. M. S.; Syamasundar, K. Synthesis, Characterization and Biological Activity of Fe(II), Co(II), Ni(II), Cu(II), Zn(II), Cd(II) and Hg(II) Complexes of Tridentate Thiosemicarbazones. *Asian J. Chem.* **2012**, *24* (1), 11–14.
129. Kousar, F.; Nosheen, S.; Zahra, S. N.; Kousar, S.; Jahan, N. Synthesis and Biological Activity of Important Phenolic Mannich Bases. *Asian J. Chem.* **2013**, *25* (1), 59–62.
130. Zhao, M.; Dong, X.; Li, G.; Yang, X. Synthesis and Antibacterial Activity of Copper(I) Complexes with Bisbenzoylthiourea. *Asian J. Chem.* **2014**, *26* (1), 277–279.

CHAPTER 17

SURFACE-MODIFIED MAGNETIC NANOPARTICLES FOR CELL LABELING

B. A. ZASONSKA[1], V. PATSULA[1], R. STOIKA[2], and D. HORÁK[1]

[1]Institute of Macromolecular Chemistry, Academy of Sciences of the Czech Republic, Heyrovskeho Sq. 2, 162 06 Prague 6, Czech Republic

[2]Institute of Cell Biology, National Academy of Science of Ukraine, Drahomanov St. 14/16, 79005 Lviv, Ukraine

CONTENTS

17.1 INTRODUCTION

A great effort has been recently devoted to the design and synthesis of new magnetic nanoparticles driven by the rapid development of the nanomedicine and nanobiotechnology.[1]Among them, iron oxide nanoparticles, in particular magnetite (Fe_3O_4) and maghemite(γ-Fe_2O_3), play a prominent role since iron is indispensable component of living organisms and has reduced toxicity.[2]Surface-modified iron oxide nanoparticles have been found very attractive for cell separation[3] and labeling,[4]cancer therapy,[5]drug delivery,[6] and as contrast agents for magnetic resonance imaging (MRI).[4]

There are many methods to obtain various types of iron oxide nanoparticles differing in shape, morphology, size, and availability of the reactive groups on the surface.[3]The oldest preparation involves the size reduction,[7]that is, grinding of bulk magnetite in the presence of large amounts of surfactant in a ball mill for 500–1000 h. Other synthetic approaches for development of magnetic nanomaterialsinclude hydrothermal process,[8] sol–gel method,[9]or spray pyrolysis.[10]However, the most popular techniquesfor preparation of such particles include coprecipitation of Fe(III) and Fe(II) salts in the presence of an aqueous base (e.g., NH_4OH or NaOH)or thermal decomposition of organometalliccomplexesin high-boiling solvents.[11]For the latter, precursors, such as Fe(III) acetylacetonate,[12]Fe*N*-nitrosophenylhydroxylamine[13] or $Fe(CO)_5$ were suggested.[14]

Iron oxide nanoparticles possess a lot of unique properties, such as small size (<100nm) allowing them to function at the cellular level, superparamagnetism, high magnetization, and large specific surface area. However, neat(uncoated) particles show high nonspecific adsorption of biomolecules,undesirablein vitroandin vivointeractions, relative toxicity, and tendency to aggregate.[15] This can be avoided by their surface modification with biocompatible polymers, which also determines ability of the nanoparticles to interact with living cells in a well-defined and controlled manner, as well as ensures immunotolerance and biocompatibility.Typical polymer shells are made from organic materials, like polyethylene glycol (PEG),[16]poly(vinyl alcohol),[17]poly(*N,N*-dimethylacrylamide) (PDMAAm),[18] or inorganic materials, for example,silica.[19]This additional layer can render the particles with colloidal stability, avoids interactions with thesurrounding environment, and introducesspecific functional groups on the surface.

In this chapter, synthesis, properties, and some applications of new poly(*N,N*-dimethylacrylamide)-coated maghemite (γ-Fe_2O_3@PDMAAm), silica-coated maghemite(γ-Fe_2O_3@SiO_2),andmethoxy polyethylene glycol-coated magnetite(Fe_3O_4@mPEG)nanoparticles aredescribed.Both PDMAAmand silica are hydrophilic,chemically inert,andbiocompatible materials, hence,they are attractivefor drug delivery systems and applications in medical diagnostics. Moreover, the polymerscan behave like transfection agents enabling efficient engulfment of the particles by the cells, for example, stem or neural cells and macrophages. Macrophages,that are formed in response to an infection and accumulate damaged or dead cells, are important in the immune system.[20]These large, specialized cells can recognize, engulf, and destroy foreignobjects.Through their ability to clear pathogens and instruct other immune cells, theyplay a pivotal role in protecting the host. They also contribute to the pathogenesis of inflammatory and degenerative diseases.[21]Labeling of macrophages with magnetic particles enables, thus their tracing in the organism using magnetic resonance imaging (MRI).

17.2 PREPARATION OF MAGNETIC NANOPARTICLES

Chemical and physical properties of magnetic nanoparticles, such as size and size distribution, morphology and surface chemistry, strongly depend on selection of the synthetic method, starting components, and their concentration.[11,22] Nanoparticles ranging in size from 1 to 100nm exhibit superparamagneticbehavior.[22] In this report, two methods of iron oxide synthesis are presented.

Coprecipitation method. Typical synthesis of magnetic nanoparticles is exemplified by formation of maghemite (γ-Fe_2O_3) duringcoprecipitation of Fe(II) and Fe(III) salts followed by oxidation of Fe_3O_4 with sodium hypochlorite.[23] Briefly,0.2 Maqueousiron(III) chloride (100ml) and 0.5 M iron(II) chloride (50 ml) were sonicated for a few minutes and mixed with 0.5 Maqueous ammonium hydroxide (100 ml). The mixture was then continuously stirred (200 rpm) at room temperature for 1h. Formed Fe_3O_4 nanoparticles were magnetically separated and seven times washed with distilled water.Subsequently, the colloid was sonicated with 5 wt.% sodium hypochlorite solution (16 ml) and again five times washed with water to obtain the final γ-Fe_2O_3 nanoparticles.

Thermal decomposition. Another possibility to produce superparamagnetic nanoparticles consists in thermal decomposition of iron organic compounds, for example, iron(III) oleate.[24]The method allows preparation of monodisperse Fe_3O_4 nanoparticles with controlled size. As an example, we describe preparation of iron(III) oleateby reaction of $FeCl_3 \cdot 6H_2O$ (10.8 g) and sodium oleate (36.5 g) in a water/ethanol/hexane mixture(60/80/140 ml) at 70°C for 4 hunder vigorous stirring. The upper organic layer was then separated, three times washed with water (30 ml each), and the volume reducedon a rotary evaporator. Obtained brown waxy product was vacuum-dried under phosphorus pentoxide for 6 h. The resulting Fe(III) oleate (7.2 g) and oleic acid (4.5 g) were then dissolved in octadec-1-ene (50 ml) andheated at 320°C for 30 min under stirring (200 rpm). The reaction mixture wascooled to room temperature, the particles precipitated by addition of ethanol (100 ml) and collected by a magnet. Obtained nanoparticles were then five times washed with ethanol (50 ml) and redispersed in toluene and stored.

<H1>17.3MODIFICATION OFTHE NANOPARTICLE SURFACE

Disadvantage of neat iron oxide colloidsis that they induceundesirable interactions, for example, adhesiontothe cells. To prevent this,it is recommended to coat theiron oxide surface with a biocompatible polymer shell. Surface of theγ-Fe_2O_3nanoparticles was, therefore, firstly modified with an initiator and *N,N*-dimethylacrylamide was then polymerized from the surface. 2,2'-Azobis(2-methylpropionamidine) dihydrochloride (AMPA)servedasuitablepolymerization initiator.In contrast, if the particles were hydrophobic, that is, obtainedfrom the thermal decomposition, they weredispersible only in organic solvents. To make them waterdispersible and suitable for biomedical applications, their surface was modified with mPEGderivatives via a ligand exchange.

17.3.1 COATING WITH POLY(N,N-DIMETHYLACRYLAMIDE) (PDMAAM)

Coating of the γ-Fe_2O_3 nanoparticles with PDMAAm via grafting from approach is schematically shown in Figure17.1. In the following section, example of this synthetic approach is described in more detail. The polymerizationwas run in a 30-ml glass reactor equipped with an anchor-type stirrer.First, the AMPAinitiator (4.8 mg) was added to 10 ml of the colloid (47 mg γ-Fe_2O_3/ml) during 5 min, DMAAm (0.3 g) was dissolved and the mixture purged with nitrogen for 10 min. The polymerization was started by heating at 70°C for 16 h under stirring(400 rpm). After completion of the polymerization, the resulting γ-Fe_2O_3@PDMAAm particles were magnetically separated and washed 10 times with distilled water until all reaction byproducts were removed.Advantage of the γ-Fe_2O_3@PDMAAm particles consists in possibility to introduce additional functional comonomer into the shell to attach a highly specific bioligand, such as antibody, peptide, or drug.

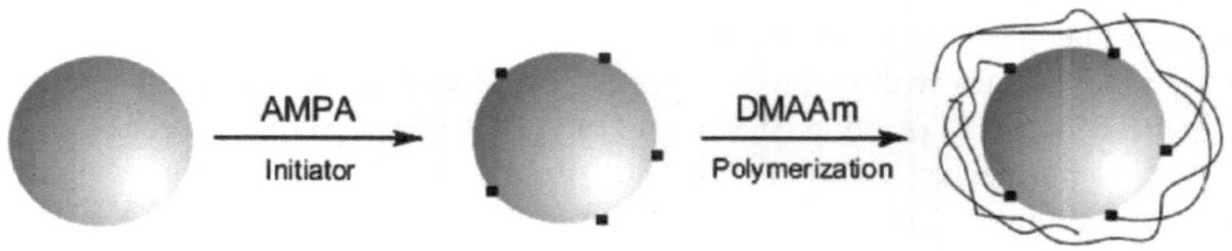

FIGURE 17.1 Scheme of preparation γ-Fe_2O_3@PDMAAm nanoparticles via grafting-from approach using 2,2'-Azobis(2-methylpropionamidine) dihydrochloride (AMPA) initiator.

17.3.2 COATING WITH TETRAMETHOXYORTOSILICATE (TMOS) AND (3-AMINOPROPYL)TRIETHOXYSILANE(APTES)

Another frequently used coating of iron oxide particles is based on silica. Silicais generally synthesized by hydrolysis and condensation of tetraethylorthosilicate (TEOS) or tetramethylorthosilicate (TMOS) (Fig.17.2). Neat silica particles are obtained by Stöber method in ethanol underthe presence of ammonia catalyst[25]or in surfactant-stabilizedreverse microemulsion containing two phases.[26]Theγ-Fe_2O_3nanoparticles were coated by a silicashell using TMOS according to earlier publishedmethod.[27]Shortly, solution containing 2-propanol (24 ml), water (6 ml),and 25 wt.%aqueous ammonia (1.5 ml) was mixed with γ-Fe_2O_3 colloid (1 ml; 50 mg γ-Fe_2O_3) for 5 min. TMOS (0.2 ml) was added and the mixture stirred (400 rpm) at 50°C for 16 h. Resulting γ-Fe_2O_3@SiO_2colloid(Fig.17.3c,g) was then five times washed with ethanol using magnetic separation. In the next step, amino groups were introduced on the particle surface using (3-aminopropyl)triethoxysilane(APTES). In a typical experiment,γ-Fe_2O_3@SiO_2 nanoparticles were dispersed in ethanol (50 ml) under sonication for 15 min and APTES (0.15 ml), ethanol (20 ml), and water (1 ml) were added. After completion of the reaction, the resulting γ-Fe_2O_3@SiO_2–NH_2 particles (Fig.17.3d,h) were washed with water.

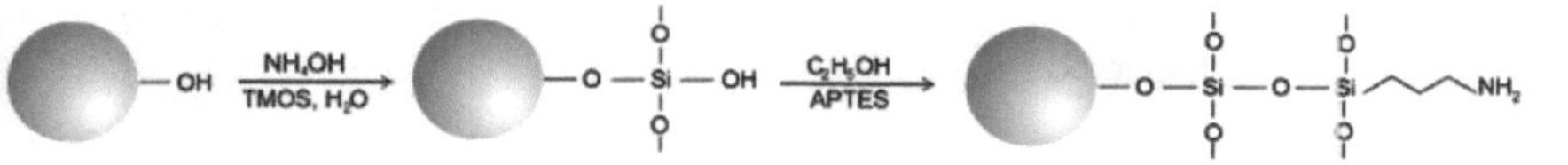

FIGURE 17.2 Scheme of silanization of γ-Fe_2O_3 with tetramethylorthosilicate (TMOS) and modification of γ-Fe_2O_3@SiO_2 nanoparticles with(3-aminopropyl)triethoxysilane(APTES).

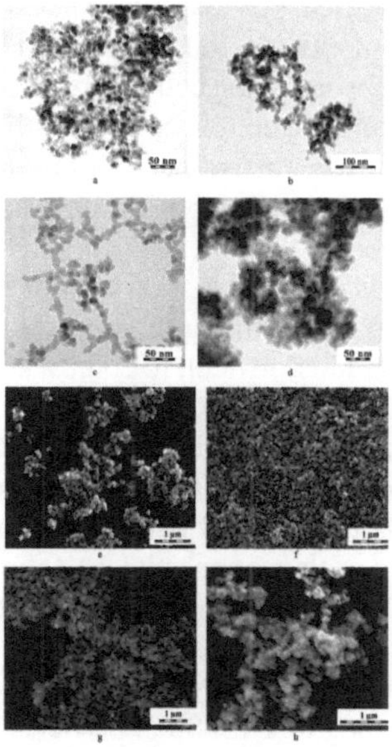

FIGURE 17.3(A–D) TEM and (e–h) SEM micrographsof (a, e) neatsuperparamagnetic γ-Fe_2O_3 nanoparticles synthesized by coprecipitation method, (b, f) γ-Fe_2O_3@PDMAAm (via grafting-from approach), (c, g) γ-Fe_2O_3@S_iO_2, and (d, h) γ-Fe_2O_3@SiO_2–NH_2 nanoparticles.

17.3.3 COATING WITH METHOXY POLYETHYLENE GLYCOL(MPEG)

In order to make the hydrophobic iron oxide particles dispersible in water, their surface was modified by a ligand exchange method.[28] As a hydrophilic ligand, mPEG was selected due to its nontoxicity, hydrophilicity, and low opsonization in biological media. mPEG was terminated with groups, such as phosphonic ($PO(OH)_2$)[29] and hydroxamic (NHOH) acid,[30] exhibiting strong interactions with the iron ions.Fe_3O_4 particles prepared by thermal decomposition were coated bymPEG,terminated with phosphonic(PA-mPEG) or hydroxamicacid (HA-mPEG).In the following section, the surface modification is described in more detail.HA- or PA-mPEG (70 mg) and hydrophobic Fe_3O_4 nanoparticles (10 mg) were added to 4 ml of tetrachloromethane/toluene mixture (1:1 v/v) and sonicated for 5 min. The mixture was then heated at 70°C for 48 h under vigorous stirring. The Fe_3O_4@PEGnanoparticles were purified by repeated precipitation with petroleum ether (3 × 30 ml) at 0°C and diethyl ether (3 × 30 ml) and redispersed in water.

17.4 PROPERTIES OF THE SURFACE-MODIFIED IRON OXIDE NANOPARTICLES

The synthesized surface-modified iron oxide particles were thoroughly characterized by a range of methods including transmission (TEM) and scanning electron microscopy (SEM), atomic absorption spectroscopy (AAS), attenuated total reflectance–Fourier transform infrared spectroscopy (ATR–FTIR),dynamic light scattering (DLS), and magnetic measurements.Shape of the iron oxideparticles, prepared by the coprecipitationand thermal decomposition methods,was sphericaland cubic, respectively. The number-average diameter (D_n)of the γ-Fe_2O_3particles prepared by precipitationwas 10 nm (TEM) andpolydispersity index PDI (D_w/D_n) = 1.24 (D_w is the weight-average diameter) suggesting a moderately broad particle size distribution (Fig. 17.3a).Since it was rather difficult to control size and particle size distribution by the precipitation method, thermal decomposition approach was investigated. Size of the Fe_3O_4 particles was controlled in the 8–25 nmrange andmonodispersity was achieved (Fig.17.4).For example, if the reaction temperature increased from 320 to 340°C, the $D_n$is increased from 8 (Fig.17.4a) to 17 nm (Fig.17.4b) due to anincrease in the growth rate of the nanoparticles. If the concentration of oleic acid stabilizer increased from 0.008 to 0.08 mmol/ml, the particle size decreased from 12 to 8 nm (Fig.17.5) because more stabilizersstabilizes more particles. However, the particles prepared by this method were hydrophobic; the organic shell formed ~80 wt.% of the total mass according toAAS. Such particles formed very stable colloids in organic solvents,such as toluene or hexane, but not in water.Magnetic properties of the nanoparticles were described earlier.[28]

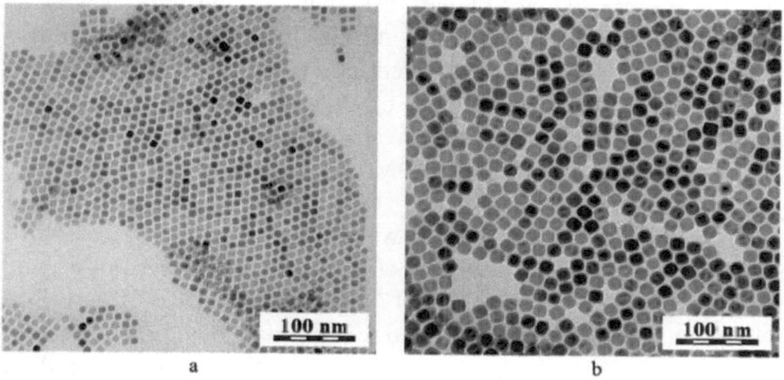

FIGURE 17.4 TEM micrographs of (a) 8 and (b)17 nm superparamagnetic Fe_3O_4 nanoparticles prepared by thermal decomposition method at (a) 320 and (b) 340°C.

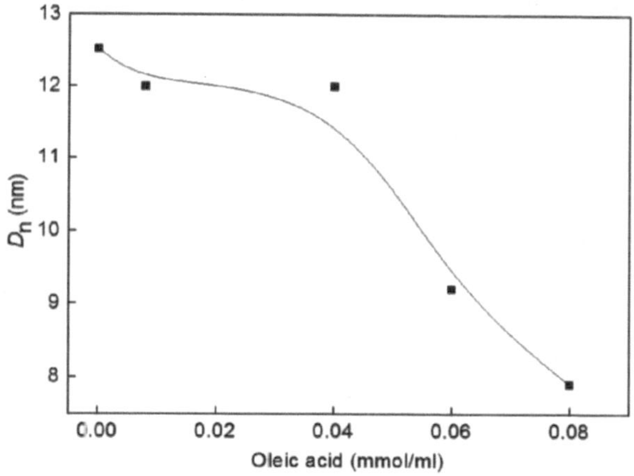

FIGURE 17.5 Dependence of number-average diameter D_n of Fe_3O_4 nanoparticles on oleic acid concentration. Particles were prepared in octadec-1-ene at 320°C for 30 min.

Compared with the neat nanoparticles, D_n of the dried γ-Fe_2O_3@PDMAAm nanoparticles was larger(12 nm) due to presence of the shell, but the polydispersity substantially did not change(PDI 1.18; Fig. 17.3b).The hydrodynamic diameterD_h of γ-Fe_2O_3@PDMAAm, PA- and HA-mPEG-coated $Fe_3O_4$was substantially larger, that is,206, 35,and 68 nm,respectively, than D_n. The reason consists in that the DLS provided information about D_h of the particle dimers and clusters in water, where hydrophilic PDMAAm chains swell. Zeta potential (ZP) of theγ-Fe_2O_3@PDMAAm, PA- and HA-mPEG-$Fe_3O_4$was −53, 26.3, and 12.4 mV, respectively. Since ZP of theγ-Fe_2O_3@PDMAAm was highly negative, the nanoparticle dispersions were very stable (up to a few months) due to the electrostatic repulsion. Regardless of the low positive ZP of theFe_3O_4@mPEGparticles, their colloidsolutions were also very stable due to steric repulsion provided by mPEG. PA-mPEG-Fe_3O_4 colloid (D_h~40 nm) was stable also at various NaCl concentrations ranging from 1 to1000 mmol/l. In contrast, HA-mPEG-coated Fe_3O_4(D_h~65 nm) demonstrated stability only at 1 and 10 mmol of NaCl/l.ATR–FTIR and Fe analysisconfirmed successful coating of the iron oxide nanoparticleswith both PDMAAm by the grafting-from methodand mPEGby ligand exchange method.[23,28]

Optionally, the γ-Fe_2O_3nanoparticles were covered with a silica shell at various γ-Fe_2O_3/TMOS ratios (0.1–0.8 w/w) to control the morphology and size of the nanoparticlesobserved by TEM (Fig.17.3c) and SEM (Fig. 17.3g). Size of the γ-Fe_2O_3@SiO_2 particles ranged from 12 to 192 nm depending on the γ-Fe_2O_3/SiO_2 ratio (Fig.17.6). With increasing amounts of silica relative to the iron oxide and with introduction of amino groups by reaction with APTES,$D_n$of theγ-Fe_2O_3@SiO_2 and γ-Fe_2O_3@SiO_2–NH_2nanoparticles increased (Fig.17.3d)due to their aggregation. This was accompanied with broadening of the particle size distribution. According to AAS, content of iron decreased from 66.1 in γ-$Fe_2O_{3}$to 27.7 and 19.8 wt.% in γ-Fe_2O_3@SiO_2 and γ-Fe_2O_3@SiO_2–NH_2 nanoparticles, respectively. This was inagreement with increasing thickness of the silica shell surrounding the γ-Fe_2O_3 particles. Nevertheless, this amount of iron was sufficient to confer the particles with good magnetic properties. Coating of the γ-Fe_2O_3 particles with a thin silica shell-hindered particles from aggregation and made them hydrophilic; as a result, the particles were well dispersible in water. Secondary coating obtained by reaction of γ-Fe_2O_3@SiO_2 particles with APTES made possible prospective attachment of a target biomolecule,for example, protein, antibody, enzyme, or drug. However, γ-Fe_2O_3@SiO_2–NH_2 nanoparticles often formed aggregates at neutral pH suggesting that the initial γ-Fe_2O_3@SiO_2 particles agglomerated during the reaction with APTES.

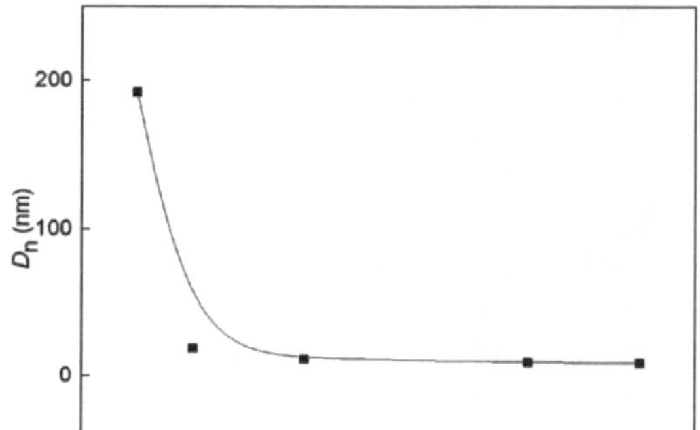

FIGURE 17.6 Dependence of number-average diameter D_n of γ-Fe_2O_3@SiO_2 particles on γ-Fe_2O_3/SiO_2 ratio.

17.5 ENGULFMENT OF THE NANOPARTICLES BY STEM CELLS AND MACROPHAGES

Labeling of the cells with surface-functionalized iron oxide nanoparticles is increasingly important for diagnostic andseparation of DNA,[31] viruses,[32] proteins,[33] and other biomolecules.[34]A great deal of attention is recentlydevoted tostem cells and their ability to differentiate in any specialized cell type. Earlier, we have developedpoly(L-lysine)-coated γ-Fe_2O_3 nanoparticles(γ-Fe_2O_3@PLL) and γ-Fe_2O_3@PDMAAm particles obtained by the solution radical polymerization in the presence of γ-Fe_2O_3.[18,35]Such particleswere found to be highly efficient forin vitrolabeling of human (hMSCs) and rat bone marrow mesenchymal stem cells (rMSCs).In this report, bothγ-Fe_2O_3@PDMAAm obtained by grafting-from approach and γ-Fe_2O_3@SiO_2nanoparticleswere investigated in termsof their engulfment by macrophages (Fig. 17.7).This is an important task from the point of view of controlling introduction, movement and overall fate of the labeled cells after their implantation in the organism.

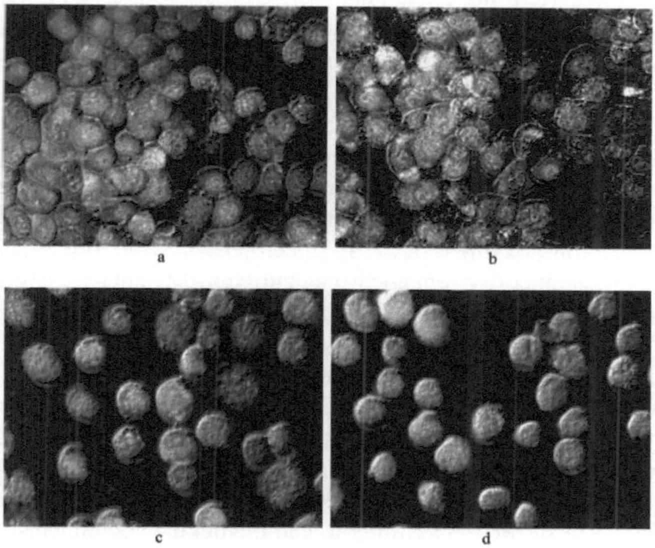

FIGURE 17.7 Fluorescence micrographs of murine J 774.2 macrophages treated with (a) γ-Fe_2O_3,(b) γ-Fe_2O_3@PDMAAm (via grafting-from approach), (c) γ-Fe_2O_3@SiO_2, and (d) γ-Fe_2O_3@SiO_2–NH_2 nanoparticles.

In a typical stem cell labeling experiment, the hMSCs or rMSCswere cultured in Dulbecco's Modified Eagle's Medium (DMEM) in a humidified 5% CO_2 incubator; the medium was replaced every 3 days until the cells grew to convergence. Uncoated,

γ-Fe_2O_3@PLL, γ-Fe_2O_3@PDMAAm particles (via the solution polymerization),and the commercial contrast agent Endorem□ (dextran-coated iron oxide) were then used for labeling of the stem cells. After 72 h of labeling, the contrast agent was stained to produce Fe(III) ferrocyanide (Prussian blue). The quantification of labeled and unlabeled cells wasperformed using TEM and inverted light microscope. Compared with Endorem□ and unmodified nanoparticles, thePDMAAm- and PLL-modified particles demonstrated high efficiency of intracellular uptake into the human cells. Optionally, the labeled rMSCs cells were intracerebrallyinjected into the rat brain and magnetic resonance(MR) images were obtained.MR images of theγ-Fe_2O_3@ PDMAAm (via the solution polymerization)- and γ-Fe_2O_3@PLL-labeled rMSCs implanted in a rat brain confirmed their better resolution compared with Endorem□-labeled cells.[18,35]

In our experiments, both γ-Fe_2O_3@PDMAAm(via grafting from approach),γ-Fe_2O_3@$SiO_2$and γ-Fe_2O_3@SiO_2–NH_2nanoparticles(4.4 mg/ml) were opsonized with FBS proteins at 37°C for 24 h. They were thenincubated with murine J774.2 macrophages and stainedwith acridine orange and Hoechst 33342. Uptakeof the particles by the cells and their morphological changeswere analyzed using fluorescence microscopy. Cytotoxicity of the γ-Fe_2O_3@PDMAAm and neat γ-Fe_2O_3 nanoparticles was estimated using a hemocytometer chamber for counting number of the cells treated in the presence of nanoparticles (0.025, 0.5, and 1 wt.%) in the culture medium for 24 h.

The efficiency of the engulfment of the γ-Fe_2O_3@ PDMAAmand neat γ-Fe_2O_3 nanoparticles by the murine J 774.2 macrophages was determined after 30 min, 1, 2, 3, and 24 h cell cultivation in the presence of the particles. Figure 17.7shows acridineorange and Hoechst 33342-stainedmacrophages treated with the nanoparticles for 3 h. After 30-min treatment of J774.2 macrophages with γ-Fe_2O_3@PDMAAm nanoparticles, their majority remained unengulfedin the culture medium. Visible engulfment of the nanoparticles appeared after 1-h treatment. After 2-h treatment, granulation of the cytoplasm was observed due to accumulation ofthe γ-Fe_2O_3@ PDMAAm nanoparticles in the peripheral region of the cytoplasm. After 3-h treatment, majority of the γ-Fe_2O_3@PDMAAm nanoparticles were engulfed by the macrophages and some cells demonstrated signs of lysosomal activation characterized by red acridin orange fluorescence. Only a minimal amount of the γ-Fe_2O_3@PDMAAm nanoparticles remained unengulfed indicating that theengulfment was very efficient.PDMAAm showed the affinity to cell membrane components facilitating, thus the endocytosis.

As a control experiment, the engulfment of the neat γ-Fe_2O_3 nanoparticles in the macrophages was investigated. Within 1–3-h treatment, the number of vacuoles, their size, as well as the number of lysosomal clusters associated with large vacuoles, increased with time. Numerous unengulfed γ-Fe_2O_3 nanoparticles were accumulated on the surface of treated macrophages, while free γ-Fe_2O_3nanoparticles were almost absent. The size of the cells treated with γ-Fe_2O_3nanoparticles was also increased.

All described superparamagneticnanoparticles were relatively nontoxic for the cultured cells. Apparently, for the efficient particle engulfment by the macrophages, the presence of positively charged amidinegroups in γ-Fe_2O_3@PDMAAm nanoparticles is beneficial. The efficiency of engulfment of the γ-Fe_2O_3@PDMAAm nanoparticles was quite high since after 2-h treatment most cells engulfed the nanoparticles and only few nanoparticlesremained in the culture medium. Fluorescence microscopy confirmed only weak activation of lysosomes, which manifested itself by a change in the color of acridine orange from green to red. Acridine orange,a weakly basic amino dye, is known to be a lysosomotropic agent. In its stacked form, that is, in lysosomes, it emits red fluorescence, while in the cell nuclei at neutral pH, it emits yellow-green fluorescence. Activation of macrophages during the engulfment of foreign extracellular materials was accompanied by an increase in the activity of digestive vacuoles, and thus it caused a red fluorescence shift due to accumulation of the dye in lysosomes. Activation of lysosomal compartments accompanied intracellular processing of the engulfed particles (microorganisms, viruses, damaged cells, and foreign macromolecules).[23] Chemical structure of uncoated γ-Fe_2O_3 nanoparticles, thus provided potential toxicity for the treated cells, which manifested itself by time-dependent evolution of vacuolesin the cell cytosol.

17.6 CONCLUSION

In summary, two different types of iron oxide nanoparticles were synthesized,maghemite(γ-Fe_2O_3)and magnetite (Fe_3O_4). The first oneswere prepared by coprecipitation of Fe(II) and Fe(III) salts with aqueous ammonia. Obtained magnetite was then oxidized with sodium hypochlorite to chemically stable maghemite. However, the particle size distribution of these particles was rather broad as determined by a range of physicochemical characterization methods including SEM, TEM, and DLS measurements. In contrast, monodispersesuperparamagnetic-Fe_3O_4nanoparticles with size controlledfrom 8 to 25 nm were produced by the thermal decomposition of Fe(III) oleate at different temperatures and oleic acid concentrations. The particles were successfully transferred in water bythe ligand exchange method. As a hydrophilic ligand, derivatives of mPEGwith specific functional groupswere used that strongly chemically bonded with iron. Optionally, γ-Fe_2O_3 particles were surface-modified with PLL, PDMAAm(both by thesolution radical polymerization and grafting-from method), or SiO_2. The successful coating of the iron oxide nanoparticle surface was confirmed by both ATR–FTIR spectroscopy and Fe analysis. The colloidal particles were stable in aqueous media for several months.

The biotargetingcharacteristics of the nanoparticles are mainly defined by the biomolecules conjugated to the particle surface.It is desirable that the particle shell contains either membranotropic molecules like phospholipids, polyethylene glycol, or macromolecules (proteins) present in biological fluids. In this work, surface of the formed nanoparticles was opsonized with proteins available in the fetal bovine blood serum. The γ-Fe_2O_3@PDMAAm and γ-Fe_2O_3@SiO_2nanoparticles, in contrast to the neat nanoparticles were shown to be noncytotoxic and intensively phagocytosed by the mammalian macrophages. Additionally, there was no cell irritation during the phagocytosis of the γ-Fe_2O_3@PDMAAm nanoparticles. In contrast, time-dependent vacuolization of neat γ-Fe_2O_3 nanoparticles in cytoplasm of the macrophages was observed suggesting cytotoxicity of the material.

Silica used as an inorganic inert coating of the γ-Fe_2O_3nanoparticles also proved to be suitable modification agentpreventing aggregation of the particles and enhancing their chemical stability. This inorganic material is also easily susceptible to chemical modifications, which make synthesis of particles for combined diagnosis and therapy possible. Biological experiments demonstrated that both γ-Fe_2O_3@ PDMAAm and γ-Fe_2O_3@SiO_2 and γ-Fe_2O_3@SiO_2–$NH_{2core\text{-}shell}$ nanoparticles were recognized and engulfed by the macrophages. The uptake of the surface-coated iron oxide nanoparticles by phagocytic monocytes and macrophages could provide a valuable in vivotool by which magnetic resonance imaging can monitor introduction, trace movement, and observe short- and long-term fate of the cells in the organism.

In conclusion, high potential of the polymer-coated magneticnanoparticles can be envisioned for many biological applications. The particles can be easily magnetically separated and redispersed in water solutions upon removing of the external magnetic field. Magnetically labeled cells can be steered and concentrated inside the body by a magnet. The iron oxide particles, modified with organic, as well as inorganic polymercoatings, seem to be very promising not only for cell imaging and tracking, but also for drug and gene delivery systems and capture of various cells and biomolecules required for diagnostics of cancer, infectious diseases, and neurodegenerative disorders.

ACKNOWLEDGMENT

The financial support of the Ministry of Education, Youth, and Sports (project LH14318) is gratefully acknowledged.

KEYWORDS

- **magnetic nanoparticles**
- **polymerization**
- **macrophages**
- **silica**
- **magnetite**
- **maghemite**

REFERENCES

1. Akbarzadeh, A.; Samiei, M.; Davaran, S. Magnetic Nanoparticles: Preparation, Physical Properties, and Applications in Biomedicine. *Nanoscale Res. Lett.* **2012**, *7*, 1–13.
2. Jeng, H.A.; Swanson, J. Toxicity of Metal Oxide Nanoparticles in Mammalian Cells. *J. Environ. Sci. Health A Tox. Hazard Subst. Environ. Eng.* 2006, *4*, 12699–12711.
3. Gupta, A.K.; Gupta, M. Synthesis and Surface Engineering of Iron Oxide Nanoparticles for Biomedical Applications. *Biomaterials* **2005**, *26*, 3995–4021.
4. Arbab, A.S.; Bashaw, L.A.; Miller, B.R.; Jordan, E.K.; Lewis, B.K.; Kalish, H.; Frank, J.A. Characterization of Biophysical and Metabolic Properties of Cells Labeled with Superparamagnetic Iron Oxide Nanoparticles and Transfection Agent for Cellular MR Imaging. *Radiology* **2003**, *229*, 838–46.
5. Schleich, N.; Sibret, P.; Danhier, P.; Ucakar, B.; Laurent, S.; Muller, R.N.; Jérôme, C.; Gallez, B.; Préat, V.; Danhier, F. Dual Anticancer Drug/Superparamagnetic Iron Oxide-Loaded PLGA-Based Nanoparticles for Cancer Therapy and Magnetic Resonance Imaging. *Int. J. Pharm.* **2013**, *15*, 94–101.
6. Alexiou, C.; Schmid, R.J.; Jurgons, R.; Kremer, M.; Wanner, G.; Bergemann, C.; Huenges, E.; Nawroth, T.; Arnold, W.; Parak, F.G. Targeting Cancer Cells: Magnetic Nanoparticles as Drug Carriers. *Eur. Biophys. J.* **2006**, *35*, 446–450.
7. Papell, S.S. Low Viscosity Magnetic Fluid Obtained by the Colloidal Suspension of Magnetic Particles, US Pat. 3,215,572, 1965.
8. Viswanathiah, M.; Tareen, K.; Krishnamurthy, V. Low Temperature Hydrothermal Synthesis of Magnetite. *J. Cryst. Growth* **1980**, *49*, 189–192.
9. Sugimoto, T.; Sakata, K. Preparation of Monodisperse Pseudocubic α-Fe_2O_3 Particles from Condensed Ferric Hydroxide Gel. *J. Colloid. Interface Sci.* **1992**, *152*, 587–590.
10. Strobel, R.; Pratsinis, S. Direct Synthesis of Maghemite, Magnetite and Wustite Nanoparticles by Flame Spray Pyrolysis. *Adv. Powder Technol.* **2009**, *20*, 190–194.
11. Cornell, R.M.; Schwertmann, U. *The Iron Oxides: Structure, Properties, Reactions, Occurrences and Uses*, 2nd ed., Wiley: Darmstadt, DE, 2000.
12. Willis, A.; Chen, Z.; He, J.; Zhu, Y.; Turro, N.; O'Brien, S. Metal Acetylacetonates as General Precursors for the Synthesis of Early Transition Metal Oxide Nanomaterials. *J. Nanomater.* **2007**, 1–7.
13. Rockenberger, J.; Scher, E.; Alivisatos, P. A New Nonhydrolytic Single-Precursor Approach to Surfactant-Capped Nanocrystals of Transition Metal Oxides. *J. Am. Chem. Soc.* **1999**, *121*, 11595–11596.
14. Woo, K.; Hong, J.; Choi, S.; Lee, H.; Ahn, J.; Kim, C.; Lee, S. Easy Synthesis and Magnetic Properties of Iron Oxide Nanoparticles. *Chem. Mater.* **2004**, *16*, 2814–2818.
15. Baalousha, M.; Manciulea, A.; Cumberland, S.; Kendall, K.; Lead, J.R. Aggregation and Surface Properties of Iron Oxide Nanoparticles: Influence of pH and Natural Organic Matter. *Environ. Toxicol. Chem.* **2008**, *27*, 1875–1882.
16. Barrera, C.; Herrera, A.P.; Rinaldi, C. Colloidal Dispersions of Monodisperse Magnetite Nanoparticles Modified with Poly(Ethylene Glycol). *J. Colloid. Interface Sci.* **2009**, *329*, 107–13.
17. Chastellain, M.; Petri, A.; Hofmann, H. Particle Size Investigations of a Multistep Synthesis of PVA-Coated Superparamagnetic Nanoparticles. *J. Colloid Interface Sci.* **2004**, *278*, 353–60.
18. Babič, M.; Horák, D.; Jendelová, P.; Glogarová, K.; Herynek, V.; Trchová, M.; Likavčanová, K.; Hájek, M.; Syková, E. Poly(*N,N*-Dimethylacrylamide)-Coated Maghemite Nanoparticles for Stem Cell Labeling. *Bioconjugate Chem.* **2009**, *20*, 283–294.
19. Lu, Y.; Yin, Y.; Mayers, B.T.; Xia, Y. Modifying the Surface Properties of Superparamagnetic Iron Oxide Nanoparticles Through a Sol–Gel Approach. *Nano Lett.* **2002**, *2*, 183–186.

20. Mosser, D.M. The Many Faces of Macrophage Activation.*J. Leukocyte Biol.* **2003**, *73*, 209–212.
21. Chawla A.; Nguyen, K.D.; Goh, Y.P.S.Macrophage-Mediated Inflammation in Metabolic Disease.*Nat. Rev. Immunol.* **2011**, *11*, 738–749.
22. Lu, A.-H.; Salabas, E.L.; Schüth, F. Magnetic Nanoparticles: Synthesis, Protection, Functionalization, and Application.*Angew. Chem. Int. Ed.* **2007**, *46*, 1222–1244.
23. Zasonska, B. A.; Boiko, N.; Horák, D.; Klyuchivska, O.; Macková, H.; Beneš, M.; Babič, M.; Trchová, M.; Hromádková, J.; Stoika, R.The Use of Hydrophilic Poly(*N,N*-Dimethylacrylamide) Grafted From Magnetic γ-Fe_2O_3Nanoparticles to Promote Engulfment by Mammalian Cells.*J. Biomed. Nanotechnol.* **2013**, *9*, 479–491.
24. Park, J.; An, K.J.; Hwang, Y.S.; Park, J.G.; Noh, H.J.; Kim, J.Y.; Park, J.H.; Hwang, N.M.; Hyeon, T.Ultra-Large-Scale Syntheses of MonodisperseNanocrystals.*Nat. Mater.* **2004**, *3*, 891–895.
25. Stöber, W.; Fink, A. Controlled Growth of Monodisperse Silica Spheres in the Micron Size Range.*J. Colloid Interface Sci.* **1968**, *26*, 62–69.
26. Finnie, K.S.; Bartlett, J.R.; Barbé, C.J.A.; Kong, L. Formation of Silica Nanoparticles in Microemulsions.*Langmuir***2007**, *23*, 3017–3024.
27. Sakka, S.*Sol-Gel Science and Technology*. Springer: San Diego, CA, 2005.
28. Patsula, V.; Petrovský, E.; Kovářová, J.; Konefal, R.; Horák, D. MonodisperseSuperparamagnetic Nanoparticles by Thermolysis of Fe(III) Oleate and Mandelate Complexes.*Colloid Polym. Sci.*, DOI:10.1007/s00396-014-3236-6.
29. Mohapatra, S.; Pramanik, P. Synthesis and Stability of Functionalized Iron Oxide Nanoparticles Using Organophosphorus Coupling Agents.*Colloids Surf. A***2009**, *339*, 35–42.
30. Ramis, G.; Larrubia, M. An FT-IR Study of the Adsorption and Oxidation of *N*-Containing Compounds Over Fe_2O_3/Al_2O_3 SCR Catalysts.*J. Mol. Catal. A Chem.* **2004**, *215*, 161–167.
31. Saiyed, Z.; Ramchand, C.; Telang, S. Isolation of Genomic DNA Using Magnetic Nanoparticles as a Solid-Phase Support.*J. Phys. Condens. Matter* **2008**, *20*, 204153.
32. Imbeault, M.; Lodge, R.; Ouellet, M.; Tremblay, M. Efficient Magnetic Bead-Based Separation of HIV-1-Infected Cells Using an Improved Reporter Virus System Reveals that p53 Up-Regulation Occurs Exclusively in the Virus-Expressing Cell Population.*Virology* **2009**, *393*, 160–167.
33. Lee, S.; Ahn, C.; Lee, J.; Lee, J.H.; Chang, J. Rapid and Selective Separation for Mixed Proteins with Thiol Functionalized Magnetic Nanoparticles.*Nanoscale Res. Lett.* **2012**, *7*, 279.
34. Haun, J.B.;Yoon, T.-J.;Lee, H.;Weissleder, R.*Magnetic Nanoparticle Biosensors. Wiley Interdiscip. Rev. Nanomed. Nanobiotechnol.***2010**, *2,* 291–304.
35. Babič, M.; Horák, D.; Trchová, M.; Jendelová, P.; Glogarová, K.; Lesný, P.; Herynek, V.; Hájek, M.; Syková, E. Poly(L-Lysine)-Modifiediron Oxide Nanoparticlesfor Stem Cell Labeling.*BioconjugateChem.* **2008**, *19*, 740–750.

CHAPTER 18

THE REINFORCEMENT OF PARTICULATE-FILLED POLYMER NANOCOMPOSITES BY NANOPARTICLES AGGREGATES

G. E. ZAIKOV[1], G. V. KOZLOV[2], and A. K. MIKITAEV[2]

[1]N. M. Emanuel Institute of Biochemical Physics, Russian Academy of Sciences, 4, Kosygin St., Moscow 119334, Russian Federation

[2]Kh.M. Berbekov Kabardino-Balkarian State University, Chernyshevsky st., 173, Nal'chik 360004, Russian Federation

CONTENTS

ABSTRACT

The applicability of irreversible aggregation model for theoretical description of nanofiller particles aggregation process in polymer nanocomposites has been shown. The main factors influencing on nanoparticles aggregation process were revealed. It has been shown that strongly expressed particulate nanofiller particles aggregation results in sharp (in about four times) formed fractal aggregates real elasticity modulus reduction. Nanofiller particles aggregation is realized by cluster–cluster mechanism and results in the formed fractal aggregates density essential reduction, which is the cause of their elasticity modulus decreasing. Distinct from microcomposites, nanocomposites require consideration of interfacial effects for elasticity modulus correct description in virtue of a well-known large fraction of phases division surfaces for them.

18.1 INTRODUCTION

In the course of technological process of preparing particulate-filled polymer composites in general[1] and nanocomposites[2–4] in particular, the initial filler powder particles aggregation in more or less large particles aggregates always occurs. The aggregation process exercises essential influence on composites (nanocomposites) macroscopic properties.[1–5] For nanocomposites, the aggregation process gains special significance, since its intensity can be such that nanofiller particles aggregates size exceeds 100 nm—the value that is assumed (although conditionally enough[6]) as upper dimensional limit for a nanoparticle. In other words, the aggregation process can be resulted in the situation when initially supposed nanocomposite ceases to be as such. Therefore, at present, a number of methods exists, allowing to suppress nanoparticles aggregation process.[4,7]

Analytically, this process is treated as follows. The authors[5] obtain the equation:

$$k(r) = 7.5 \times 10^{-3} S_u, \tag{18.1}$$

where $k(r)$ is aggregation parameter, S_u is specific surface of nanofiller initial particles, which is given in m^2/g.

In turn, the value S_u is determined as follows[8]:

$$S_u = \frac{6}{\rho_n D_p} \tag{18.2}$$

where ρ_n is nanofiller density, D_p is diameter of its initial particles.

From the eqs 18.1 and 18.2, it follows that D_p reduction results in S_u growth that, in turn, reflects in the aggregation intensification, characterized by the parameter $k(r)$ increasing. Therefore, in polymer nanocomposites, strengthening (reinforcing) element are not nanofiller initial particles themselves, but their aggregates.[9] This results in essential changes of nanofiller elasticity modulus, the value of which is determined with the aid of the equation[9]:

$$E_{agr} = E_{nan}\left(\frac{a}{R_{agr}}\right)^{3+d_1}, \qquad (18.3)$$

where E_{agr} is nanofiller particles aggregate elasticity modulus, E_{nan} is elasticity modulus of material, from which the nanofiller was obtained, a is initial nanoparticles size, R_{agr} is a nanoparticles aggregate radius, d_1 is chemical dimension of the indicated aggregate, which is equal to ~1.1.[9]

As it follows from the eq 18.3, the initial nanoparticles aggregation degree enhancement, expressed by R_{agr} growth, results in E_{agr} decrease (rest of the parameters in the eq 18.3 are constant) and, as consequence, in nanocomposite elasticity modulus reduction.

Very often, the elasticity modulus (or reinforcement degree) of polymer composites (nanocomposites) is described within the frameworks of numerous micromechanical models, which proceed from elasticity modulus of matrix polymer and filler (nanofiller) and the latter volume contents.[10] Additionally, it is supposed that the above-mentioned characteristics of a filler are approximately equal to the corresponding parameters of compact material, from which a filler is prepared. This practice is absolutely inapplicable in case of polymer nanocomposites with fine-grained nanofiller, since in this case, a polymer is reinforced by nanofiller fractal aggregates, whose elasticity modulus and density differ essentially from compact material characteristics (see the eq 18.3).[5,9] Therefore, the microcomposite models application, as a rule, gives a large error at polymer composites elasticity modulus evaluation that, in turn, results in the appearance of an indicated models modifications large number.[10]

Proceeding from the above-mentioned information, the present work purpose is the theoretical treatment of particulate nanofiller aggregation process and elasticity modulus (reinforcement degree) particulate-filled polymer nanocomposites with due regard for the indicated effect within the framework of irreversible aggregation models and fractal analysis.

18.2 EXPERIMENTAL

Polypropylene (PP) "Kaplen" of mark 01030 with average weight molecular mass of ~(2–3) × 10^3 and polydispersity index 4.5 was used as matrix polymer. Nanodimensional calcium carbonate ($CaCO_3$) in compound form of mark Nano-Cal P-1014 (production of China) with particles size of 80 nm and mass contents of 1–7 mass % and globular nanocarbon (GNC) (production of corporations group "United Systems," Moscow, Russian Federation) with particles size of 5–6 nm, specific surface of 1400 m^2/g and mass contents of 0.25–3.0 mass % were applied as nanofiller.

Nanocomposites PP/$CaCO_3$ and PP/GNC were prepared by components mixing in melt on a twin screw extruder Thermo Haake, model Reomex RTW 25/42, production of German Federal Republic. Mixing was performed at temperature 463–503 K and screw speed of 50 rpm during 5 min. Testing samples were prepared by casting under pressure method on a casting machine, Test Sample Molding Apparatus RR/TS MP of firm Ray-Ran (Taiwan) at temperature 483 K and pressure 43 MPa.

The nanocomposites melt viscosity was characterized by a melt flow index (MFI). MFI measurements were performed on an extrusion-type plastometer, Noselab ATS A-MeP (production of Italy), with capillary diameter of 2.095 ± 0.005 mm at temperature 513 K and load of 2.16 kg. The sample was maintained at the indicated temperature during 4.5 ± 0.5 min.

Uniaxial tension mechanical tests have been performed on the samples in the shape of a two-sided spade with sizes according to GOST 112 62-80. The tests have been conducted on a universal testing apparatus, Gotech Testing Machine CT-TCS 2000, production of German Federal Republic, at temperature 293 K and strain rate $\sim 2 \times 10^{-3}$ s^{-1}.

18.3 RESULTS AND DISCUSSION

The particulate nanofiller aggregation degree can be evaluated and aggregates diameter D_{agr} quantitative estimation can be performed within the framework of strength dispersive theory,[11] where shear yield stress of nanocomposite τ_n is determined as follows:

$$\tau_n = \tau_m' + \frac{G_n b_B}{\lambda}, \tag{18.4}$$

where τ_m is shear yield stress of polymer matrix, b_B is Burgers vector, G_n is nanocomposite shear modulus, λ is distance between nanofiller particles.

In case of nanofiller particles aggregation, eq 18.4 has the look[11]:

$$\tau_n = \tau_m' + \frac{G_n b_B}{k(r)\lambda}, \tag{18.5}$$

where $k(r)$ is aggregation parameter.

The parameters included in the eqs 18.4 and 18.5 are determined as follows. The general relationship between normal stress σ and shear stress τ has the look[12]:

$$\tau = \frac{\sigma}{\sqrt{3}} \tag{18.6}$$

The intercommunication of matrix polymer τ_m and nanocomposite polymer matrix τ_m' shear yield stresses is given as follows[5]:

$$\tau_{m}' = \tau_{m}\left(1-\phi_{n}^{2/3}\right), \tag{18.7}$$

where φ_n is nanofiller volume content, which can be determined according to the well-known formula[5]:

$$\phi_{n} = \frac{W_{n}}{\rho_{n}}, \tag{18.8}$$

where W_n is nanofiller mass contents, ρ_n is its density, which for nanoparticles is determined according to the equation[5]:

$$\rho_{n} = 188\left(D_{p}\right)^{1/3}, \text{kg/m}^3, \tag{18.9}$$

where D_p is given in nm.

The value of Burgers vector b_B for polymeric materials is determined as follows[13]:

$$b_{B} = \left(\frac{60.5}{C_{\infty}}\right)^{1/2}, \text{Å}, \tag{18.10}$$

where C_∞ is characteristic ratio, connected with nanocomposite structure dimension d_f by the equation[13]:

$$C_{\infty} = \frac{2d_{f}}{d(d-1)(d-d_{f})} + \frac{4}{3}, \tag{18.11}$$

where d is the dimension of Euclidean space, in which a fractal is considered (it is obvious, that in our case $d = 3$).

The value d_f can be calculated according to the equation[14]:

$$d_{f} = (d-1)(1+\nu), \tag{18.12}$$

where ν is Poisson's ratio, estimated according to the mechanical tests results with the aid of the equation[15]:

$$G_{n} = \frac{E_{n}}{d_{f}}, \tag{18.13}$$

where σ_Y and E_n are yield stress and elasticity modulus of nanocomposite, respectively.

Nanocomposite moduli E_n and G_n are connected between themselves by the equation[14]:

$$G_{\text{n}} = \frac{E_{\text{n}}}{d_{\text{f}}} \tag{18.14}$$

And at last, the distance λ between nanofiller non-aggregated particles is determined according to the equation[11]:

$$\lambda = \left[\left(\frac{4\pi}{3f_n} \right)^{1/3} - 2 \right] \frac{D_p}{2}. \tag{18.15}$$

From the eqs 18.5 and 18.15, $k(r)$ growth from 5.65 up to 43.70 within the range of $W_{\text{n}} = 0.25$–3.0 mass% for nanocomposites PP/GNC and from 1.0 up to 2.87 within the range of $W_{\text{n}} = 1$–7 mass% for nanocomposites PP/$CaCO_3$ follows. Let us note that the indicated variation $k(r)$ for the considered nanocomposites corresponds completely to the eqs 18.1 and 18.2. Let us consider how such $k(r)$ growth is reflected on nanofiller particles aggregates diameter D_{agr}. The eqs 18.8, 18.9, and 18.15 combination gives the following equation:

$$k(r)\lambda = \left[\left(\frac{0.251\pi D_{\text{agr}}^{1/3}}{W_{\text{n}}} \right)^{1/3} - 2 \right] \frac{D_{\text{agr}}}{2}, \tag{18.16}$$

allowing at D_{p} replacement on D_{agr} to determine real, that is, with accounting of nanofiller particles aggregation, nanoparticles aggregates diameter of the used nanofiller. Calculation according to the eq 18.16 shows D_{agr} increasing (corresponding to $k(r)$ growth) from 25 up to 125 nm within the range of $W_{\text{n}} = 0.25$–3.0 mass% for GNC and from 80 up to 190 nm within the range of 1–7 mass% for $CaCO_3$. Further nanofiller particles aggregates density can be calculated according to the eq 18.9 at the condition of D_{p} replacement by D_{agr}.

Within the framework of irreversible aggregation model, D_{agr} value is given by the following relationship[16]:

$$D_{\text{agr}} \sim \left(\frac{4c_0 kT}{3\eta m_0} \right)^{1/d_{\text{f}}^{\text{agr}}} t^{1/d_{\text{f}}^{\text{agr}}}, \tag{18.17}$$

where c_0 is nanoparticles initial concentration, k is Boltzmann constant, T is temperature, η is medium viscosity, m_0 is mass of initial nanoparticle, $d_{\text{f}}^{\text{agr}}$ is fractal dimension of particles aggregate, and t is aggregation process duration.

Let us consider estimation methods of the parameters included in the eq 18.17. In the simplest case, it can be accepted that all particles of nanofiller initial powder have the same size and mass. In this case, $c_0 \approx \varphi_{\text{n}}$, where φ_{n} value is determined according to the eq 18.8 with using nanofiller particles aggregates diameter D_{agr}. η value is accepted equal to reciprocal of MFI value and m_0 magnitude was calculated as follows. In supposition of nanofiller initial particles spherical shape, the nanoparticle volume was calculated according to the known values of their diameter D_p and then, using ρ_n value, calculated according to the eq 18.8, their mass m_0 can be estimated. T value is accepted as constant and equal to nanocomposites processing duration, that is, 300 s.

The fractal dimension of nanofiller particles aggregates structure d_f^{agr} was calculated with the aid of the equation[17]:

$$\rho_n = \rho_{dens}\left(\frac{D_{agr}}{2a}\right)^{d_f^{agr}-d}, \tag{18.18}$$

where ρ_{dens} is density of compact material of nanofiller particles, a is self-similarity (fractality) lower scale of nanofiller particles aggregates.

ρ_{dens} value for carbon is accepted equal to 2700 kg/m^2, for $CaCO_3$—2000 kg/m^{25} and a value is accepted equal to the initial GNC particle radius, that is, 2.5 nm. d_f^{agr} values, calculated according to the eq 18.18, are equal to 2.09–2.67 and 2.47–2.75 for GNC and $CaCO_3$ nanoparticles aggregates, respectively.

In Figure 18.1, the dependences $D_{agr}(W_n)$, plotted according to the eqs 18.16 and 18.17, comparison is adduced. As one can see, the good enough correspondence of estimations according to both indicated methods was obtained (the average discrepancy of D_{agr} values, calculated with the usage of these relationships, makes up ~16%). This circumstance indicates that irreversible aggregation models can be used for the theoretical description of particulate nanofiller particles aggregation processes. Besides, eq 18.17 analysis demonstrates various factors influence on nanofiller particles aggregates size (or their aggregation degree). So, c_0, T, and t increasing results in aggregation processes intensification and η, m_0, and d_f^{agr} enhancement—to their weakening.

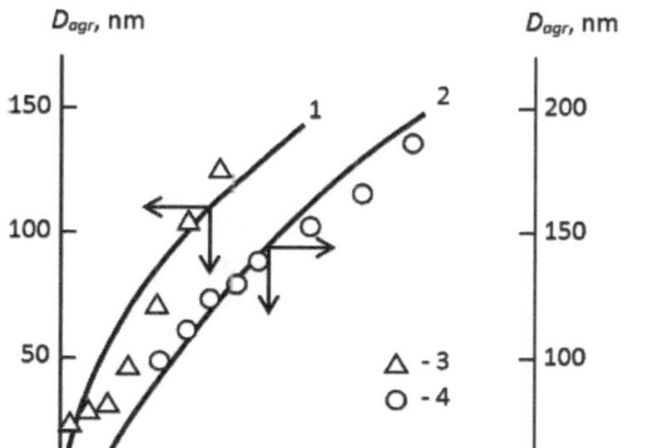

FIGURE 18.1 The dependences of nanofiller particles aggregates diameter D_{agr} on nanofiller mass contents W_n for nanocomposites PP/GNC (1, 3) and PP/$CaCO_3$ (2, 4). 1, 2—calculation according to the eq 18.16; 3, 4—calculation according to the eq 18.17.

Let us note in conclusion that proportionality coefficient in the eq 18.17 for GNC and $CaCO_3$ (c_{GNC} and d_f^{agr}, respectively) can be approximated by the following equation:

$$\frac{c_{CaCO_3}}{c_{GNC}} = \left(\frac{m_0^{CaCO_3}}{m_0^{GNC}} \right)^{1/d_f^{av}}, \tag{18.19}$$

where $m_0^{CaCO_3}$ and d_f^{av} are masses of the initial particles of $CaCO_3$ and GNC, respectively, d_f^{av} is average fractal dimension of the indicated nanoparticles aggregates.

Further elasticity modulus E_{agr} of nanofiller particles aggregates according to the eq 18.3 can be determined. Let us consider the concrete conditions of this equation usage in reference to nanocomposites PP/GNC. Two possible variants exist at parameter a choice in the indicated equation. The first from them supposes that the value a is equal to GNC initial particles diameter,[9] that is, 5.5 nm. Such supposition means that GNC nanoparticles aggregates are formed by particle–cluster (P–Cl) mechanism, that is, by separate particles GNC joining to a growing aggregate.[18] However, such supposition gives unreal high E_{agr} values of order of 5×10^5 GPa. The other variant assumes that nanofiller aggregation is realized by a cluster–cluster (Cl–Cl) mechanism, that is, small clusters association in larger ones.[18] In such model, aggregate radius R_{agr}^{i-1} on the previous (i–1)th aggregation stage is accepted as a and then the eq 18.3 can be rewritten as follows:

$$E_n^T . \tag{18.20}$$

The elasticity modulus E_{agr} real values within the range of 21.3–5.0 GPa were obtained at such calculation method. Further, the simplest microcomposite models can be used for nanocomposite elasticity modulus E_n estimation. For the case of uniform strain in nanocomposite phases, the theoretical value E_n (E_n^T) is given by a parallel model[10]:

$$E_n^T , \tag{18.21}$$

where E_m is elasticity modulus of matrix polymer.

For the case of uniform stress in nanocomposite phases, the lower theoretical boundary E_n^T is determined according to the serial model[10]:

$$E_n^T = \frac{E_{agr} E_m}{E_{agr}\left(1-\phi_n\right) + E_m \phi_n} \tag{18.22}$$

In Figure 18.2, the comparison of the experimentally received E_n and calculated according to the eqs 18.21 and 18.22 E_n^T elasticity modulus values of the considered nanocomposites PP/GNC is adduced. As one can see, the experimental data correspond better to the determined, according to the eq 18.21 E_n^T upper boundary (in this case, average discrepancy of E_n and E_n^T makes up ~8%). The indicated discrepancy is due to objective causes. As it is known,[10] at the eqs 18.21 and 18.22 derivation, the equality of Poisson's ratio for nanocomposite both phases was supposed. In practice, this condition nonfulfillment defines discrepancy between experimental and theoretical data.

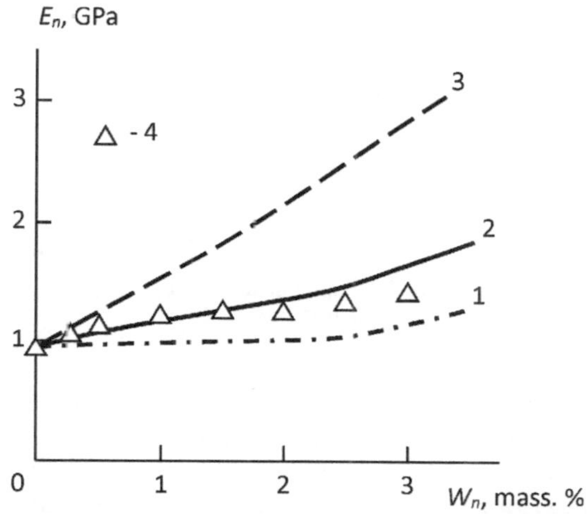

FIGURE 18.2 The dependences of elasticity modulus E_n on nanofiller mass contents W_n for nanocomposites PP/GNC. 1—calculation according to the eq 18.22; 2—according to the eq 18.21 at E_{agr} = variant; 3—according to the eq 18.21 at E_{agr} = const = 21.3 GPa; 4—experimental data.

In Figure 18.2, the dependence $E_n^T(W_n)$, calculated according to the eq 18.21 in supposition E_{agr} = const = 21.3 GPa, is also adduced. As one can see, in this case, the theoretical values of elasticity modulus E_n^T exceed the ones received essentially experimentally E_n. Hence, the good correspondence of experiment and calculation according to the eq 18.21 is due to real values E_{agr} usage only.

It is obvious that nanoparticles aggregates elasticity modulus reduction is due to the indicated aggregates diameter growth and, as consequence, their density ρ_n reduction, which can be calculated according to the eq 18.18. In Figure 18.3, the dependence $E_{agr}(\rho_n)$ is adduced, which, as was expected, proves to be linear, passing through coordinates origin and is described analytically by the following empirical equation:

$$E_n = E_m(1-\phi_n)+bE_{nan}\phi_n, \text{ GPa,} \tag{18.23}$$

where ρ_n is given in kg/m^3.

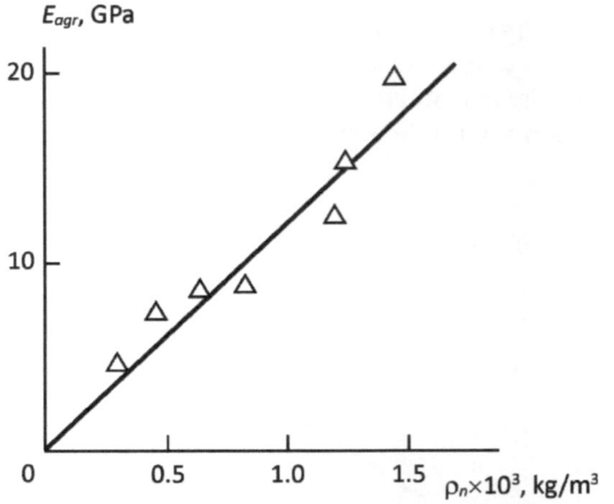

FIGURE 18.3 The dependence of GNC nanoparticles fractal aggregates elasticity modulus E_{agr} on their density ρ_n for nanocomposites PP/GNC.

The limiting magnitude $\rho_n = \rho_{dens}$ allows to obtain the greatest value $E_{agr} \approx 34$ GPa for GNC aggregates, which is the real value of this parameter.[1]

The authors[19] proposed to use a modified mixtures rule for nanocomposites elasticity modulus E_n determination, which in original variant gives upper limiting value of composites elasticity modulus[10]:

$$E_n = E_m(1-\phi_n) + bE_{nan}\phi_n, \tag{18.24}$$

where $b < 1$ is coefficient, reflected nanofiller properties realization degree in polymer nanocomposite. In the present work context, the parameter bE_{nan} as a matter of fact presents nanofiller effective modulus or, more precisely, its aggregates modulus E_{agr} (compared with the eq 18.21).

In Figure 18.4, the dependence of parameter b in the eq 18.24 on nanofiller particles aggregates diameter D_{agr}, calculated according to the eq 18.16, for the studied nanocomposites is adduced. As one can see, this dependence disintegrates on two linear parts: at small D_{agr}, fast decay of b at D_{agr} growth is observed and at large enough D_{agr}, the value b ≈ const ≈ 0.175. Let us note that dimensional interval of the indicated transition showed in Figure 18.4 by a shaded area makes up $D_{agr} \approx$ 70–100 nm, that is, it coincides approximately with upper dimensional boundary of nanoparticles interval (although conditional enough[6]), which is equal to about 100 nm. As a matter of fact, the indicated dimensional interval defines the transition from nanocomposites to microcomposites, the dependence $b(D_{agr})$ for which differs actually qualitatively. The adduced in Figure 18.4 dependence $b(D_{agr})$ can be described analytically by the following integrated equation:

$$b = 0.67 - 6.7\times10^{-3} D_{agr}\text{, for } D_{agr} \leq 70 \text{ nm,}$$

$$b = \text{const} = 0.175\text{, for } D_{agr} > 70 \text{ nm.} \quad (18.25)$$

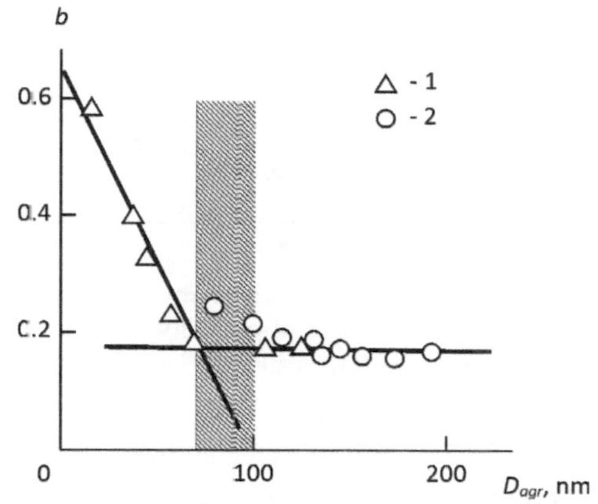

FIGURE 18.4 The dependence of parameter b on nanofiller particles aggregates diameter D_{agr} for nanocomposites PP/GNC (1) and PP/$CaCO_3$ (2). The shaded region indicates transition of nanofiller particles aggregates from nano- to microbehavior.

In Figure 18.5, the comparison of experimentally obtained and calculated according to the eq 18.25 dependences $E_n(\varphi_n)$ is adduced for the studied nanocomposites. In this case, the parameter b value was estimated according to the eq 18.25 and values E_{nan} were accepted equal to 30 GPa for GNC and 15 GPa for $CaCO_3$. As one can see, the good correspondence of theory and experiment is obtained (their mean discrepancy makes up 3% that approximately equals to the experimental error of E_n determination). Higher values E_n for nanocomposites PP/GNC in comparison with PP/$CaCO_3$, even at D_{agr} > 100 nm, are due to two factors: the initial nanoparticles a smaller size, which gives higher values φ_n at the same W_n values (see the eq 18.8 and 18.9) and higher value E_{nan}. It is important to note close values E_{agr} for nanocomposites PP/GNC, determined according to the eq 18.20 and as bE_{nan}.

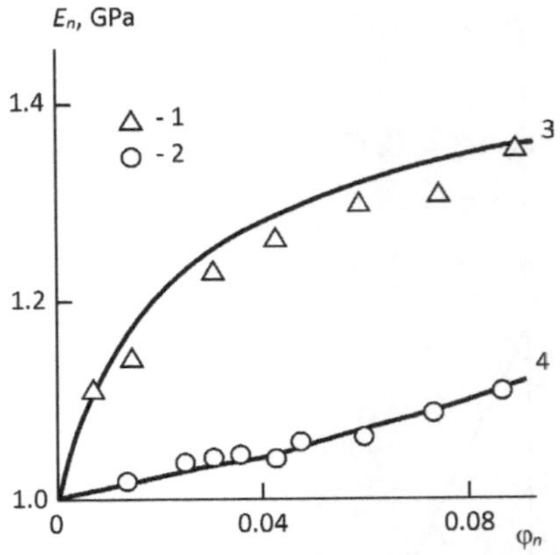

FIGURE 18.5 The comparison of experimentally received (1, 2) and calculated according to the eqs 18.24 and 18.25 (3, 4) dependences of elasticity modulus E_n on nanofiller volume contents ϕ_n for nanocomposites PP/GNC (1, 3) and PP/$CaCO_3$ (2, 4).

The authors[8] proposed the following percolation relationship for polymer microcomposites reinforcement degree E_c/E_m description:

$$\frac{E_c}{E_m} = 1 + 11(\phi_n)^{1.7}, \tag{18.26}$$

where E_c is elasticity modulus of microcomposite.

Later, the eq 18.26 was modified in reference to the polymer nanocomposites case[20]:

$$\frac{E_n}{E_m} = 1 + 11(\phi_n + \phi_{if})^{1.7}, \tag{18.27}$$

where φ_{if} is relative fraction of interfacial regions.

It is easy to see that the modified eq 18.27 considers a factor of sharp increase of division surfaces polymer matrix-nanofiller.[21] In Figure 18.6, the comparison of experimentally obtained and calculated according to the eq 18.26 dependences $E_n(\varphi_n)$, for the considered nanocomposites, is adduced. As it follows from this figure data, eq 18.26 describes the experimental data for nanocomposites PP/$CaCO_3$ well, but the corresponding data for nanocomposites PP/GNC set essentially higher than theoretical curve. This discrepancy cause is obvious from the eqs 18.26 and 18.27 comparison—for nanocomposites PP/GNC, interfacial effects accounting is necessary, that is, parameter φ_{if} accounting. Hence, in the considered case, only compositions PP/GNC are true nanocomposites.

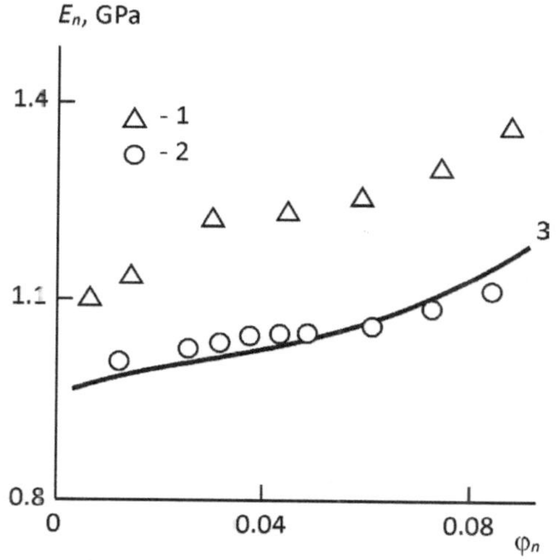

FIGURE 18.6 The comparison of experimentally received (1, 2) and calculated according to the eq 18.26 (3) dependences of elasticity modulus E_n on nanofiller volume contents ϕ_n for nanocomposites PP/GNC (1, 3) and PP/$CaCO_3$ (2, 4).

18.4 CONCLUSION

The applicability of irreversible aggregation models for theoretical description of particulate nanofiller particles aggregation processes in polymer nanocomposites has been shown. Analysis within the framework of the indicated models allows to reveal either factors influence on aggregation degree.

Strongly expressed aggregation of particulate nanofiller particles results in sharp (in about four times) formed fractal aggregates real elasticity modulus reduction. In turn, this process defines nanocomposites as the whole elasticity modulus reduction. Nanofiller particles aggregation is realized by a cluster–cluster mechanism and results in the formed fractal aggregates density essential reduction, which is the cause of their elasticity modulus decreasing.

A nanofiller elastic properties realization degree is defined by the aggregation of its initial particles level. Unlike microcomposites, nanocomposites require interfacial effects accounting for elasticity modulus correct description in virtue of well-known large fraction of phases division surfaces for them.

KEYWORDS

- **nanocomposite**
- **globular nanocarbon**
- **calcium carbonate**
- **aggregation**
- **interfacial effects**
- **reinforcement**

REFERENCES

1. Kozlov, G. V.; Yanovskii, Yu. G.; Zaikov, G. E. *Structure and Properties of Particulate-Filled Polymer Composites: The Farctal Analysis*, Nova Science Publishers, Inc.: New York, 2010; p 282.
2. Kozlov, G. V.; Zaikov, G. E. *Structure and Properties of Particulate-Filled Polymer Nanocomposites,* Lambert Academic Publishing: Saarbrücken, DE, 2012; p 112.
3. Kozlov, G. V.; Yanovskii, Yu. G.; Zaikov, G. E. Particulate-Filled Polymer Nanocomposites. *Structure, Properties, Perspectives,* Nova Science Publishers, Inc.: New York, 2014; p 273.
4. Edwards, D. C. *J. Mater. Sci.* **1990**, *25* (12), 4175–4185.
5. Mikitaev, A. K.; Kozlov, G. V.; Zaikov, G. E. *Polymer Nanocomposites: Variety of Structural Forms and Applications,* Nova Science Publishers, Inc.: New York, 2008; p 319.
6. Buchachenko, A. L. *Adv.Chem.* **2003**, *72* (5), 419–437.
7. Kozlov, G. V.; Yanovsky, Yu. G.; Zaikov, G. E. *Synergetics and Fractal Analysis of Polymer Composites Filled with Short Fibers,* Nova Science Publishers, Inc.: New York, 2011; p 223.
8. Bobryshev, A. N.; Kozomazov, V. N.; Babin, L. O.; Solomatov, V. I. Synergetics of Composite Materials. *Lipetsk NPO ORIUS* **1994**, *47*, 154.
9. Witten, T. A.; Rubinstein, M.; Colby, R. H. *J. Phys. II France* **1993**, *3* (3) 367–383.
10. Ahmed, S.; Jones, F. R. *J. Mater. Sci.* **1990**, *25* (12) 4933–4942.
11. Sumita, M.; Tsukumo, Y.; Miyasaka, K.; Ishikawa, K. *J. Mater. Sci.* **1983**, *18* (5) 1758–1764.
12. Honeycombe, R. W. K. *The Plastic Deformation of Metals,* Edwards Arnold Publishers: Cambridge, UK, 1968; p 402.
13. Kozlov, G. V.; Zaikov, G. E. *Structure of the Polymer Amorphous State,* Brill Academic Publishers: Utrecht, Boston, 2004; p 465.
14. Balankin, A. S. *Synergetics of Deformable Body,* Publishers of Ministry Defence SSSR: Moscow, RU, 1991; p 404.
15. Kozlov, G. V.; Sanditov, D. S. *Anharmonic Effects and Physical-Mechanical Properties of Polymers*; Nauka: Novosibirs, RU, 1994; Vol. 29, p 261.
16. Weitz, D. A.; Huang, J. S.; Lin, M. Y.; Sung, J. *Phys. Rev. Lett.* **1984**, *53* (17), 1657–1660.
17. Brady, L. M.; Ball, R. C. *Nature*, **1984**, *309* (5965), 225–229.
18. Shogenov, V. N.; Kozlov, G. V. Fractal Clusters in Physics-Chemistry of Polymers. *Nal'chik, Polygraphservice T* **2002**, 18, 268.
19. Komarov, B. A.; Dzhavadyan, E. A.; Irzhak, V. I.; Ryabenko, A. G.; Lesnichaya, V. A.; Zvereva, G. I.; Krestinin, A. V. *Polym. Sci., Ser. A* **2001**, *53* (6), 897–905.
20. Malamatov, A. Kh.; Kozlov, G. V.; Mikitaev, M. A. Reinforcement Mechanisms of Polymer Nanocomposites. Publishers of D.I. Mendeleev RKhTU: Moscow, RU, 2006; p. 240.
21. Andrievsky, R. A. *Russ. Chem. J.* **2002**, *46* (5), 50–56.

CHAPTER 19

A NEW CONCEPT OF PHOTOSYNTHESIS

G.G. KOMISSAROV

N. N. Semenov Institute for Chemical Physics, Russian Academy of Sciences, Kosygin St. 4, Moscow 119991, Russia. E-mail: komiss@chph.ras.ru; gkomiss@yandex.ru

CONTENTS

ABSTRACT

A history of the formation of a new concept of photosynthesis proposed by the author is considered ranging from 1966 to 2013. Its essence is as follows: the photosynthetic oxygen (hydrogen) source is not water, but exogenous and endogenous hydrogen peroxide; thermal energy is a necessary part of the photosynthetic process; along with carbon dioxide, air (oxygen, inert gases) is included in the photosynthetic equation. Here, the mechanism of the photovoltaic (Becquerel) effect in films of chlorophyll and its synthetic analogue—phthalocyanine—is briefly touched. There are presented works on artificial photosynthesis performed in the laboratory of Photobionics of the Semenov Institute of Chemical Physics RAS.

I know Gennady Efremovich Zaikov from the first day of my stay at the Institute of Chemical Physics RAS (since 1967). I was always amazed and still am at his lively mind, a fantastic performance, and a strong sense of humor. His gait is always swift, handshake is vigorous, and at meetings I always see a smiling face.

Taking this opportunity, I wish dear Gennady Efremovich good health and new successes, happiness.

Photosynthesis is the "global," "fundamental," and "unique" biological process. Its mechanism study is one of the central tasks of the modern natural science. Investigations into photosynthesis started in 1771, when the outstanding English chemist, Joseph Priestley, discovered the capacity of plants to "repair air, distorted by the burning of candles", that is, release oxygen. A significant contribution to the development of the concept was introduced by Ingenhousz, who showed the necessity of solar light for the occurrence of photosynthesis: the release of oxygen occurs only if the plants are illuminated (in darkness, they lose this capacity). He also established that photosynthesis is accompanied by the buildup of organic products. The experiments carried out by J. Senebier and N.Th. Saussure revealed initial substances of photosynthesis (carbon dioxide and water). The energy aspect of the problem was discussed for the first time by J.R. Mayer. In 1941, A.P. Vinogradov and R.V. Teis in USSR and S. Ruben with his colleagues in the USA established that oxygen is released from water and not from carbon dioxide.[1] During the last 70 years, the main equation of photosynthesis does not change and is written as follows:

$$CO_2 + H_2O \xrightarrow{\text{Light}} \text{carbohydrates} + O_2 \quad (19.1)$$

During breathing (reaction (1) taking place from the left to the right), the energy stored in the final products by the plants is released. It should also be mentioned that all living substances are constructed from molecules that initially formed in plants. Examination of the unique biological process of storage of solar energy has been continuing for more than two centuries, but the final mechanism has not as yet been completely explained.

19.1 FUNCTIONAL MODELING OF PHOTOSYNTHESIS

Regardless of the complexity of photosynthesis, it is possible to define two main stages in this process: the light stage, whose occurrence requires the direct effect of light, and the dark stage, which follows the light stage. It is assumed that the first, light stage, is characterized by the occurrence of photosplitting of water with the generation of molecular oxygen, which is released into the atmosphere as a secondary product:

$$H_2O \xrightarrow[\text{Chlorophyll}]{\text{Light}} [H] + O_2 \quad (19.2)$$

The resultant hydrogen subsequently enters into the thermal cycle responsible for the fixation of carbon dioxide, which is completed with the formation of carbohydrates, conventionally denoted by $\{CH_2O\}$:

Ferments

$$[H] + CO_2 \longrightarrow \{CH_2O\} \quad (19.3)$$

This stage, including a large number of fermentation reactions, has been examined in detail in studies by the outstanding American chemist M. Calvin who was awarded the Nobel Prize in 1961.

Thus, "photosynthesis" consists of two main stages: photolysis of water and dark synthesis of carbohydrates. In the second stage, there are no specific photosynthesis processes; the fixation of carbon dioxide in dark may be carried out by the liver cells of a rat, if a suitable hydrogen donor is available.

In this chapter, we shall approach the problem of photosynthesis from the physical–chemical position. Can artificial physical–chemical systems reproduce the light stage of photosynthesis?

The sequence of transformation of energy during photosynthesis, as described in Refs 1–3, may be written in the following form:

$$E_L \rightarrow E_E \rightarrow E_C \ (19.4)$$

where E is the energy with the appropriate indexes "light," "electrical," and "chemical.". Equation 19.4 greatly simplified the problem of functional modeling of photosynthesis because it makes it possible to solve the problem in two independent stages:

$$E_L \rightarrow E_E \quad (19.5)$$

$$E_E \rightarrow E_C \quad (19.6)$$

and transformation (eq 19.6). Electrolysis of water with the release of gaseous oxygen has been solved a long time ago. On an industrial scale, the process is carried out with high efficiency (current efficiency is 95–99%). Thus, the initial task is to find a device capable of generating electrical energy under the influence of light.

In 1839, E. Becquerel[4] first described the occasionally detected photoeffects. The principle of the phenomenon, subsequently referred to as the Becquerel effect or photovoltaic effect, is the formation of a difference of the potentials between two metallic electrodes placed in an electrolyte; one of the electrodes is coated with a layer of a light-sensitive substance.

In 1874, Becquerel observed that the illumination of a silver-coated platinum electrode carrying a chlorophyll film, with red light, changes its potential.[5] We believe that this date should be regarded as the basic date in the physical–chemical modeling of photosynthesis, although Becquerel never mentioned this approach. His work was directed to confirm the phenomenon of optical sensitization in photography, discovered by H.W. Vogel.[6]

Prior to starting modeling experiments, we thought that it was essential to evaluate (at least approximately) the intensity of the photoflux in chloroplast. Using Faraday's law, it can easily be shown that the total current in chloroplasts, containing 0.1 mg of chlorophyll, is equal to 10^{-5} A. However, in a single chloroplast, the current is evidently many times smaller (10^{-13} A).

In 1966, as a functional model of chloroplast, we proposed a photovoltaic battery.[1–3] Its first in the world variant was realized at the Institute of Chemical Physics of the Academy of Sciences of the USSR in 1968.[1,7,8] Then a similar battery was built in the United States.[9] The battery consisted of four electrodes coated with a synthetic analogue of chlorophyll, that is, phthalocyanine. It was characterized by the following parameters[8]: the light potential reached 2.4 V, which is fully sufficient for electrolysis of water with the generation of molecular oxygen (the dissociation potential of water 1.23 V); the current taken from the battery was 5.6×10^{-5} A so that it was possible to record oxygen by the conventional methods. The quantum yield of the photocurrent was not high, 0.01–0.1%. *The proposed device is capable of releasing oxygen from water under the effect of visible light, that is, reproduce one of the main functions of the chloroplast, and the process, taking place in the battery, is of photocatalytic nature.*

19.2 THE BECQUEREL EFFECT IN PHTALOCIANINE FILMS

The maximum value of the quantum yield of the photocurrent is detected in the case of monomolecular coating of the electrode with the pigment (Fig. 19.1).[10] In the chloroplast, the membranes are also covered on average by a monomolecular layer of chlorophyll.

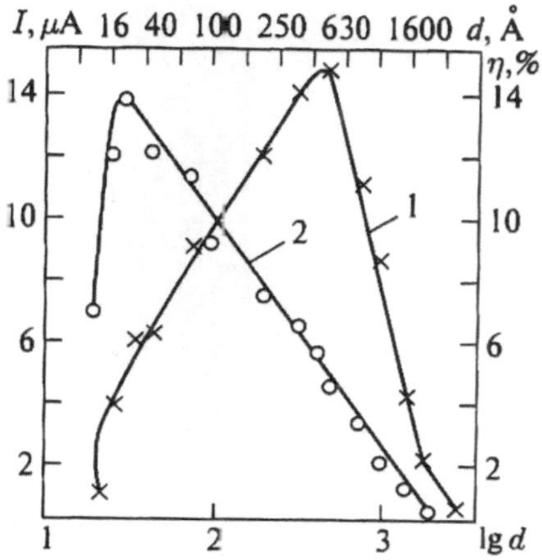

FIGURE 19.1 Dependence of the short circuit current (curve1) and the quantum yield (curve 2) on the thickness of the pigment film.

To increase the quantum yield, it was necessary to carry out a systematic examination of the mechanism of the Becquerel effect. This was carried out in our laboratory. As a result, the quantum yield was increased to units and, subsequently, to tens of percent.[1,10–14] In other words, it was possible to show that with regards to efficiency, the modern photovoltaic systems are not inferior to photosynthetic structures.

Examination of the dependence of the magnitude of current and potential of pigmented electrodes on the pH value of the electrolyte gave an unexpected result. Fixed value of pH in darkness was observed, these values are almost identical with the values in platinum electrodes in the absence of the pigment in them. This indicates that the pigment film was penetrated by a large number of pores (Fig. 19.2).[1,13]

The methods of the preparation of films, the values of current, potentials, and the dynamics of these values unambiguously show that they used porous films, but the interpretation of the observed relationships was based on the considerations regarding the monolithic pigmented layer of the electrode. This was followed by the development of methods of production of pore-free films and at the present time, it is possible to specify the main types of photovoltaic effects in the films of semiconductor pigments contacting, on the one side, with the electrolyte and, on the other side, with the metal (Fig. 19.3).[1,12,13,15,16]

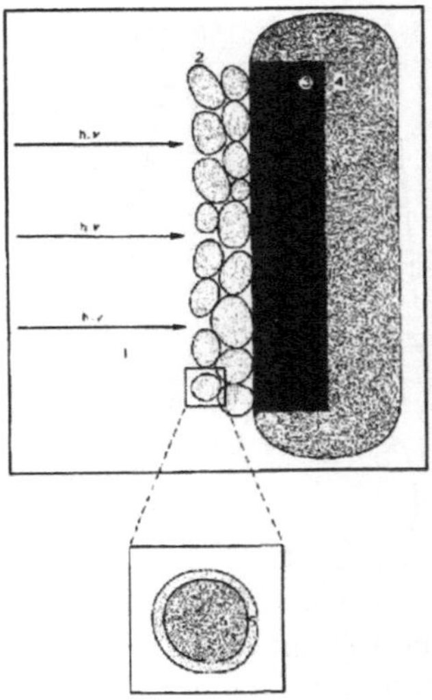

FIGURE 19.2 Scheme of an electrode coated with a layer of pigment: 1—electrolyte, 2—pigment layer, 3—platinum electrode, 4—insulating layer, 5—thickness of layer penetrated by incident light, 6—main body of the pigment suspension.

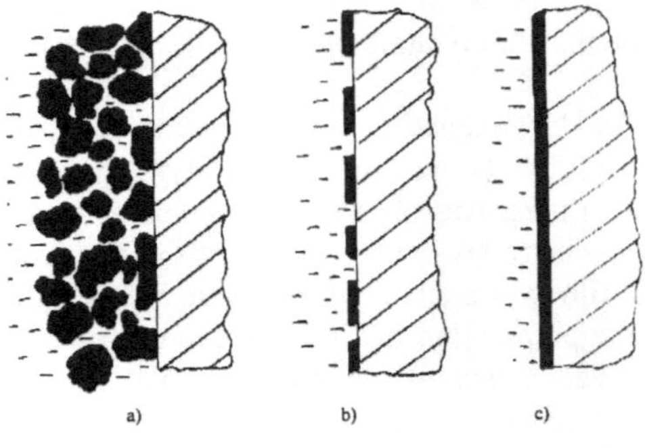

FIGURE 19.3 Schematic of three types of pigment films (photovoltaic effects). Dark areas show the pigment, dashes indicate electrolyte, the metal is shaded by solid inclined lines. (a) A thick porous film (light does not reach the pigment–metal contact). Light-induced potential arises due to a change in concentration of potential-governing ions at the electrode surface ($\phi = A(B - \ln C)$; ϕ is light-induced potential; $A = RT/nF$. R, T, n, and F have the commonly accepted meaning). The time needed to reach the steady photo-induced potential is of photo-induced potential in minutes, the dark current $\approx 10^{-6}$ A/cm^3. (b) Intermediate type. (c) A thin nonporous film (light penetration depth exceeds the film thickness). The response to illumination is a result of semiconductor processes in the bulk of the film and its boundaries. The time taken to reach the steady photo-induced potential is a fraction of second; the dark current is 10^{-12} A/cm^3.

The nature of the generation of photoresponse in the films differs principally. The values of current and their kinetic parameters differ tens, hundred, or more times. The previously described considerations enabled us to start investigations in the area of structural–functional modeling, the development of artificial systems reproducing the structure (composition) and function of natural photosynthetic formations. The analysis of literature data on the generation of oxygen from water in physical–chemical systems (electrolysis, photolysis, and radiolysis of water) and also the data obtained in our laboratory on the generation of oxygen in natural and modeling systems have made it possible to propose an original schema of photosynthetic release of oxygen.[1,12]

According to the schema, the generation of a single molecule of oxygen requires at least four light quanta, each of which generates an "electron–hole" couple, and the electron is used in the reaction $H^+ + e \rightarrow H$, required for subsequent fixation of CO_2. The surface of lamellae (sheets from which chloroplasts is produced) is a unique photoelectrode, consisting of alternating anodic and cathodic microregions (Fig. 19.4).[1,12,13] On the whole, the proposed schema of the dissociation of water has made it possible to explain several relationships in the photosynthetic generation of oxygen.

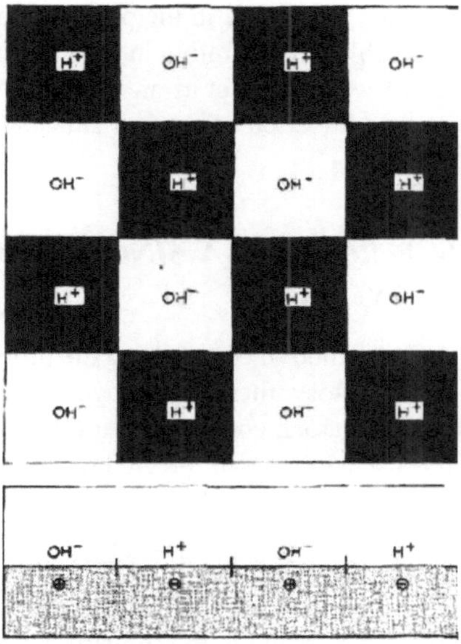

FIGURE 19.4 Structural scheme of the absorbed layer on the surface of the pigment (view from above and from side).

The experimental confirmation required many years of work resulting in the development of a new concept of photosynthesis.

19.3 HYDROGEN PEROXIDE—SOURCE OF PHOTOSYNTHETIC OXYGEN (HYDROGEN)

19.3.1 THE EFFECT OF HYDROGEN PEROXIDE ON THE KINETICS OF GENERATION OF OXYGEN DURING PHOTOSYNTHESIS

The sequence of dissociation of water during photosynthesis, proposed by us in 1973, may be represented in the form: $H_2O \rightarrow H_2O_2 \rightarrow HO_2 \rightarrow O_2$, that is, water is oxidized to hydrogen peroxide, which, in the final analysis, results in the generation of molecular oxygen (see the previous chapter). At present, there is a large number of literature data indicating the participation of hydrogen peroxide as an intermediate product in the course of formation of oxygen. We attempted to use the kinetic methods for confirming the participation of hydrogen peroxide in the course of photosynthetic generation of oxygen. For this purpose, we proposed a kinetic model based on the representation of vector algebra and projection geometry.[1,17] The results of kinetic analyses and the experimentally obtained changes of the kinetics of photosynthetic generation of oxygen in the presence of exogenous hydrogen peroxide indicate that it is capable of penetrating in light into the oxygen-generating complex of growth instead of water and act as an independent source of electrons and an independent source of generation of oxygen situated outside the four-stage oxygen cycle.

19.3.2 HYDROGEN PEROXIDE—A SINGLE SOURCE OF PHOTOSYNTHETIC OXYGEN

As already reported, the main method of overcoming the problem of photosynthesis is based on the light stage of photosynthesis (the stage of photodissociation of water). The work of current investigators, concerned with the examination of the photosynthesis mechanism, has been and is still directed to finding approaches capable of explaining the mechanism by which the chlorophyll (the photosynthetic oxygen-generating reactions center) is capable of storing the energy of several light quanta in order to use this energy for the formation of molecular oxygen. Many original studies have been published in the course of these investigations.

Already in our first study, concerned with the justification of the photoelectrochemical hypothesis of photosynthesis,[3] we paid attention to the need to take into account the changes of the properties of water in the chloroplast where it is situated between the lamellae with the distance between them not exceeding 100 Å. It is well known that the properties of structured liquid (in particular, and/or situated at the boundary with the solid phase) greatly differ from the properties in the volume. The viscosity of water in capillaries is an order of magnitude higher than the viscosity of water in the volume, and the heat conductivity of water in the layers increases tens of times, dielectric permittivity decreases from 81 (water in the volume) to 3–4 (in the interlayers with a thickness of 0.5–0.6 nm). Similar examples make it possible to assume that the puzzle of biological oxidation of water with the generation of oxygen is "hidden not only in the properties of the chloroplast but also of the water itself."[3]

The extensive and long-term examination of the physical–chemical properties of water resulted in an unambiguous conclusion: *in nature, there is no pure water, and water always contains an impurity, that is, hydrogen peroxide.* In a thridistillate, the concentration of hydrogen peroxide is 10^{-9} M,[18] in water of natural reservoirs (seas, rivers, lakes), it reaches 10^{-6} M, in rainwater is 10^{-5} M.[1,19] This makes it possible to supplement the equation of biological oxidation of water (eq 19.2) by another term, that is, hydrogen peroxide, which was discovered in 1818 by L.J. Thernard and referred to as "oxidized water"[20]:

$$H_2O_2,\ H_2O \xrightarrow[\text{Chlorophyll}]{\text{Light}} [H] + O_2 \qquad (19.7)$$

At first sight, it may appear that in this case, we are talking about some negligibly small impurity, with no relationship to photosynthesis; however, this is not so. By evaporating the aqueous solution of the peroxide, it is possible to increase its concentration tens of times, because its volatility is considerably smaller than in the case of water. The method of concentration of the aqueous solutions of hydrogen peroxide by evaporation of water has been used in the chemical practice for a long time now.[20] Transpiration (evaporation of water by plants) evidently plays the same function in addition to the protection of plants against overheating. For each kg of water absorbed by the roots from soil, only 1 g (1/1000 part of) is used by the plant for the construction of tissue. Thus, the green leaves may be regarded as a unique concentrator of hydrogen peroxide. It should also be mentioned that the second initial substance in photosynthesis is CO_2 whose content in air is only 0.03% (less than the content of inert gases). Its concentration is higher than the concentration of hydrogen peroxide in the initial water. However, the solubility of hydrogen peroxide in water is seven times higher than that for CO_2 (2×10^5 and 4.5×10^{-2} mole/L atm, respectively). This results in a serious consequence. Regardless of the area of formation of H_2O_2 (in air, in air bubbles in soil), its concentration in contacting water will be higher than in CO_2. Naturally, in this case, it is necessary to take into account the initial concentration of these substances. In addition to exogenous hydrogen peroxide, the photosynthesized cell also contains endogenous hydrogen peroxide. Its source in the cytoplasm is the mitochondrin (in high-intensity photosynthesis, it converges to chloroplasts), peroxisomes, and so forth. For example, up to 40%

of hydrogen peroxide, generated in peroxisomes, is transferred into the cytosol. In other words, the hydrogen peroxide *in vivo* is completely sufficient for explaining the observed intensity of the generation of molecular oxygen from this hydrogen peroxide.[1]

The data, obtained in Ref 21, are most interesting from the viewpoint of examining the role of hydrogen peroxide in the generation of photosynthetic oxygen. The authors carried out mass spectrometric examination of photosynthetic generation of oxygen using hydrogen peroxide, marked with respect to oxygen ($H_2\ ^{18}O_2$). The results show that $H_2\ ^{18}O_2$ is the source of the entire amount of generated oxygen. It is well known[18] that the hydrogen peroxide in chemical systems rapidly exchanges hydrogen (deuterium) with water. This was already established in 1934. However, when using the oxide marked with respect to oxygen, the situation is completely different. As reported in the monograph,[20] "the oxygen of water, used as a solvent, does not take part in the dissociation or reaction of hydrogen peroxide; no exchange was found between the water and the resultant molecular oxygen or oxidized products." Consequently, if the photosynthesized systems contain water and hydrogen peroxide, it is evident that the dissociation with the generation of oxygen explained case of the latter compound.

In the photoelectrochemical mechanism of formation of oxygen *in vivo* from two possible sources (hydrogen peroxide and water), the first mechanism is obviously preferred. In our laboratory, using the method of a spinning disk electrode with a ring (the spinning speed up to 3000 rpm), we showed the catalase activity of chlorophyll films that were in contact with the aqueous solution of hydrogen peroxide. Under the effect of light, the rate of dissociation of hydrogen peroxide increased 2–3 times in comparison with the darkness values.[1]

The results obtained in our laboratory and also the critical analysis of the literature data on the photosynthetic generation of oxygen have made it possible to propose a completely new viewpoint. According to the viewpoint, the source of photosynthetic oxygen (hydrogen) is not water but the exogenous and endogenous hydrogen peroxide.[1,22–26]

Naturally, it is difficult for a conventional photosynthetist to understand our position, because in the current literature, water is regarded as a source of oxygen. It is useful to mention that up to the studies carried out by A.P. Vinogradov[27] and S. Ruben,[28] which appeared 70 years ago, the majority of investigators of photosynthesis had assumed that CO_2 is the source of oxygen in photosynthesis.

At the Fifth International Biochemical Congress (Moscow, 1961), A.P. Vinogradov and V.M. Kutyurin[29] attempted to evaluate the variants of the methods of dehydration of water during photosynthesis. We believe that A.P. Vinogradov had experimental justification for proposing the hydrogen peroxide as a source of oxygen in photosynthesis. We shall discuss the results. The studies[27,28] show convincingly that CO_2 cannot be the source of photosynthetic oxygen as assumed at that time by the majority of researchers. This completed the revolutionary break in the examination of the photosynthesis mechanism. Since it was not possible during the experiments to achieve the equality of the isotope composition of the oxygen generated during photosynthesis and the oxygen in water, it would be necessary to introduce an assumption on the effect of breathing on the investigated process.[29] The situation existing in 1961 and this problem was characterized by R. Wurmser [30] as "it is almost evident (bold face by me, *G.K.*) that the generated oxygen comes from water." In 1975, H.J. Metzner [31] published a large article: "The dissociation of water during photosynthesis? Critical review." Analyzing the literature data and his own results, the author concluded that they reject the hypothesis on the oxidation of water in photosynthesis. The study ended with the words: "If we take together the data of published isotope experiments as a confirmation against the splitting of O–H bonds, we should postulate the rapture of another oxygen-containing bond, that is, C–O bond or **O–O** bond in the peroxide precursor of oxygen" (bold face by me, *G.K.*).

We have assumed (and we shall remain on these positions) that the pigment system of the chloroplast is a highly autonomous structure, designed for the generation of protons and molecular oxygen.

19.3.3 THE ROLE OF THERMAL ENERGY IN PHOTOSYNTHESIS AND CORRECTION OF THE FUNDAMENTAL PHOTOSYNTHESIS EQUATION

It is generally accepted (see any textbook on photosynthesis) that the thermal energy is a waste production of the photosynthetic process. In the case of a high-intensity process, only 0.5–5% of light energy is used up for photosynthesis, whereas ~95% of energy "degrades into heat." It is difficult to assume that in billions of years of their existence on earth, the plants have not adapted themselves to a more efficient utilization of light energy. In 1973, the author assumed that the thermal energy is not only an important but also essential participant of the photosynthetic process.[1,12,32] This viewpoint is presented in the most complete form in the study "Photosynthesis as a thermal process."[33] According to this concept, in the regions with the size of the order of the chlorophyll molecule, the local temperature may greatly exceed (by several tens of degrees) the temperature of the surrounding medium. According to our estimates, this temperature reaches 70°C as a result of the recombination of charge carriers in the reaction center in which the adsorption of initial substances has not been completed at the given moment. Increase of the temperature inside the chloroplasts accelerates at the diffusion of both the products of photosynthesis and initial substances. An increase of temperature also facilitates the transport of ions through the membrane. According to calculations, the energy required for the transport of anion from the electrolyte into a lipid membrane is 250 kJ/mole. However, the energy of transport of the ion (e.g., sodium, potassium) through a membrane channel is considerably lower (~20 kJ/mole).[34]

Already at the start of previous century, K.A. Timiryazev[35] proposed a hypothesis regarding the effect of thermal energy in photosynthesis. He assumed that the heating of chloroplasts, caused by sunlight, may be sufficient for the occurrence, in the chloroplasts, of a process thermodynamically reversed in relation to combustion, and, consequently, this may be used to explain the principle of photosynthesis. This viewpoint is at present of only historic interest because it was criticized. However, in the light of our considerations regarding hydrogen peroxide as a source of molecular oxygen in photosynthesis and local heating of the chloroplasts, it is possible that this hypothesis will be developed further.

It may be assumed that in photosynthesis, together with the photoelectrochemical mechanism, there is also the possibility of the thermal dissociation of hydrogen peroxide as the release of molecular oxygen. The thermal stability of water is incomparably higher than that of hydrogen peroxide.[1,20]

An additional confirmation of the possibility of thermal dissociation of H_2O_2 in photosynthesis may be the data of modeling systems—the films of phthalocyanine (synthetic analogue of chlorophyll) on a Pt electrode, which is in contact with the electrolyte, where the dependence of the photopotential of the films of phthalocyanine on heating of the electrolyte may be examined. At a temperature higher than 80°C, the photoresponse cannot be recorded. This can be naturally explained by the thermal dissociation of hydrogen peroxide formed on the surface of the platinum electrode.[1,22]

It should be mentioned that the assumptions regarding the local heating of chloroplast, introduced by us in 1973, were initially met with excessive criticism, but in recent years, these concepts in the sphere of biophysics do not lead to any dispute.[36] In addition, a number of studies have been published recently on polymers, the theory of the photographic process, media for optical memory, glasses, where the assumption on the local temperature is used successfully for explaining the observed relationships. In the monograph by S.F. Timashev "Physics and chemistry of membrane processes,"[34] the role of the local heating is treated in a special section.

On the basis of all these considerations, it may be assumed that the local equation of photosynthesis should include not only light but also thermal energy.[1,22–26,33] The point is that in the implicit form, the thermal energy, from our viewpoint, has been included in the general equation of photosynthesis for a long time. We shall pay attention to the thermodynamic potential of dissociation of water. The minimum difference of the potentials, required for the electrolysis of water at 25°C, 1 atm, and on the condition of the supply of thermal energy, is 1.23 V. The thermoneutral potential of dissociation of water at 25°C is 1.47 V. In all investigations into photosynthesis, only the first magnitude (1.23 V) has been used, but the necessity for supplying thermal energy in this case is simply not mentioned.

Thus, the previously presented data on the exogenous and endogenous hydrogen peroxide as a source of oxygen (hydrogen) in photosynthesis and the role of thermal energy make it possible to write the basic equation of photosynthesis in the following form[1,22–26]:

$$CO_2\text{ (air)} + H_2O_2\text{ (water)} \xrightarrow[\text{Thermal energy (+/-)}]{\text{Light energy}} \text{carbohydrates} + O_2 \qquad (19.8)$$

The main difference between this equation and eq 19.1 is the replacement of H_2O by H_2O_2, and water plays the role of the reaction medium for CO_2 and H_2O_2. It should be mentioned that an identical situation was recorded previously with CO_2. Initially, it was assumed that "*air*" (my italics, the author) damaged by the combustion of candles takes place in photosynthesis, and, subsequently, the modification- CO_2 (G. Senebier, 1782) was introduced. The concentration of CO_2 in air is only 0.03%. It is also possible that in future, new components of both air and natural water, taking part in photosynthesis, will be detected. For example, reports have appeared according to which the inert gases affect the rate of splitting of cells. Therefore, we regard it as useful to write in the main photosynthesis equation, air in addition to CO_2 and water together with H_2O_2 and, naturally, photosynthesis, like any other life processes, is not possible without water.

The sign (+/−) in eq19.8 indicates that at high densities of the light the leaf (chloroplast) transfers the energy to the surrounding medium, and at low densities, it takes the energy from the surrounding medium. In the latter case, the coefficient of transformation of solar energy may be higher than 100% because the contribution of thermal energy is not taken into account. A similar situation was found in the early stages of examination of the efficiency of operation of fuel elements.

Thus, we shall make a conclusion. A new concept, according to which the source of oxygen (hydrogen) in photosynthesis is the exogenous and endogenous hydrogen peroxide, and not water, has been proposed. The dissociation of hydrogen peroxide with the generation of molecular oxygen is possible either by photoelectrochemical and/or thermochemical mechanism.

19.4 SOME CONSEQUENCES RESULTING FROM THE NEW CONCEPT OF PHOTOSYNTHESIS

19.4.1 THE EFFECT OF HYDROGEN PEROXIDE ON THE GROWTH OF PLANTS

Within the framework of the proposed concept, it was reasonable to propose that the variation of the concentration of peroxide in water has a significant effect on the rate of growth of plants because photosynthesis, in particular, determines the rate of the physiological processes. The authors of Ref 37 presented results of experimental verification of this assumption. Naturally, the effect of different amounts of exogenous hydrogen peroxide may have an appropriate effect also on the rate of generation of endogenous peroxide. The mechanism of participation of hydrogen peroxide in the physiological processes of growth is being studied intensively at the present time. Peroxide determines the intensity of photophosphorylation, photobreathing, fungitotoxicity of the surface of leaves, and so on.

In order to record the growth of plants in the presence of hydrogen peroxide, in addition to the traditional biological approach, we have also used the recently proposed method of laser interference auxanometry.[1,24] It should be mentioned that the problem of the growth of plants is one of the central problems in current phytophysiology.

Analysis of the literature shows that the stimulating effect of hydrogen peroxide on the growth and development of plants has been known for a long period of time. Already at the beginning of this century, it was known that the solutions of hydrogen peroxide stimulate the growth of seeds. The introduction of hydrogen peroxide into soil increases the yield of corn, soya beans, accelerates the growth of seeds, but the reason for this effect, as reported in the above investigations, was not clear.[20]

USING PLANTS FOR THE GENERATION OF AIR IN A CLOSED VOLUME

If we compare the traditional form of the main equation of photosynthesis (eq 19.1) with eq 19.8 proposed by us, it may be seen that in addition to the differences mentioned previously, there is one important consequence in our view. The classic equation of photosynthesis assumes that there is complete agreement with respect to the initial and final products between photosynthesis and breathing (eq 19.1, read from right to left, is the equation of breathing). In the concepts proposed by us, the relationship between the fundamental biological process and breathing is complicated because the final product in breathing is water which, in our view, does not split during photosynthesis. This is illustrated by the diagram of formation of oxygen and its usage during breathing (Fig. 19.5).

It is well known that in addition to photosynthesis, the leaf is characterized by the occurrence of an opposite process stimulated by light, that is, the generation of carbon dioxide and absorption of oxygen (photobreathing). In the dark conditions, the intensity of the process, opposite photosynthesis (dark breathing), is 5–7% of photosynthetic gas exchange (plants, like living creatures, require oxygen). However, the intensity of photobreathing is an order of magnitude higher in comparison with dark breathing and equals ~50% of photosynthetic gas exchange.[38,39] Since the photobreathing greatly reduces the rate of real photosynthesis and, consequently, the productivity of the plants, a larger number of attempts have been made to liquidate this process (or, at least, reduce its intensity). These attempts have not been successful. According to V.I. Chikov,[38] "none of the directions in the search for the methods of decreasing the intensity of photobreathing have resulted in a positive result for increasing the productivity of the plants. In addition, it is believed that the conditions, suppressing photobreathing, also decrease the productivity of the plants."

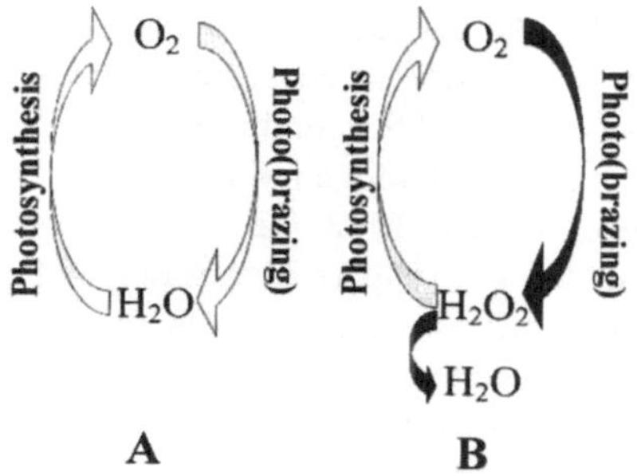

FIGURE 19.5 The relationship between photosynthesis and (photo) breathing within the framework of the conventional considerations regarding photosynthesis (A) and in accordance with the concept proposed by the author of the article (B).

Previously, it was mentioned that breathing is accompanied by the formation of endogenous hydrogen peroxide. In accordance with the proposed concept of photosynthesis, the endogenous hydrogen peroxide is a source of not only oxygen, released into the atmosphere, but also of hydrogen used in the synthetic processes of growth. It should be mentioned that photobreathing is most active when the plant is supplied with abundance of mineral nitrogen, that is, favorable conditions are created for the growth of biomass. The stimulating effect of the exogenous hydrogen peroxide (at the optimum concentration of the latter) was examined in the previous section.

Thus, it may be assumed that the activity of synthetic processes in the plants is determined in approximately equal parts by the intensity of photosynthesis and photobreathing. Naturally, photobreathing uses the products stored as a result of photosynthesis plus atmospheric oxygen. A further examination of this process by physiologists of plants would enable, in our opinion, the development of the optimum technology to cultivate agricultural plants.

The considerations described here cast doubts on the exclusive role of plants in the formation of the current oxygen atmosphere of the earth, widely cited in the biological literature. We believe that in the geochemical investigations, which, in our view, are most important in the given problem, it is easy to find an opposite viewpoint.

Explanation of the mechanism of generation of photosynthesizing oxygen is very important in the current period because of the development of space technology, the development of essential conditions for the life of man in a closed space. If the plants ensure the supply of oxygen to our atmosphere, it is natural that it should be attempted to use them for the generation of air in a closed system, for example, in a space station. Our concept of photosynthesis is directly related to this problem.

In September 1991, eight volunteers started two-year tests in the Biosphere 2—complex isolated from the outer world. The complex is a prototype of future cosmic stations on the planets of the solar system.[40] The Biosphere 2 is a large experimental system generated for the examination of ecological processes taking place on the earth, and also for the development of the conditions of life activity in future cosmic stations, which will be constructed mainly on Mars. The experiment was planned for two years. However, already in June 1992, the oxygen content inside the complex was greatly reduced (from 20.94 to 16.4%). The subsequent decrease of the oxygen concentration was 0.25–0.3% per month and it was, therefore, necessary to supply oxygen into the complex.[41] A special commission has been formed for investigating the reasons resulting in a decrease of the oxygen concentration.

We believe that one of the reasons resulting in the unsatisfactory dynamics of oxygen in the Biosphere 2 is associated with the fact that the developers of the object used the conventional assumptions according to which the water is a source of oxygen in photosynthesis. In accordance with our concept of photosynthesis (see eq 19.8 and Fig. 19.5), for the normal functioning of a hermetically sealed complex, containing plants, it is necessary to ensure the occurrence in the complex of the processes resulting in the generation of hydrogen peroxide. On the earth, the hydrogen peroxide forms as a result of storm discharges, the radiolysis of underground water, and so on.

19.4.2 ARTIFICIAL PHOTOSYNTHESIS AND A PROBLEM OF THE LIFE ORIGIN

The attempts to model natural photosynthesis have been made since long ago. Back in the early 20th century, there were studies aimed to reproduce the fundamental biological process or its separate steps by means of simple photochemical systems.[1] Currently, the number of publications dealing with this topic increases in an avalanche-like fashion (Fig. 19.6). However, the attempts made did not result in the design of artificial physicochemical systems able to form organic products from carbon dioxide and water with simultaneous evolution of molecular oxygen under the action of visible light. It was not until 1969 that only one of the key functions of chloroplast was reproduced for this first time, in particular, photocatalytic evolution of molecular oxygen from water under the action of light absorbed by synthetic analogs of chlorophyll.[8,9]

What is the main obstacle that precluded implementation of artificial photosynthesis despite numerous attempts? Probably, the traditional equation of photosynthesis bears some error, which prevents implementation of the process in vitro.

In 1993, a new photosynthesis equation was proposed at the Institute of Chemical Physics of the RAS and substantiated in detail in a monograph.[1] According to this equation, exogenic and endogenic hydrogen peroxide rather than water serves as the source of oxygen (hydrogen) in the photosynthesis. In chloroplast, there is one hydrogen peroxide molecule per chlorophyll molecule.[1] As shown by quantum chemical calculations, the bond energy of peroxide with chlorophyll dimers is higher than that for water. Depending on the configuration of the dimeric complex, the difference may reach 3.8 kcal/mol.

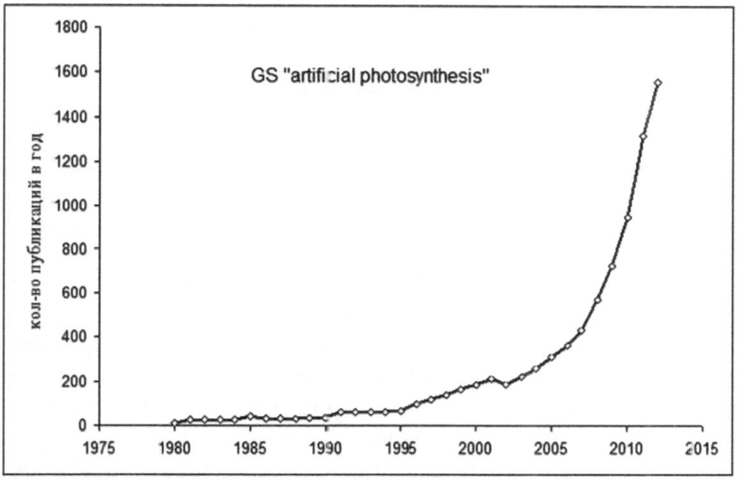

FIGURE 19.6 Annual number of publications mentioning artificial photosynthesis (1980–2012).

In 1994, we proposed to organize the international project "Artificial photosynthesis."[42] It should be noted that the developed concept allows us to see a new approach to the problem artificial photosynthesis. Thermodynamic estimates of the enthalpy ($\Delta H°$) and the Gibbs energy ($\Delta G°$) for reactions of CO_2 both with water and with hydrogen peroxide to give formaldehyde, formic acid, glucose, and other products were reported.[43] The results demonstrated unambiguously that syntheses of organic compounds from CO_2 and H_2O_2 require less energy than the corresponding reactions using water. Indeed, replacement of H_2O by H_2O_2 decreases the $\Delta G°$ of product formation by 30% for formaldehyde, by 34% for methanol, and by 42% for formic acid.

On the basis of the above, a successful attempt was made in 2004 to detect the formation of organic compounds upon the action of light on an aqueous suspension of adsorbed phthalocyanines, which were used as synthetic analogs of chlorophyll.[43,44] The suspension contained 0.2 mol/L of hydrogen peroxide and 0.4 mol/L of $NaHCO_3$. By means of chemical analysis and spectrophotometry, photocatalytic formation of formaldehyde was detected (Fig. 19.7).

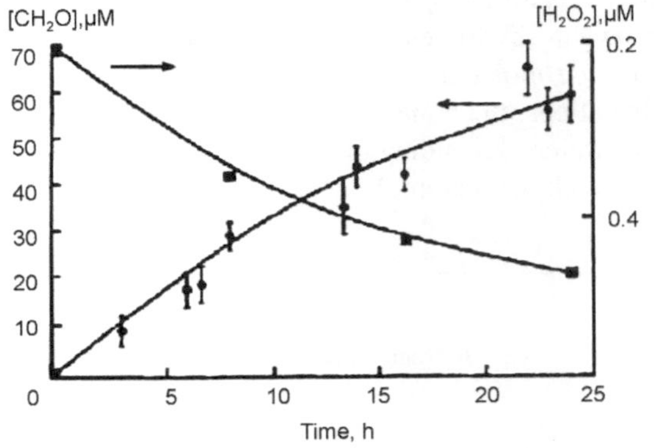

FIGURE 19.7 Kinetics of formaldehyde accumulation and hydrogen peroxide consumption in the photocatalytic reduction bicarbonate anion ($[NaHCO_3]_0$ = 0.4 M, 20°C, photocatalyst is aluminum phtalocyanine) (time, hours).

Analysis of the IR spectra of the irradiated reaction mixture attested to possible formation of organic products of various nature in such systems.[43–45]

The work presented in Ref 46 deals with a GC/MS detection of photogenerated organic products in an aqueous suspension of adsorbed aluminum phthalocyanine, H_2O_2, and HCO_3 and in the system containing H_2O_2 and CO_2, which we considered as a new step toward artificial photosynthesis.

The photophysical and photochemical properties of phthalocyanine are similar to those of chlorophyll but the stability against destructive impacts is much higher. This is why phthalocyanine was chosen for the experiment.[1] We used spectral-grade chlorinated aluminum phthalocyanine (Kodak). Since chlorophyll is linked in vivo to a support (protein), we used phthalocyanine also in the adsorbed state. As a support with developed surface, we took silica gel L 40/100, which allowed easy separation of the pigment from the reaction products by mere centrifugation. Aluminum phthalocyanine was adsorbed from a solution in DMF. The adsorbed amount was measured as the decrease in the pigment concentration in solution by spectrophotometry on a DR/4000V instrument (HACH-Lange, USA). As found from the adsorption isotherm, a monomolecular layer of the phthalocyanine metal complex is formed on the support surface when the phthalocyanine concentration in the initial solution is in the range 10^{-4} mol/L.[45] The preliminary experiments showed that the highest photocatalytic activity of the adsorbed aluminum phthalocyanine in decomposition of H_2O_2 is observed in the region of pre-monomolecular coverage of the support surface. Considering these data, the catalysts were prepared by adding 10 ml of a 5×10^{-4} M solution of aluminum phthalocyanine to 1 g of silica gel. The adsorbates, thus obtained contained 5.4×10^{-7} mol of the pigment per g of silica gel.

Kinetic experiments were carried out in 10 ml of a reaction mixture containing H_2O_2 (0.2 mol/L), special purity grade $NaHCO_3$ (0.4 mol/L) (Reakhim), and supported catalyst (200 mg) in distilled water. The suspension was distilled in a visible light with intense stirring. A halogen lamp (150 W) fitted with lenses, a condenser, and a KS-13 light filter cutting off the radiation below 630 nm was used as the source of visible light. The light flux power was 10 mW/cm^2. A quantitative determination of formaldehyde showed that its concentration reaches 10^{-5} mol/L after 24 h of irradiation. By this moment, more than 70% of the initial amount of H_2O_2 has been consumed, mainly via (photo) catalytic disproportionation. Under these conditions, we did not observe destruction or poisoning of the catalyst. The conclusion about the stability of the metal complex is also confirmed by the results of control runs carried out without $NaHCO_3$ in which no formaldehyde was detected. This implies that only CO_2 (HCO_3^-) can serve as the source of CH_2O.

Other products were analyzed by the GC/MS method. The GC/MS facility used to analyze the samples comprised a Thermo Focus GC gas chromatograph and a Thermo DSQ II mass spectrometer with a 60 m × 0.25 mm capillary glass column (0.25 m thick 100% dimethtylpolysiloxane as the stationary phase). The temperature program of the chromatograph included 10 min heating from 35 to 80°C (heating rate 1°C/min), then to 110°C (heating rate 5°C/min), and to 210°C (heating rate 10°C/min). The injector temperature was 200°C and the interface temperature was 250°C. Electron impact ionization was used, the electron energy being 70 eV and the mass spectrum being scanned in the range of 20–270 amu. Helium served as the carrier gas; the flow rate was 1 ml/min. Compounds were identified using the NIST library of mass spectra. Under the conditions used, the GC/MS method revealed organic compounds of alcohol and ketone classes in the reaction mixture (Fig. 19.8). If one of the components (hydrogen peroxide, hydrocarbonate, or phthalocyanine) is removed from the reaction mixture, organic products cannot be detected.

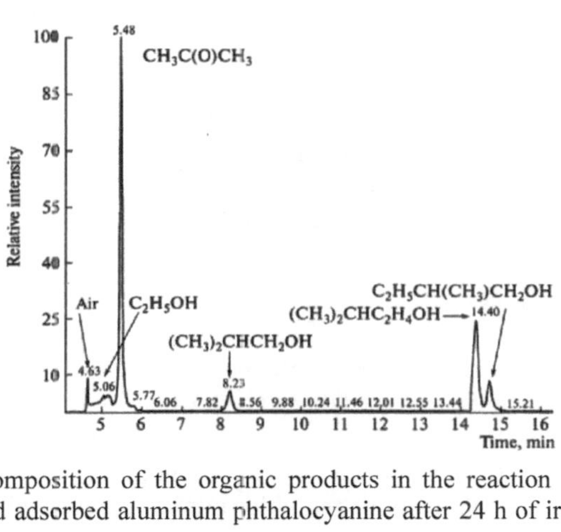

FIGURE 19.8 Composition of the organic products in the reaction mixture containing H_2O_2, $NaHCO_3$, and adsorbed aluminum phthalocyanine after 24 h of irradiation according to GC/MS data.

In the next experiment, $NaHCO_3$ was replaced by gaseous CO_2, which was passed continuously through the suspension during the irradiation (24 h). All other conditions were the same as in the previous experiment. The GC/MS analysis of the reaction mixture after irradiation showed the presence of formic acid (Fig. 19.9).

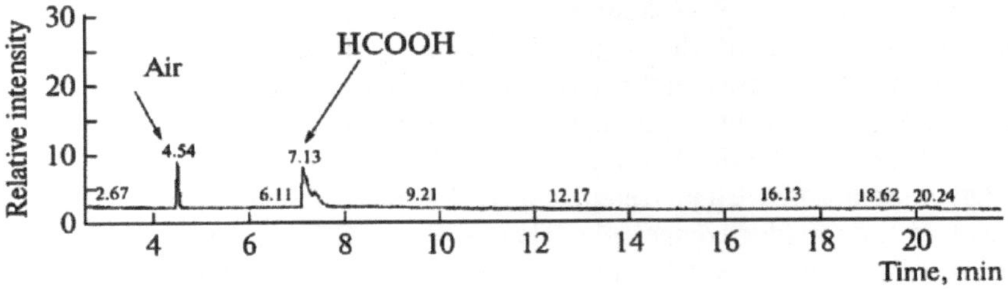

FIGURE 19.9 Formation of formic acid in the reaction mixture containing H_2O_2 and adsorbed aluminum phthalocyanine after 24 h of irradiation with continuous purging with CO_2 according to GC/MS data.

Note that in this run too, GC/MS analysis did not detect organic compounds when either hydrogen peroxide or carbon dioxide was missing.

Positive results of experiments using hydrogen peroxide are of considerable interest. They open up attractive prospects: selection of the most appropriate pigment (variation of the ligands and the central atom), support, and reactant ratio, which may provide significant results. At the current stage, we can say with confidence that a new step was made toward artificial photosynthesis in a purely abiogenic system. The value of these results is beyond the framework of photosynthesis. Particular paths to biofuel from carbon dioxide and hydrogen peroxide have been outlined, which is of paramount importance for modern ecology. According to various estimates, the period when mankind will face the problem of exhaustion of the resources of fossil combustibles is near at hand.

The obtained results can also be used in considering the problem of cosmic origin of life, as hydrogen peroxide was detected in space objects.[47–50]

19.5 CONCLUSION

In the justification of the new concept of photosynthesis resulting from all our previous investigations, carried out in the laboratory, we use the following results. Photosynthesis (or, more accurately, its light stage—the generation of oxygen) is basically a relatively simple physical–chemical process taking place in a highly complicated biological system. The pigments of the chloroplasts represented a highly autonomous system whose function is mainly the absorption of light (therefore, the intensive color of the pigments) and transformation of the light to chemical energy. The effect of light in the chloroplasts results in the formation of an excess number of protons ensuring the possibility of occurrence of fermentation reactions leading to the synthesis of hydrocarbons. The results of our literature analysis and the data obtained in our laboratory convincingly indicate that the source of oxygen (hydrogen) in photosynthesis is the exogenous and endogenous hydrogen peroxide. Water plays the role given to the solvent in conventional chemical reactions. Naturally, the water protons may take part in the formation of endogenous hydrogen peroxide. However, the problem of dissociation (oxidation) of water is not so important for us at the moment because the amount of hydrogen peroxide ("oxidized water," the initial name of hydrogen peroxide) in the chloroplasts (cytoplasm) is such that it is possible to explain the experimental detected intensity of generation of oxygen (for each molecule of chlorophyll, including the molecule of the antennae, there is one molecule of hydrogen peroxide). Evidently, if the system contains water and hydrogen peroxide, only hydrogen peroxide will undergo photodissociation under the effect of visible light. Naturally, the development of detailed schema of the dissociation of hydrogen peroxide in photosynthesis requires a certain period of time. As an example, in current catalysis, hydrogen peroxide is used widely in heavy and light organic synthesis. This is one of the oldest variants of catalysis. However, according to the descriptive expression by Academician I.I. Moiseev "Hydrogen peroxide is widely used in organic synthesis industry, although the mechanism of its action is not elucidated fully" (Report of the seminar on Catalysis in ICP RAS).

The participation of hydrogen peroxide in the photosynthetic generation oxygen is far more likely than the participation of water, not only from the physical–chemical viewpoint (comparison of the dissociation potentials, thermal stability, etc.). It makes it possible to explain the existence of a large number of physiological processes accompanying photosynthesis. In the literature on physiology of plants, transpiration is treated as an "unavoidable evil," with attempts being made to eliminate it, although these attempts have not been successful. Within the framework of our considerations, transpiration is essential for the concentration of exogenous hydrogen peroxide. The identical situation is also characteristic of the process of photobreathing with which the physiologists "fight" without success and cannot get rid of it. Our considerations, presented in this chapter, indicate that the plants accumulate not only light but also thermal energy (unique thermal pump). The significance of the latter is especially large in the plant associations where the illumination of leaves because of mutual shading is relatively low (in the individually standing trees, the mutual screening of the leaves is also high). This makes it possible to use a new approach to explain the fact that the intensity of photosynthesis in the case of light illumination is almost an order of magnitude higher in comparison with high-intensity illumination.

In conclusion, it should be mentioned that the long-term investigations into modeling and examination of the photosynthesis mechanism, described in the chapter, have been used by us in the formulation and computer analysis of the model in which the process of search for the solution of a scientific problem by the investigator is examined in the generalized form.[51,52]

ACKNOWLEDGMENT

I am very grateful to the colleagues and graduates of the Laboratory of Photobionics of the Institute of Chemical Physics. I am also grateful to a large number of undergraduates and graduates (in most cases, of the Physical Department of the Moscow State University and Moscow Physico-Technical Institute) for solving the given problems.

I am especially grateful to Academician A.L. Buchachenko, the Head of the Department of the Dynamics of Chemical and Biological Processes of the Institute of Chemical Physics of the Russian Academy of Sciences, for supporting investigations carried out in our laboratory. The experiments in recent years have been supported by continuing financial support of the Russian Fund of Fundamental investigations to which we are grateful (Grants No. 94-02-04972a, No. 95-03-08982a, No. 96-0334064a, No. 98-0332061a, No. 04-03-32890a and No. 00-15-97404a, 08-03-00875a; the Presidential program "Leading Scientific Schools (NSh–2003–2013)"—Coordinator Academician A.L.Buchachenko.

2005–2007 Basic Research Program of the Presidium of RAS "Organic and hybrid nanostructured materials for photonics"—Program Coordinator Academician M.V. Alfimov.

2005–2006 Program of Presidium of RAS No. 7P-05 "Hydrogen Energy"—Program Coordinator Academician I.I. Moiseev.

2006–2014 Program of the Presidium of RAS No.18 "The problem of origin of the Earth's biosphere and its evolution"—Program Coordinator Academician E.M.Galimov.

2009–2011 Program of the Presidium of RAS "Chemical aspects of energy"—Program coordinator of Academician I.I. Moiseev.

2004–2007 ISTS project No. 2876 "Research and development of photoelectrochemical light energy converters based on organic semiconductors using the principles of photobionics."

2009–2013 ISTS project No. 3910 "Modeling of primary stages of photosynthesis on the basis of nano-sized supramolecular systems."

KEYWORDS

- **photosynthesis**
- **artificial photosynthesis**
- **photovoltaic effect (Becquerel effect)**
- **photocells**
- **quantum yield on a photocurrent**
- **the types of structures of pigment films on electrodes (three kinds of Becquerel effect)**
- **new equation of photosynthesis**
- **chlorophyll**
- **phthalocyanine**
- **chloroplast**
- **transpiration**
- **water**
- **hydrogen peroxide**
- **carbon dioxide**
- **oxygen**

REFERENCES

1. Komissarov, G. G. *Photosynthesis (Physicochemical Approach)*; Ed. URSS: Moscow, RU, 2003; p 223 (in Rus.); Komissarov, G.G. Fotosintesis: um enfoque fisicoquimico Ed. URSS, 2005; p 258 (in Spain).
2. Komissarov, G.G. *Abstr. Second Intern. Biophys. Congr.* **1966,** 234.
3. Komissarov, G. G. *Biophysics* **1967,** *12* (3), 558–561.
4. Becquerel, E. C. R. *Academy of Sciences,* Paris **1839**, *9,* 561.
5. Becquerel, E. C. R. *Academy of Sciences,* Paris **1874**, *79,* 185.
6. Chibisov, K. V. Comments on history of photography, Art, Moscow (1987), 2_8 (in Rus.).
7. Komissarov, G. G.; Shumov, Yu. S. The Reports of USSR. **1968,** *182,* 1226–1229.
8. Komissarov, G. G.; Shumov, Yu. S.; Borisevich, Yu.E. The Reports of USSR. **1969,** *187,* 670–673.
9. Wang, J. H. *Proc. Natl. Acad. Sci. U.S.A.* **1969,** *62,* 653–660.
10. Ilatovsky, V. A.; Dmitriev, I. B.; Komissarov, G. G. *Russ. J. Phys. Chem.* **1978,** *52,* 66–68.
11. Ilatovsky, V. A.; Apresyan, E. S.; Komissarov, G. G. *Russ. J. Phys. Chem.* **1989,** *63,* 2242–2244.
12. Komissarov, G. G. *Russ. J. Phys. Chem.* **1973,** *47,* 927–932.
13. Komissarov, G. G. *Sov. Sci. Rev.* **1971,** 285–290.
14. Ilatovsky, V. A.; Komissarov, G.G. *The Reports of Russian Academy of Sciences.* **2008,** *420,* 66–69.

15. Komissarov, G. G. *UPAC Abstr. 5th Int.. Symp. Macromoleculare Complexes, Bremen, Germany,* **1993,** 420.
16. Komissarov, G. G. *UPAC Abstr. XVIth Int Symp. Photochemistry, Helsinki, Finland,* **1996,** 332.
17. Ptitsyn, G. A.; Komissarov, G. G. *Sov. J. Chem. Phys.* **1994,** 2137–2147.
18. Das, T. N., et al. *J. Indian Chem. Soc.* **1982,** *59,* 85–87.
19. Stamm, E. V.; Purmal, A. P.; Skurlatov, Yu. I. Successes of Chemistry (in Rus.), 1991, 60, 2373–2398.
20. Schumb, W. C.; Satterfield, C. N.; Wentworth, R. L. *Hydrogen Peroxide, Reinold Publishing Corporation*: New York. 1955; p 578.
21. Mano, J.; Takahashi, M. A.; Asada, K. *Biochemistry* **1987,** *26,* 2495–2497.
22. Komissarov, G. G. *Chem. Phys. Rep.* **1995,** *14* (11), 1723–1732.
23. Komissarov, G. G. *Sci. Russia* (in Rus.) **1994,** *5,* 52–55.
24. Komissarov, G. G. *J. Advanc. Chem. Phys.* **2003,** *2* (1), 28–61.
25. Komissarov, G. G. *Optics Spectrosc.* **1997,** *83,* 607–610.
26. Komissarov, G. G. A New Concept of Photosynthesis Mechanism in Book Problems of Ecological Security in Agriculture Moscow. *Sergiev Posad* **2003,** *6,* 5–25.
27. Vinogradov, A. P., Teis R.V. *The Reports of USSR.* **1941,** *33,* 497–499.
28. Ruben, S., et al. *J. Amer. Chem. Soc.* **1941,** *63,* 877–879.
29. Vinogradov, A. P.; Kutyurin, V.M. *5th Int. Biochem. Congr. Sci. (in Rus.), Moscow* **1962,** 264–274.
30. Vyurmser, Z. *5th Int. Biochem. Congr. Sci. (in Rus.), Moscow* **1962,** 21.
31. Metzner, H. *J. Theoret. Biol.* **1975,** *51,* 201–216.
32. Komissarov, G. G. Chemistry and Physics of Photosynthesis. *Knowledge (in Rus.), Moscow* **1980**, 63.
33. Komissarov, G. G. *Current Research in Photosynthesis;* Baltscheffsky, M., Ed.; Kluwer Academic Publishers, 1990; Vol. IV, pp 107–110.
34. Timashev, S. V. Physico-chemistry of Membrane Processes. *Chemistry (in Rus.), Moscow* **1988,** 237.
35. Rabinovich, E. Photosynthesis. *Foreign Lit. (in Rus.) Moscow* **1959,** *3,* 936.
36. Kucheva, N. S., et al. *Biophysics* **1997,** *42,* 628–632.
37. Apasheva, L. M.; Komissarov, G. G. *Biol. Bull.* **1996,** *23,* 518–519.
38. Chikov, V. I. *Soros Educ. J. Chem* **1996,** *11,* 2–6. (in Rus.).
39. Golovko, T. K. Breathing of Plants (Physiological Aspects). *Science (in Rus.)* **1999,** SPb, 190.
40. Allen, J.; Nelson, M. *Space Biospheres;* Synergetic Press: Arizona, 1989; p 108.
41. Nelson, M., et al, *Her. Russ. Acad. Sci.* **1993,** *63,* 1024–1036
42. Komissarov, G. G. Artificial Photosynthesis: When? *10th Int. Conf. Photochem. Conversion and Storage Solar Energy (IPS-10) Book of Abstracts;* Calzafferri, G, Ed.; 1994.
43. Lobanov, A. V.; Kholuiskaya, S. N.; Komissarov, G. G. *Rep. Phys. Chem.* (in Rus.) **2004,** *399,* Part 1. 266–268.
44. Lobanov, A. V.; Kholuiskaya, S. N.; Komissarov, G. G. *Chem. Phy.* (in Rus.) **2004,** *23* (5), 44–47.
45. Nevrova, O. V.; Lobanov, A. V.; Komissarov, G. G. J. *Charact. Develop. Novel Mater.* **2011,** *3* (3), 172–176.
46. Komissarov, G. G.; Lobanov, A. V.; Nevrova, O. V. *Rep. Phys. Chem.* (in Rus.) **2013,** *453,* Part 2, 275–278.
47. Houtkooper, J. M.; Schulze-Makuch, D. Planet. Space Sci. **2009,** *57,* 449–453.
48. Bergman, P.; Parise B.; Liseau, R. et al. *A&A* **2011,** *531* (L8), 1–4.
49. Du, F., Parise B.; Bergman, P. *A&A* **2012,** 544 (C4), 1–2.

50. Encreannaz, T.; Greathouse, T. K.; Lefevre, F.; Atreya, S. K. *Planet. Space Sci.* **2012,** *68*, 3–17.
51. Komissarov, G. G.; Avakyanz, G. S.; Mazo, M. A. *Pros. Int. Conf. Nonlinear Word,* Astrakhan 2000, 110.
52. Komissarov, G. G.; Petrenko, U. M.; Avakyants, G. S.; Rubtsova, N. A. *Abstr. XX Intern Scien. Conf. Math. Methods in Technics and Technology Yaroslavl,* 2007, Vol. 2, pp 253–256.

CHAPTER 20

APPLICATION OF ORGANIC PARAMAGNETS IN BIOLOGICAL SYSTEMS

M. D. GOLDFEIN and E. G. ROZANTSEV

Saratov State University, Saratov, Russia. E-mail: goldfeinmd@mail.ru

CONTENTS

ABSTRACT

Research of the condensed phases containing stable radicals, by means of radio-spectroscopy, represents a method of paramagnetic sounding. This method (using iminoxyl radicals) allowed to set the mechanism of interaction of antigens with antibodies in case of study of immune gamma globulins, particularly structural transitions in biological membranes, to study the structure of some model systems. It is shown that certain derivatives of iminoxyl radicals have low toxicity and exhibit a relatively high antileukemic activity, greatest quantities were observed for the inhibition ratios hemocytoblasts in peripheral blood, bone marrow, and during chemotherapy of certain cancers.

20.1 INTRODUCTION

The presence of paramagnetic particles in liquid or solid objects opens new opportunities to study it by the EPR technique. Ready free radicals and substances forming paramagnetic solutions due to spontaneous homolization of their molecules in liquid and solidmedia (such as triphenylmethyl dimer, Frémy's salt or 4,8-diazaadamantan-4,8-dioxide) can act as sources of paramagnetic particles.

The experimental technique of radio spectroscopic examination of condensed phases with theaid of paramagnetic impurities is usually calledparamagnetic probe method. Though iminoxyl radicals have found broadest applications for probing of biomolecules, nevertheless, the first application of the paramagnetic probe technique to study a biological system is associated with a quite unstable aminazine radical cation:

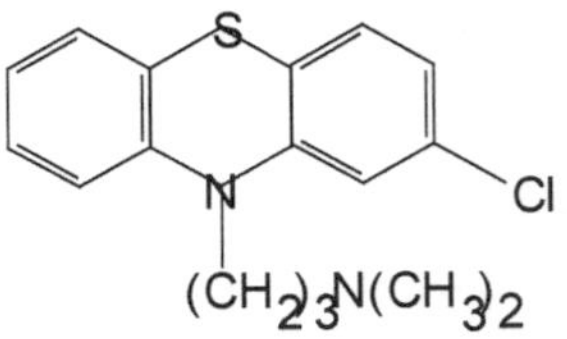

The progress in the theory and practice of EPR usage in biological research isrestrained bythe narrow framework of chemical reactivity ofnonfunctionalized stable radicals with alocalized paramagnetic center as follows:

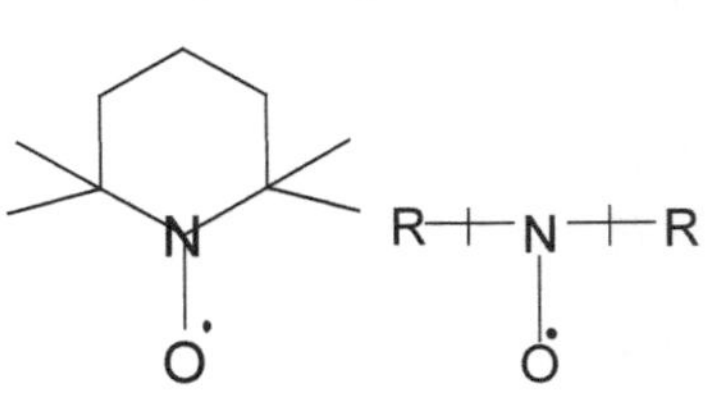

The substances of this class only enter intocommon, well-known free radical reactions, namely, recombination, disproportionation, addition to multiple bonds, isomerization, and β-splitting.[1] All these reactions proceed with the indispensable participation of a radical center and steadily lead to full paramagneticloss,though the synthesis ofnonfunctionalized stable radicalsplaysa very important role. No expressed delocalization of anuncoupled electronover amultiple-bond system has been shown to beobligatory for a paramagnet to be stable.

Despite of the basic importance of thediscovery of stable radicals of anonaromatic type,[2] this event has not changed contemporary ideas on the reactivity of stable radicals.

In the early 1960s, one of the authors of this book laid the foundation of a new lead in the chemistry of free radicals, namely, the synthesis and reactivity of functionalized stable radicals with an expressed localized paramagnetic center.[3]

The opportunity toobtain and studya wide range of such compounds with various functional substituents arose in connection with thediscovery of free radicalreactions with their paramagnetic center unaffected.

Functionalized free radicals have found broad applicationslikeparamagnetic probes for exploring molecular motion in condensed phases of various natures. The introduction of a spin-labeled technique (covalently bound paramagnetic probe) is associated with itsusage;this idea isnot new and is based on the dependence of the EPR spectrumshape of the free radical on the properties of its immediate atomic environment and the way of interaction of the paramagnetic fragment with the medium. The reactions of free radicals withtheir paramagnetic center unaffected (Neumann–Rozantsev's reactions)[4]have become the chemical basis forobtaining spin-labeled compounds.

The concept of the usage of non-radical reactions tostudy macromolecules was formulated at the Institute of Chemical Physics (USSR Academy of Sciences) by G.I. Lichtenstein[5]in 1961, and the theoretical bases of this method, calculationof algorithmsfor thecorrelation times of rotary mobility of a paramagnetic particle from their EPR spectrum shape were developed byMcConnell,[6]Freed and Fraenkel,[7]Kivelson,[8] and Stryukov[9]have investigated the behaviorofiminoxyl radicals in various systems and obtained important and interesting results.[10]

Let us cite several important aspects of the application of organic paramagnets to researching biological systems.

20.2 APPLICATION OF IMINOXYL-FREE RADICALS FORSTUDYING OF IMMUNE GAMMA GLOBULINS

From the physicochemical viewpoint, the mechanism of various immunological reactions is determined by changesinthe phase state of a system. Despite of the wide use of these reactions in medical practice, the nature of interaction of antigens with antibodies isnot quite clear. Tostudy this process,gamma globulins labeled with iminoxyl radicals were used.[11]The molecule of gamma globulin is known to consist four polypeptide chains bound with each other with disulfide bridges. When the interchain disulfide bonds are split, the polypeptide chains continue to keep together. A spin label (an iminoxyl derivative ofmaleimide)

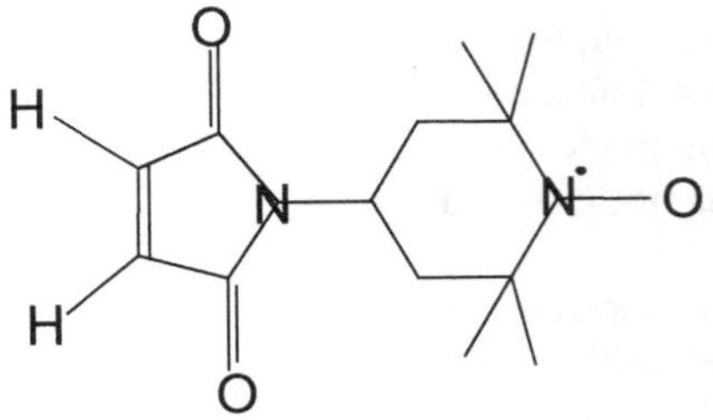

was attached to the sulfhydryl groupsobtained by restoration of disulfide bonds with β-mercaptoethanol. Experiments were made on the rabbit and human gamma globulins. The EPR spectra in both cases corresponded to rather high mobility of free radicals (correlation times=$1.1 \cdot 10^{-9}$s for human gamma globulin and $7.43 \cdot 10^{-9}$s for rabbit'sone). By comparing the correlation times in these proteins with the values obtained in experiments with serum albumin labeled withsulfhydryl groups, treatedwith urea ($\tau = 1.09 \cdot 10^{-9}$s) and dioxane ($2.04 \cdot 10^{-9}$s), it is possible to conclude that the fragments of polypeptide chains bearing free radicalspossess no ordered secondary structure. This is accordingto the data on very low contents of α-helical structures in gamma globulins. Such a character of the EPR spectrum of immune gamma globulinwith preservation of its specific activity opens the possibility to explore conformational and phase transitions at specific antigen–antibody reactions. Sharp distinctions in the mobility of spin labels were revealed at sedimentation of the rabbit antibodies by salting-outwithammonium sulfate and precipitation with a specific antigen (egg albumin). Precipitation of antibodies with a specific antigen led only to a small reduction of the paramagnetic labelmobility,whereas sedimentation withammonium sulfate caused strong retardation of the rotary mobility of free radicals (Fig. 20.1).

These results can be considered as direct confirmation of the alternative theory[12] according to which theprecipitate formation is associated with the immunological polyvalency of theantigen and antibody relative to each other. Really, exceptingthelocation of spinlabels inside the antibody's active center, it is possible to conclude that the rather mobile condition of spinlabels in theantigen–antibody precipitatemay remain if only there is no strong dehydration of the antibodies due to intermolecular interactions. Unlike gamma globulin precipitatedwithammonium sulfate, the specific precipitate, according to the lattice theory, has a microcellular structure. At long storage of theantigen–antibody precipitate with no addition of stabilizers, the mobility degree of iminoxyl radicals sharply decreased. This is apparently a result of secondary dehydration of the antibodies owing to protein molecule interactions in theprecipitate.

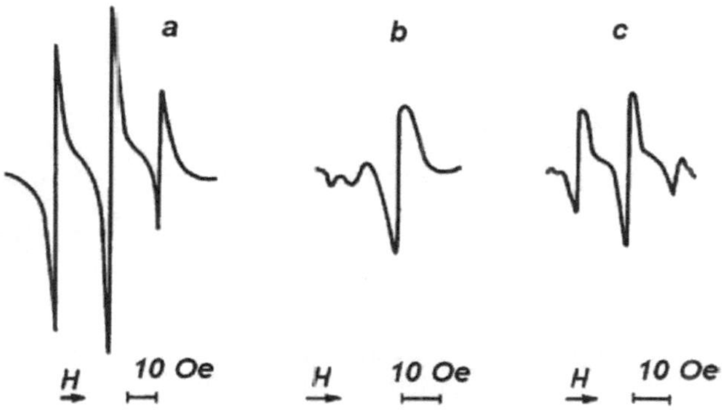

FIGURE 20.1 EPR spectra ofgamma globulin labeled with animinoxylradical with its free valency unaffected: (a)in solution; (b)in theprecipitateobtained bysalting-out with ammonium sulfate;(c)in thespecificprecipitate.

20.3 EXPLORING STRUCTURAL TRANSITIONS IN BIOLOGICAL MEMBRANES

Biological membranes, in particular,mitochondrial membranesare known to play a huge role in redox processes in the cell, being theonly place of respiratory chain enzyme localization. Baum and Riske[13] have found essential distinctions in the properties of one of the mitochondrial membrane fragments (Complex III of Electron Transport Chain) upon transition from the oxidized form to the reduced one: the sulfhydryl groups, easily titrated in the oxidized form of this fragment, become inaccessible in the reduced form. The nature of trypsin digestion of this complexalso strongly changes.

Earlier, changes in the repeating structural units of mitochondria in conditions leading to the formation of macroergic intermediate products or their provision, for example, active ion transfer, wererevealed by electron microscopy.[14]All this has allowed us to assume that anyredox reaction catalyzed by anenzymatic chain of electron transfer is accompanied by some kind of "conformational wave"probably covering not only the protein component of the membrane but also a higher level of its organization, namely, fragments of the membrane of more or lesscomplexity degree, including its lipidic part. To verify this assumption,a modification of thespin-labeledmethod (a method of noncovalentlybound paramagnetic probe) was used. The radical is kept by the matrix (membrane) involving only weak hydrophobic bonds. Such an approach allows studying of weak interactions in the system without essential disturbance of the biochemical functions of the biomembrane and its structure. The paramagnetic probe was 2,2,6,6-tetramethylpiperidine-1-oxyl caprylic ester:

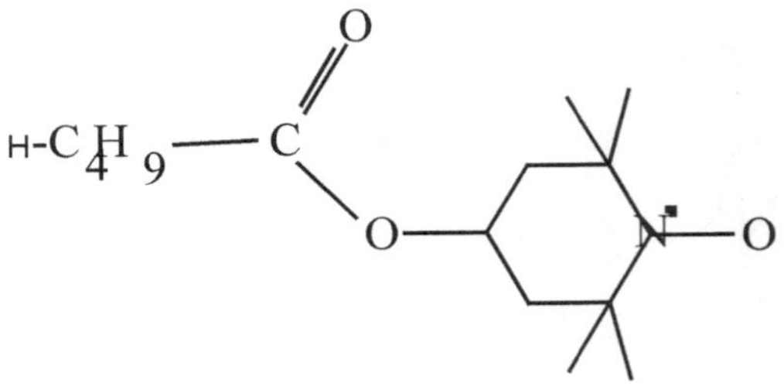

This compound was prepared from caprylic acid chloranhydride and 2,2,6,6-tetramethyl-4-oxpiperidin-1-oxyl in a triethylamine mediumbya radical reaction withfree valency unaffected. The paramagnetic probe was introduced into a suspension of electrontransport particles (ETP) isolated from thebull heart mitochondriabythetechnique described in Ref.at theLaboratory of Bioorganic Chemistry, Moscow State University. These fragments of the mitochondrial membrane are characterized by arather fullset of enzymes of the respiratory chain with the same molar ratio as in the intact mitochondria.[14]The ability of oxidizing phosphorylation, however, is lost under the used way of isolation.

The paramagnetic probe is insoluble in water but solubilized by ETPsuspended in a buffer solution. Owing to this, the observed EPR spectrum is free fromany backgrounddue to the radicalsnot attached with the object under study. The presence of a voluminous hydrocarbonic chain provides "embedding" of a molecule of the probe into the lipidic part of ETP. Therefore, the EPR spectrum reflects the condition of exactly this fraction of the membrane. To detect conformational transitions, EPR spectra were recorded before and after the introduction of oxidation substrata (succinate and NAD-N), and after oxidation of the earlier reduced respiratory chain with potassium ferricyanide. Typical results are shown in Figure 20.2.[14]

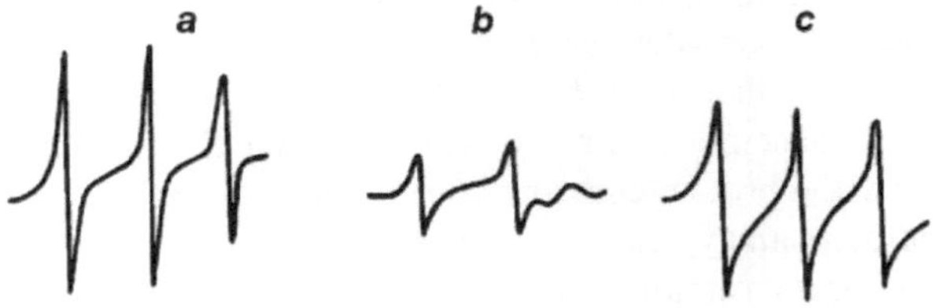

FIGURE 20.2 Changes in theEPRspectrumanisotropyof thehydrophobiciminoxylradicalin asuspension ofelectron transportparticlesfrom bovine heart mitochondriawith oxidation substrates added.

The enhanced anisotropy of the EPR spectrumof iminoxyl after substratum introduction is clearly seen. The spectrum in the ferricyanide-oxidized ETPalmost doesnot differ from that in the intact ETP.

ETP inactivation by long storage at room temperature or cyanide inhibition eliminated this effect. Comparison of theshape of signals and correlation times shows that the EPR spectrum ofiminoxyl in the intact ETP consists of two signals differing by anisotropy. The radical localized in that part of the membrane where the effective free volume available for radical motion is rather large gives aweakly anisotropic signal. The strongly anisotropic (retarded) spectrum belongs to the radicals localized in other sites of the system with a smaller effective free volume. Reduction of the respiratory chain with substrata leads, owing to cooperative-type conformational transitions, to a reducedfraction of sites with large free volume (i.e., to an increased microviscosity of the immediate environment of the radical).The correlation time of thewholespectrum changes from$20 \cdot 10^{-10}$s in the oxidized ETP to$4 \cdot 10^{-10}$s in the reduced ETP.

Concurrently with enhancingthe anisotropy of the signal, its intensity decreases as well: iminoxyl reduces, apparently, to hydroxylamine derivatives. Potassium ferricyanide inverts this process. It is necessary to consider that oxidation substrata,themselves, donot interact considerably with iminoxyls. Obviously, the conformational transition not only leads to achanged microviscosity but also eliminatesany steric obstacles complicating reduction of the radical. In principle, this circumstance points to possiblya new, actually chemical, aspect of application of the paramagnetic probe technique.

20.4 EXPLORING THE STRUCTURE OF SOME MODEL SYSTEMS

The application of the paramagnetic probe technique in systems like biological membranes poses a number of questions concerning the behavior of hydrophobic labels in media with an ordered arrangement of hydrophobic chains. As a first stage, mixtures of a nonionic detergent(Tween80) and water were studied. Tween80 originates from polyethoxylated sorbitan and oleic acid (polyoxyethylene (20) sorbitan monooleate) and classifies as anonionic detergent based on polyethylene oxide.[14]Our choice of this object was determined by some methodical conveniences, and also some literature data on the structure of aqueous solutions of Tween, obtained by classical methods (viscometry, refractometry, etc.) The properties of this detergent are also interesting in themselves since it finds quite broad applicationsfor biological membrane fragmentation.

The esters of 2,2,6,6-tetramethyl-4-oxypiperidin-1-oxyl and saturated acids of the normal structure with a hydrocarbonic chain length of 4,7,or 17 carbon atoms or the corresponding amides were used as a paramagnetic probe. For comparison, the behavior of hydrophobic labels IV and V in these systems was also studied

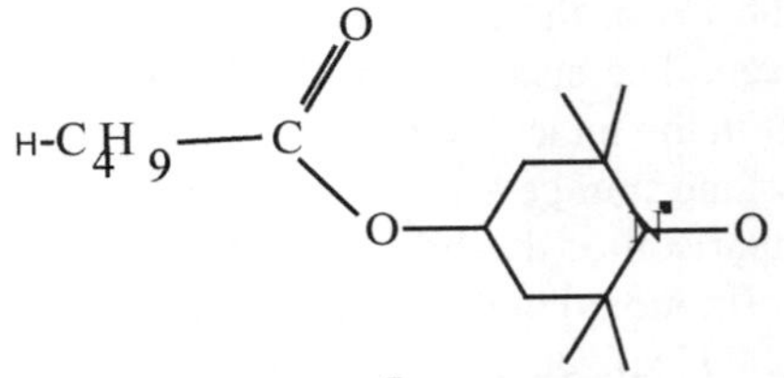

I

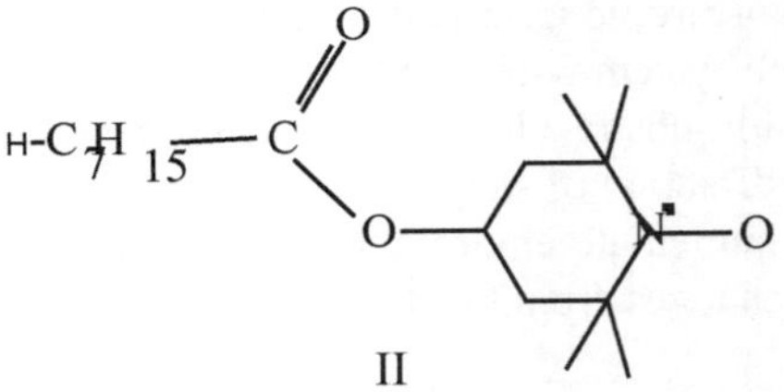

II

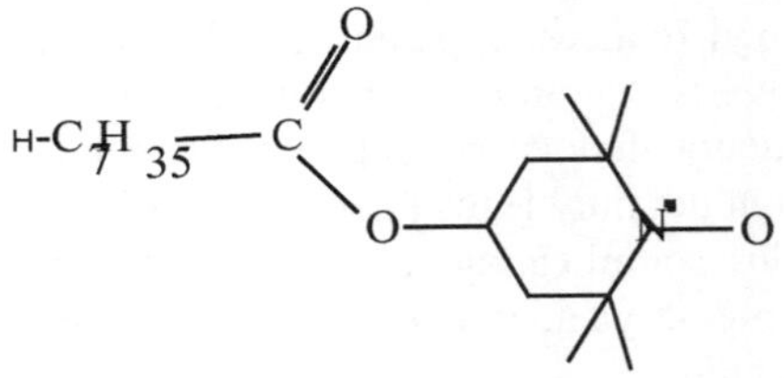

III

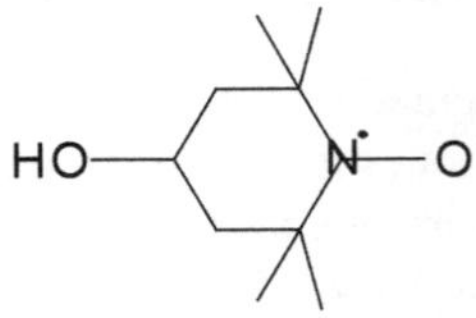

IV

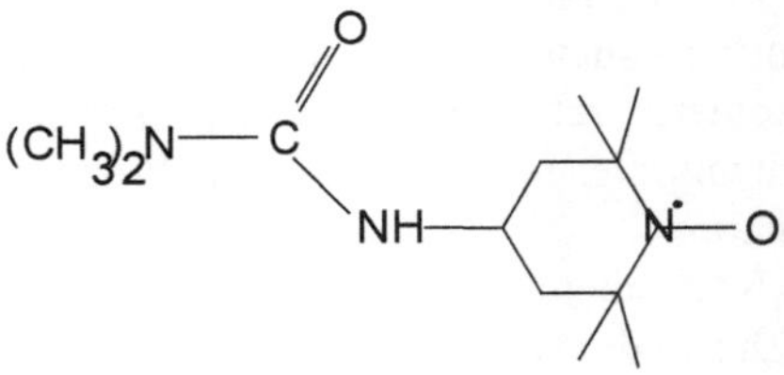

V

The course of changesin thecorrelation time of radical rotation as a function of the Tween concentration is shown in Figure 20.3.

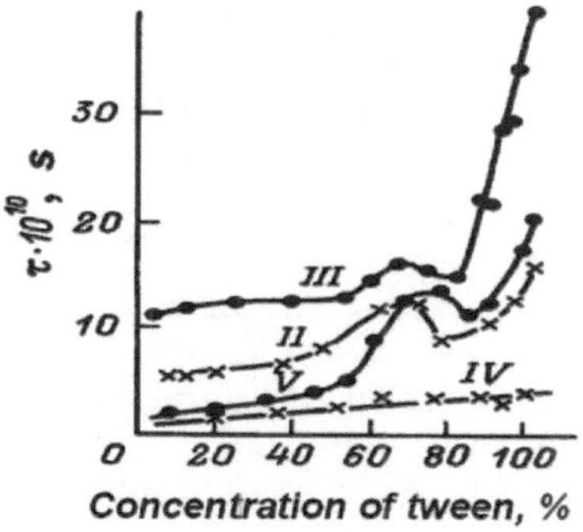

FIGURE 20.3 Changes of the correlation timein the detergent–water system, for nitroxyl radicals with astrongly localized paramagnetic center.

It is possible to resolve several areas, apparently, corresponding to various structure types. The initial fragment, only distinguishable for the easiest radicals, corresponds to an unsaturated Tween solution in water. This regionis better revealed for thedetergents with ahigher critical micelle concentration (CMC), for example, for sodium dodecyl sulfate (Fig. 20.4),[15,16] after which micelle formation occurs. Thecorrelation time of water-insoluble labels increases and comes to a plateau; at further increasingdetergent concentration, it passes through a maximumand then monotonously increases up to its value in pure Tween. It is useful to compare these data with the results of viscosity measurements by usual macromethods. Figure 20.5 shows that viscosity has one extremum about 60% of Tween. Therefore, at a high Tween concentration, the effective volume available for probe molecule rotation and macroviscosity donot correlate.

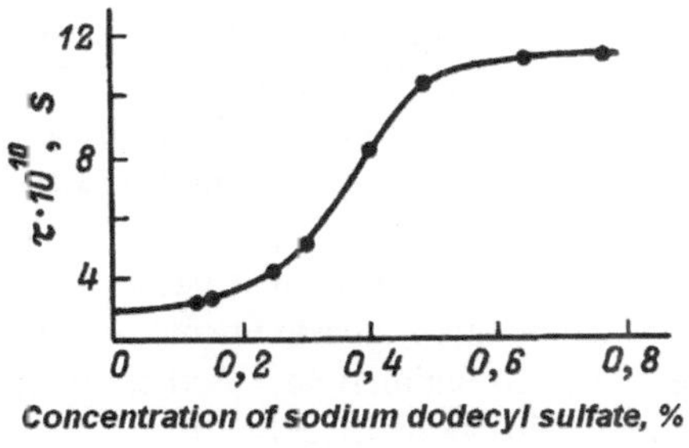

FIGURE 20.4 Estimation of the critical micelleconcentrationof sodiumdodecylsulfatewith the aid of 2,2,6,6-tertamethyl-4-hydroxypiperidyl-1-oxyl varerate.

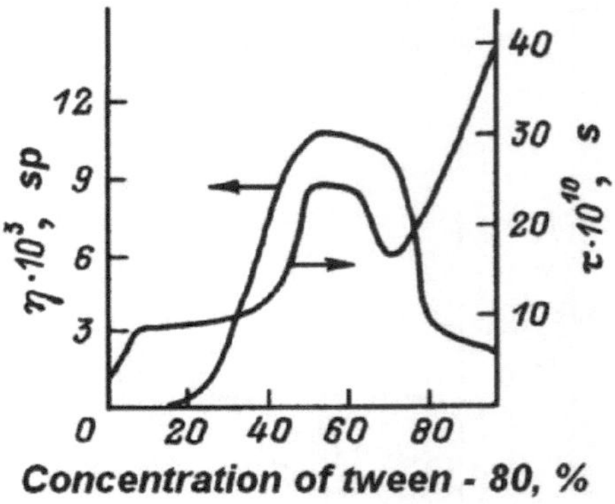

FIGURE 20.5 Changes in themicroviscosityand macroviscosity of thewater–Tween 80 system.

These results suggest the following interpretation. In pure Tween,the lamellar structure provides easy layer-by-layer sliding,hence, the macroviscosity of the system is low. However,in the absence of water, the interaction of the polar groups is strong, the hydrocarbonic chains are ordered, and the effective free volume in the field of radical localization is small. Small water amounts lead to the formation of defects in the layered structure. Sliding is hindered,and the viscosity increases. However, moistening breaks the close interaction of the polar groups inTween. Thesegroups are deformed, at the same time,the hydrocarbonic chains are disordered. Hence, the microviscosity of the hydrocarbonic layerdecreases. Upon termination of hydration of the polar groups,water-filled cavities are formed. They are a structural element (micelle) from which the system is built, for example, a hexagonal P-lattice is formed, byLuzzatti. Structure formation manifests itself as increasing microbiscosity and macroviscosity. In the field of the maximum, phase inversion is possible. Structural units of a new type (Tweenmicelles in water)are formed, passing into colloidal solution upon further dilution. The course of microviscosity changes at high Tweenconcentrations, amazingly resembles the change incorrelation time when some lyophilized cellular organellesare moistened. This similarity confirms that in the field of τmaximum, where restoration of the biochemical activity of chloroplasts begins, a phase transition occurs of the same type as in theLC"detergent–water" systems.

Some information on the behavior of radical particles in colloidal systems is provided by the results of temperature measurements. Figure 20.6 illustrates thetemperature dependence of the correlation time in Arrhenius' coordinates for several iminoxyl radicals of various hydrophobicity degrees.

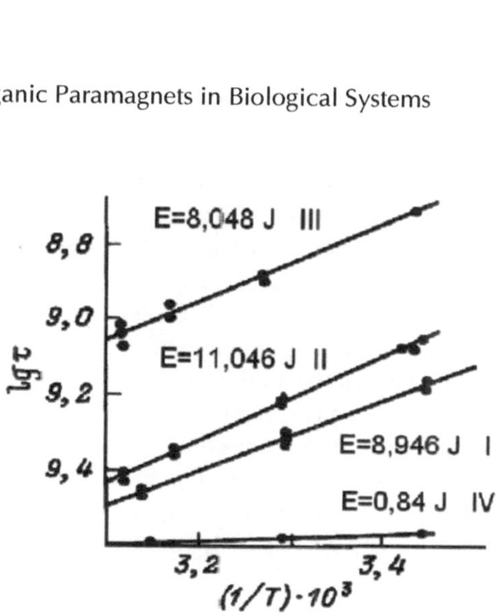

FIGURE 20.6 Activatyion energy ofrotationaldiffusion offreeiminoxyl (nitroxyl) radicalsin theTween–water (30% of Tween 80) system.

The strong dependence of the pre-exponential factoron chain length, apparently, confirms that the observed correlation time really reflects rotation of theradical molecule as a whole. However, this result also allows another interpretation, namely, depending on the hydrocarbonic chain length;the radical introducesitselfinto a detergent micelle more or less deeply. This may lead to a changedrotation frequency of iminoxyl groups round theordinarybonds in the molecule, depending on the environment of the polar end of the radical. If the polar group of the radical is on themicelle–solvent interface, the measured frequency should depend on the surface charge (potential) of the micelle. In our opinion, this circle of colloidchemical problems, closely connected with questions of transmembrane transfer in biological systems, will provide one more application field of the paramagnetic probe technique.

20.5 SOLVING OTHER BIOLOGICAL PROBLEMS WITH STABLE RADICALS

Further progress in the field of the usage of stable paramagnets to solve various biological problemsis reflected in numerous reviews and monographs.[17–20]

For dynamic biochemistry, of undoubted interest are local conformational changes of protein molecules in solutions. The distances between certain loci of biomacromolecules, in principle, can be estimated quantitatively by means of stable paramagnets. Upon introduction of iminoxyl fragments into certain sites of native protein (NRR-method),the distance between the neighboring paramagnetic centers can be calculatedfrom the efficiency of their dipole–dipole interaction in vitrified solutions of aspin-labeled preparation (the EPRmethod).

The first attempt to estimate the distances between paramagnetic centers in spin-labeledmesozyme and hemoglobin was undertaken by Liechtenstein,[21] who indicated prospects of such an approach. In this regard, there appeared a need of identification of a simple empirical parameter in EPR spectra for quantitative assessment of the dipole–dipole interaction of paramagnetic centers.

A convenient empirical parameter was found while studying vitrified solutions of iminoxyl radicals. It was the ratio of the total intensity of the extreme components of a spectrum to the intensity of the central one (Fig. 20.7). To establish a correlation of the d_1/dvalue with the average distance between the localized paramagnetic centers, the corresponding calibration plots are drawn (Fig. 20.8). Calculations have shown that the d_1/dparameter depends on the value of dipole–dipole broadening,being in fair agreement with independently obtained experimental results.

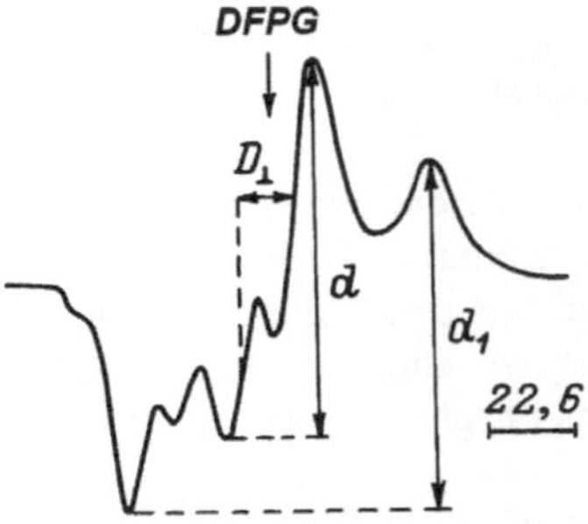

FIGURE 20.7 EPR spectrum of theiminoxyl-freebiradical (2,2,6,6-tetramethyl-4-hydroxypiperidine-1-oxyl phthalate) vitrified in tolueneat 77K.

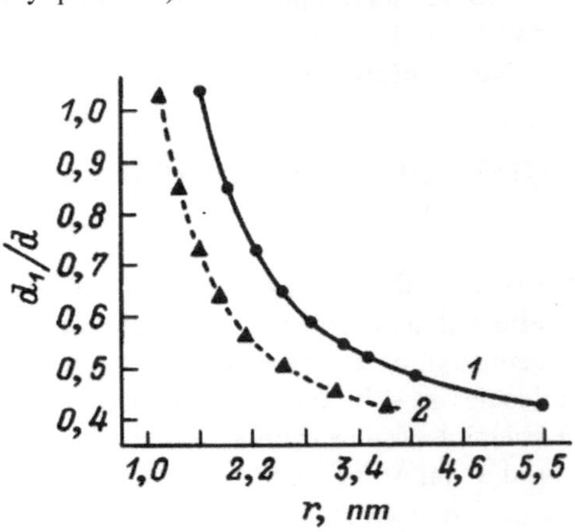

FIGURE 20.8 Dependence of thed_1/dparemeterof an EPR spectrumon themeandistance$r$betweeninteractionparamagneticcenters ofiminoxylradicals (1) andbiradicals (2) at 77 K.

Subsequently, the methods of quantitative assessment of the distances between paramagnetic centers in biradicals and spin-labeled biomolecules became reliable toolsfor structural research.

When determining relaxation rate constants of various paramagnetic centers in solution,the values of these constants have appeared to be significantly dependent on the chemical nature of the functional groups in iminoxyl radicals.

Sign inversion of the electrostatic charge of the substituent and itsdistance from the paramagnetic iminoxyl group has the strongest impact on the values of the constants. The electrostatic effect caused by the value and sign of the charge of paramagnetic particles interacting in solution is a more essential factor. An increased ionic strength leads to a change in the k value in qualitative agreement with Debye's theory. The substituent'smass is another factor considerably influencing the value of the constant. If the substituent is a protein macromolecule, the value of k decreases twice. The possibilities of application of the paramagnetic probe technique for detection of anion and cation groups and estimation of distances toexplore the microstructure of protein were analyzed. The dependences obtained in experiment show that Debye's equation with $D = 80$ can be successfully applied to experimental data analysis, and, in particular, to estimatingthe distance from the iminoxyl group of aspin label on protein to the nearest charged group, if this distance doesnot exceed 1.0–1.2 nm.

Considering the role of free radical processesin radiation cancer therapy, it was offered to investigate the influence of iminoxyl radicals upon the organism of laboratory animals. Pharmacokinetic studies[22] have shown that the elementary functionalized derivatives of 2,2,6,6-tetramethylpiperidin-1-oxyl possess rather low toxicity and show an expressed antileukemic activity. The highest values of retardationcoefficients were observed for hematopoietic stem cells in peripheral blood and bone marrow. This circumstance stimulated further structural and synthetic research directed toobtaining more effective and less toxic cancerolytics and sensibilizers for radiation cancer therapy.

After the first publication, for a rather short time,many potential cancerolytics were synthesizedin the Laboratory of Stable Radicals, Institute of Chemical Physics, USSR Academy of Sciences, among which of greatest interest from biologists wereso-called paramagnetic analogs of some known antitumor preparations, for example, a paramagnetic analog of Thiotepa[22].

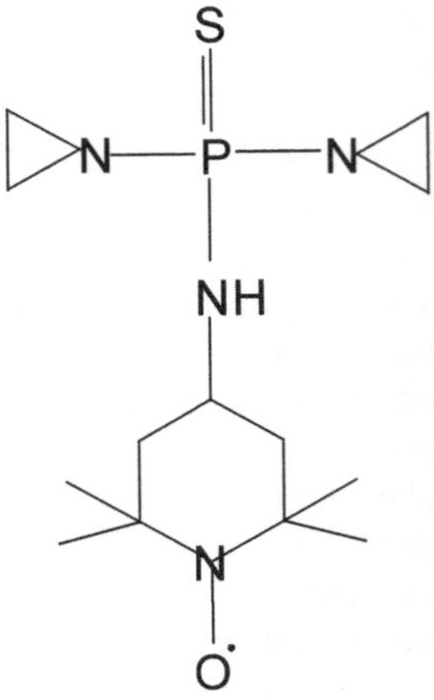

A briefreviewon theusage ofstableparamagnets of theiminoxylseriesintumor chemotherapy is presented bySuskina.[23]

KEYWORDS

- **biology**
- **organic paramagnetic method**
- **study**
- **radio spectroscopy**
- **sensing**
- **mechanism**

REFERENCES

1. Rozantsev, E.G.*Chem. Encyclopedic Dictionary*. M. SE.**1983**,p489 [Russ].
2. Rozantsev, E.G.;Lebedev, O.L.;Kazarnovskii, S.N. Diploma for the opening number 248 of 05.10.1983. TS. 1982,*6*,p. 6. [Russ].
3. Rozantsev, E.G.*Doctor Dissertation*; Institute of Chemical Physics, USSR Academy of Sciences:Moscow, RU,1965 [Russ].
4. Zhdanov, R.I. *Nitroxyl Radicals and Non-Radical Reactions of Free Radicals Bioactive Spin labels*;Zhdanov R.I., Ed.; Springer: Berlin, Heidelberg, NY, 1991,pp 24.
5. Liechtenstein, G.I. *Method of spin labels in molecular biology*;Nauka: Moscow, RU,1974;pp 255. [Russ].
6. McConnell,N. J.*Chem. Phys*. **1956**,*25*,709.
7. Freed, J.;Fraenkel, G.J. *Chem. Phys*. **1963**,*39*,326.

8. Kivelson. D. *J. Chem. Phys*. **1960**,*33*,4094.
9. Stryukov. V.B. *Stable Radicals in Chemical Physics;*Znanie: Moscow, RU, 1971 [Russ].
10. Goldfeld, M.G.;Grigoryan, G.L.;Rozantsev, E.G.*Polymer* (Gath. of preprints M. ICP AS SSSR.1979,p 269 [Russ].
11. Grigoryan, G.L.;Tatarinov, S.G.;Cullberg, L.Y.;Kalmanson, A.E.;Rozantsev, E.G.;Suskina. V.I.Abst.of USSR Academy of Sciences. 1968;*178*,p 230.*31*,p 768 [Russ].
12. Pressman, D. *Molecular Structure and Biological Specificity;*Nova: Washington D. C. 1957.
13. Baum, H.; Riske, J.;Silman, H.; Lipton, S.*Froc.Nature Acad. Sci. US.*1967,*57*,798.
14. Rozantsev, E.G.Biochemistry of Meat and Meat Products (General Part). *Manual for Students*. Depi.Print: Moscow, RU, 2006;p 240 [Russ].
15. Goldfeld, M.G.;Koltover, V.K.;Rozantsev, E.G.;Suskina, V.I.;Kolloid.ZS. *Bioactive Spin Labels*; Zhdanov, R. I., Ed.; SPRINGER,1970.
16. Schenfeld H. *Nonionic Detergents*. Marcel Dekker: New York,1963.
17. Smith, J. ShrierMuchillo, Sh.;Murch, D. Method of Spin Labels.*Free Radicals in Biology;* Mir: Moscow, RU, 1979;1,pp 179.
18. *Method of Spin Labels and Probes, Problems and Prospects;* Nauka:Moscow, RU, 1986 [Russ].
19. NitroxylRadicals. *Synthesis, Chemistry, Applications*; Nauka:Moscow, RU, 1987 [Russ].
20. Rozantsev, E.G.;Goldfein, M.D.;Pulin, V.F. *Organic Paramagnetic*. Saratov Gov. Univ.:Saratov, RU, 2000;pp 340 [Russ].
21. Liechtenstein, G.I. *J. Mol.Biol.***1968**,*2*,234.
22. Konovalova, N.P.;Bogdanov, G.N.; Miller, V.B.Abst.of USSR Academy of Sciences. T. 1964,*157*(3),pp 707. [Russ].
23. Suskina, V.I. *Candidate Dissertation*. Academy of Sciences: Moscow, RU, ICP USSR. 1970 [Russ].

CHAPTER 21

QUANTUM-CHEMICAL CALCULATION OF THE MODELS OF DEKACENE AND EICOSACENE BY METHOD MNDO WITHIN THE FRAMEWORK OF MOLECULAR GRAPHENE MODEL

V. A. BABKIN[1], V. V. TRIFONOV[1], A. P. KNYAZEV[1], V. YU. DMITRIEV[1], D. S. ANDREEV[1], A. V. IGNATOV[1], E. S. TITOVA[2], O. V. STOYANOV[3], and G. E. ZAIKOV[4]

[1]Sebrykov Department, Volgograd State Architect-build University, Akademicheskaya ul., 1, Volgograd 400074, Volgograd Oblast, Russia

[2]Volgograd State Technical University, Volgograd, Russia

[3]Kazan State Technological University, Kazan, Tatarstan, Russia

[4]N. M. Emanuel Institute of Biochemical Physics, Russian Academy of Sciences, 4, Kosygin St., Moscow 119334, Russian Federation

CONTENTS

Quantum-chemical calculation of molecules dekacene, eicosacene was done by method MNDO. Optimized by all parameters geometric and electronic structures of these compounds was received. Each of these molecular models has a universal factor of acidity equal to 33 (pKa = 33). They all pertain to class of very weak H-acids (pKa > 14).

21.1 AIMS AND BACKGROUNDS

The aim of this work is a study of electronic structure of molecules dekacene, eicosacene, and theoretical estimation of its acid power by quantum-chemical method MNDO within the framework of molecular graphene model, which was discovered by Novoselov and Game in 2004.[1]

21.2 METHODICAL PART

The calculation was done with optimization of all parameters by standard gradient method built-in in PC GAMESS.[2] The calculation was executed in approach the insulated molecule in gas phase. Program MacMolPlt was used for visual presentation of the model of the molecule.[3]

21.3 THE RESULTS OF THE CALCULATION AND DISCUSSION

Geometric and electronic structures, general and electronic energies of molecules dekacene, eicosane were received by MNDO method and are shown in Figure 21.1, Figure 21.2, and in Tables 21.1–21.3. The universal factor of acidity was calculated by formula: $pKa = 49.4–134.61*q_{max}H^{+}$ [4] (where, $q_{max}H^{+}$—a maximum positive charge on atom of the hydrogen (by Milliken[1]) $R = 0.97$, R–*a* coefficient of correlations, $q_{max}^{H+} = +0.06$). pKa = 33. This formula was successfully used in the following articles[5,6,7]:

Quantum-chemical calculation of molecules dekacene, eicosacene by MNDO method was executed for the first time. Optimized geometric and electronic structures of these compounds were received. Acid power of molecules dekacene, eicosacene was theoretically evaluated (pKa = 33). These compounds pertain to class of very weak H-acids (pKa > 14).

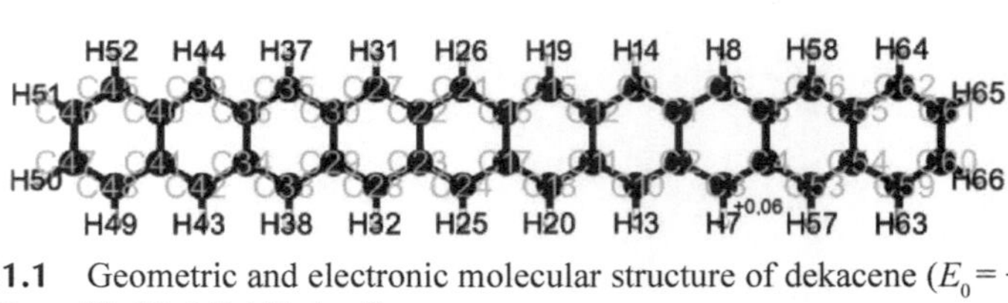

FIGURE 21.1 Geometric and electronic molecular structure of dekacene (E_0 = −5, 50, 105 kDg/mol, E_{el} = −48, 50, 841 kDg/mol).

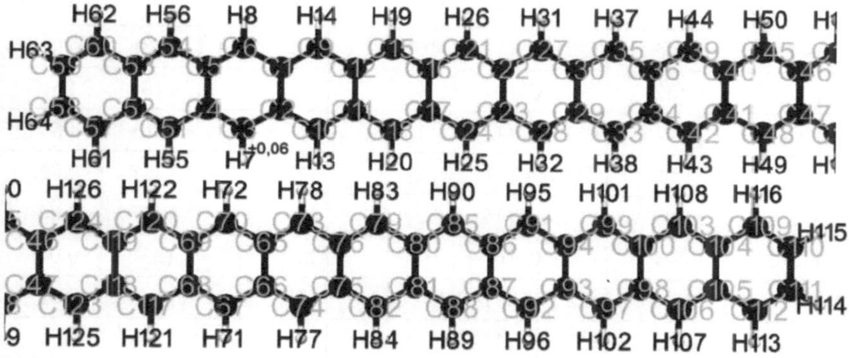

FIGURE 21.2 Geometric and electronic molecular structure of eicosacene (E_0 = −10, 69. 853 kDg/mol, E_{el} = −1, 17, 19, 827 kDg/mol).

TABLE 21.1 Optimized Bond Lengths, Valent Corners, and Atom Charges of Dekacene.

Bond lengths	R,A	Valent corners	Grad	Atom	Charge (by Milliken)
C(1)–C(2)	1.46	C(3)–C(2)–C(1)	118	C(1)	−0.04
C(2)–C(3)	1.45	C(12)–C(9)–C(1)	122	C(2)	−0.04
C(3)–C(4)	1.38	C(4)–C(3)–C(2)	123	C(3)	−0.02
C(4)–C(5)	1.47	C(9)–C(1)–C(2)	119	C(4)	−0.06
C(5)–C(6)	1.38	C(5)–C(4)–C(3)	119	C(5)	−0.06
C(6)–C(1)	1.45	C(10)–C(2)–C(3)	123	C(6)	−0.02
H(7)–C(3)	1.09	C(6)–C(5)–C(4)	119	H(7)	0.06
H(8)–C(6)	1.09	C(54)–C(53)–C(4)	123	H(8)	0.06
C(9)–C(12)	1.44	C(1)–C(6)–C(5)	123	C(9)	−0.02
C(9)–C(1)	1.39	C(53)–C(4)–C(5)	118	C(10)	−0.02
C(10)–C(2)	1.39	C(2)–C(1)–C(6)	118	C(11)	−0.04
C(11)–C(10)	1.44	C(9)–C(1)–C(6)	123	C(12)	−0.04
C(12)–C(11)	1.46	C(4)–C(3)–H(7)	120	H(13)	0.06
H(13)–C(10)	1.38	C(1)–C(6)–H(8)	117	H(14)	0.06
H(14)–C(9)	1.45	C(11)–C(12)–C(9)	118	C(15)	−0.02
C(15)–C(16)	1.42	C(15)–C(12)–C(9)	123	C(16)	−0.04

C(15)–C(12)	1.41	C(1)–C(2)–C(10)	119	C(17)	−0.04
C(16)–C(17)	1.45	C(18)–C(11)–C(10)	123	C(18)	−0.02
C(17)–C(18)	1.42	C(2)–C(10)–C(11)	122	H(19)	0.06
C(18)–C(11)	1.41	C(15)–C(12)–C(11)	119	H(20)	0.06
H(19)–C(15)	1.09	C(10)–C(11)–C(12)	118	C(21)	−0.02
H(20)–C(18)	1.09	C(16)–C(15)–C(12)	122	C(22)	−0.04
C(21)–C(22)	1.41	C(2)–C(10)–H(13)	120	C(23)	−0.04
C(21)–C(16)	1.42	C(12)–C(9)–H(14)	118	C(24)	−0.02
C(22)–C(23)	1.46	C(17)–C(16)–C(15)	119	H(25)	0.06
C(23)–C(24)	1.40	C(21)–C(16)–C(15)	123	H(26)	0.06
C(24)–C(17)	1.42	C(18)–C(17)–C(16)	119	C(27)	−0.02
H(25)–C(24)	1.09	C(22)–C(21)–C(16)	122	C(28)	−0.02
H(26)–C(21)	1.09	C(11)–C(18)–C(17)	122	C(29)	−0.04
C(27)–C(30)	1.39	C(21)–C(16)–C(17)	119	C(30)	−0.04
C(27)–C(22)	1.44	C(12)–C(11)–C(18)	119	H(31)	0.06
C(28)–C(23)	1.44	C(24)–C(17)–C(18)	123	H(32)	0.06
C(29)–C(28)	1.39	C(16)–C(15)–H(19)	118	C(33)	−0.02
C(30)–C(29)	1.46	C(11)–C(18)–H(20)	119	C(34)	−0.04
H(31)–C(27)	1.09	C(23)–C(22)–C(21)	119	C(35)	−0.02
H(32)–C(28)	1.09	C(27)–C(22)–C(21)	123	C(36)	−0.04
C(33)–C(29)	1.45	C(24)–C(23)–C(22)	119	H(37)	0.06
C(34)–C(33)	1.38	C(30)–C(27)–C(22)	122	H(38)	0.06
C(35)–C(36)	1.38	C(17)–C(24)–C(23)	122	C(39)	−0.02
C(35)–C(30)	1.45	C(27)–C(22)–C(23)	118	C(40)	−0.04
C(36)–C(34)	1.47	C(16)–C(17)–C(24)	119	C(41)	−0.04
H(37)–C(35)	1.09	C(28)–C(23)–C(24)	123	C(42)	−0.02
H(38)–C(33)	1.09	C(17)–C(24)–H(25)	119	H(43)	0.06
C(39)–C(40)	1.38	C(22)–C(21)–H(26)	119	H(44)	0.06
C(39)–C(36)	1.46	C(29)–C(30)–C(27)	119	C(45)	−0.04
C(40)–C(41)	1.48	C(35)–C(30)–C(27)	123	C(46)	−0.06
C(41)–C(42)	1.38	C(22)–C(23)–C(28)	118	C(47)	−0.06

C(42)–C(34)	1.46	C(33)–C(29)–C(28)	123	C(48)	−0.04
H(43)–C(42)	1.09	C(23)–C(28)–C(29)	122	H(49)	0.06
H(44)–C(39)	1.09	C(35)–C(30)–C(29)	118	H(50)	0.06
C(45)–C(46)	1.36	C(28)–C(29)–C(30)	119	H(51)	0.06
C(45)–C(40)	1.47	C(36)–C(35)–C(30)	124	H(52)	0.06
C(46)–C(47)	1.45	C(30)–C(27)–H(31)	120	C(53)	−0.02
C(47)–C(48)	1.36	C(23)–C(28)–H(32)	118	C(54)	−0.04
C(48)–C(41)	1.47	C(30)–C(29)–C(33)	118	C(55)	−0.04
H(49)–C(48)	1.09	C(42)–C(34)–C(33)	123	C(56)	−0.02
H(50)–C(47)	1.09	C(29)–C(33)–C(34)	123	H(57)	0.06
H(51)–C(46)	1.09	C(39)–C(36)–C(34)	118	H(58)	0.06
H(52)–C(45)	1.09	C(34)–C(36)–C(35)	119	C(59)	−0.04
C(53)–C(54)	1.38	C(39)–C(36)–C(35)	123	C(60)	−0.06
C(53)–C(4)	1.46	C(33)–C(34)–C(36)	119	C(61)	−0.06
C(54)–C(55)	1.48	C(40)–C(39)–C(36)	123	C(62)	−0.04
C(55)–C(56)	1.38	C(36)–C(35)–H(37)	120	H(63)	0.06
C(56)–C(5)	1.46	C(29)–C(33)–H(38)	117	H(64)	0.06
H(57)–C(53)	1.09	C(41)–C(40)–C(39)	119	H(65)	0.06
H(58)–C(56)	1.09	C(45)–C(40)–C(39)	123	H(66)	0.06
C(59)–C(60)	1.36	C(42)–C(41)–C(40)	119		
C(59)–C(54)	1.47	C(46)–C(45)–C(40)	122		
C(60)–C(61)	1.45	C(34)–C(42)–C(41)	123		
C(61)–C(62)	1.36	C(45)–C(40)–C(41)	118		
C(62)–C(55)	1.47	C(36)–C(34)–C(42)	118		
H(63)–C(59)	1.09	C(48)–C(41)–C(42)	123		
H(64)–C(62)	1.09	C(34)–C(42)–H(43)	117		
H(65)–C(61)	1.09	C(40)–C(39)–H(44)	120		
H(66)–C(60)	1.09	C(47)–C(46)–C(45)	121		
		C(48)–C(47)–C(46)	121		
		C(41)–C(48)–C(47)	122		
		C(40)–C(41)–C(48)	118		

		C(41)–C(48)–H(49)	118		
		C(48)–C(47)–H(50)	121		
		C(47)–C(46)–H(51)	118		
		C(46)–C(45)–H(52)	120		
		C(55)–C(54)–C(53)	119		
		C(3)–C(4)–C(53)	123		
		C(56)–C(55)–C(54)	119		
		C(60)–C(59)–C(54)	122		
		C(5)–C(56)–C(55)	123		
		C(59)–C(54)–C(55)	118		
		C(4)–C(5)–C(56)	118		
		C(6)–C(5)–C(56)	123		
		C(54)–C(53)–H(57)	120		
		C(5)–C(56)–H(58)	117		
		C(61)–C(60)–C(59)	121		
		C(53)–C(54)–C(59)	123		
		C(62)–C(61)–C(60)	121		
		C(55)–C(62)–C(61)	122		
		C(54)–C(55)–C(62)	118		
		C(56)–C(55)–C(62)	123		
		C(60)–C(59)–H(63)	120		
		C(55)–C(62)–H(64)	118		
		C(62)–C(61)–H(65)	121		
		C(61)–C(60)–H(66)	118		

TABLE 21.2 Optimized Bond Lengths, Valent Corners, and Atom Charges of Eicosacene.

Bond lengths	R,A	Valence corners	Grad	Atom	Charge (by Milliken)
C(2)–C(1)	1.46	C(5)–C(6)–C(1)	122	C(1)	−0.03
C(3)–C(2)	1.44	C(11)–C(10)–C(2)	122	C(2)	−0.04
C(4)–C(3)	1.39	C(1)–C(2)–C(3)	118	C(3)	−0.02
C(4)–C(5)	1.47	C(10)–C(2)–C(3)	123	C(4)	−0.04

C(5)–C(54)	1.46	C(5)–C(4)–C(3)	119	C(5)	−0.04
C(6)–C(5)	1.39	C(2)–C(3)–C(4)	122	C(6)	−0.02
C(6)–C(1)	1.44	C(54)–C(5)–C(4)	118	H(7)	+0.06
H(7)–C(3)	1.09	C(6)–C(5)–C(4)	119	H(8)	+0.06
H(8)–C(6)	1.09	C(53)–C(54)–C(5)	123	C(9)	−0.02
C(9)–C(1)	1.40	C(54)–C(5)–C(6)	123	C(10)	−0.02
C(10)–C(2)	1.40	C(2)–C(1)–C(6)	118	C(11)	−0.03
C(10)–C(11)	1.43	C(2)–C(3)–H(7)	118	C(12)	−0.03
C(11)–C(18)	1.42	C(5)–C(6)–H(8)	120	H(13)	+0.06
C(11)–C(12)	1.45	C(1)–C(6)–H(8)	118	H(14)	+0.06
C(12)–C(9)	1.43	C(2)–C(1)–C(9)	119	C(15)	−0.02
H(13)–C(10)	1.09	C(1)–C(2)–C(10)	119	C(16)	−0.04
H(14)–C(9)	1.09	C(18)–C(11)–C(10)	123	C(17)	−0.04
C(15)–C(12)	1.42	C(12)–C(11)–C(10)	119	C(18)	−0.02
C(16)–C(15)	1.41	C(17)–C(18)–C(11)	122	H(19)	+0.06
C(16)–C(17)	1.45	C(9)–C(12)–C(11)	119	H(20)	+0.06
C(17)–C(24)	1.44	C(15)–C(12)–C(11)	119	C(21)	−0.02
C(18)–C(17)	1.41	C(1)–C(9)–C(12)	122	C(22)	−0.04
H(19)–C(15)	1.09	C(18)–C(11)–C(12)	119	C(23)	−0.04
H(20)–C(18)	1.09	C(2)–C(10)–H(13)	119	C(24)	−0.02
C(21)–C(16)	1.44	C(11)–C(10)–H(13)	118	H(25)	+0.06
C(22)–C(21)	1.39	C(1)–C(9)–H(14)	119	H(26)	+0.06
C(23)–C(22)	1.46	C(9)–C(12)–C(15)	123	C(27)	−0.02
C(24)–C(23)	1.39	C(17)–C(16)–C(15)	119	C(28)	−0.02
H(25)–C(24)	1.09	C(12)–C(15)–C(16)	122	C(29)	−0.04
H(26)–C(21)	1.09	C(24)–C(17)–C(16)	118	C(30)	−0.04
C(27)–C(22)	1.45	C(18)–C(17)–C(16)	119	H(31)	+0.06
C(28)–C(23)	1.45	C(23)–C(24)–C(17)	122	H(32)	+0.06
C(29)–C(28)	1.38	C(24)–C(17)–C(18)	123	C(33)	−0.02
C(29)–C(30)	1.47	C(12)–C(15)–H(19)	119	C(34)	−0.04

C(30)–C(27)	1.38	C(17)–C(18)–H(20)	119	C(35)	−0.02
H(31)–C(27)	1.09	C(15)–C(16)–C(21)	123	C(36)	−0.04
H(32)–C(28)	1.09	C(17)–C(16)–C(21)	118	H(37)	+0.06
C(33)–C(29)	1.46	C(16)–C(21)–C(22)	122	H(38)	+0.06
C(34)–C(33)	1.38	C(21)–C(22)–C(23)	119	C(39)	−0.02
C(34)–C(36)	1.48	C(27)–C(22)–C(23)	118	C(40)	−0.04
C(34)–C(42)	1.46	C(22)–C(23)–C(24)	119	C(41)	−0.04
C(35)–C(30)	1.46	C(28)–C(23)–C(24)	123	C(42)	−0.02
C(36)–C(35)	1.38	C(23)–C(24)–H(25)	120	H(43)	+0.06
H(37)–C(35)	1.09	C(16)–C(21)–H(26)	118	H(44)	+0.06
H(38)–C(33)	1.09	C(21)–C(22)–C(27)	123	C(45)	−0.02
C(39)–C(36)	1.46	C(22)–C(23)–C(28)	118	C(46)	−0.03
C(40)–C(39)	1.38	C(30)–C(29)–C(28)	119	C(47)	−0.03
C(40)–C(41)	1.48	C(23)–C(28)–C(29)	123	C(48)	−0.02
C(41)–C(48)	1.45	C(27)–C(30)–C(29)	119	H(49)	+0.05
C(42)–C(41)	1.38	C(35)–C(30)–C(29)	118	H(50)	+0.05
H(43)–C(42)	1.09	C(22)–C(27)–C(30)	123	C(51)	−0.02
H(44)–C(39)	1.09	C(22)–C(27)–H(31)	117	C(52)	−0.04
C(45)–C(40)	1.45	C(23)–C(28)–H(32)	117	C(53)	−0.04
C(46)–C(45)	1.39	C(28)–C(29)–C(33)	123	C(54)	−0.02
C(47)–C(46)	1.49	C(30)–C(29)–C(33)	118	H(55)	+0.06
C(48)–C(47)	1.39	C(36)–C(34)–C(33)	120	H(56)	+0.06
H(49)–C(48)	1.09	C(42)–C(34)–C(33)	123	C(57)	−0.04
H(50)–C(45)	1.09	C(29)–C(33)–C(34)	123	C(58)	−0.06
C(51)–C(4)	1.46	C(35)–C(36)–C(34)	120	C(59)	−0.06
C(52)–C(51)	1.38	C(41)–C(42)–C(34)	123	C(60)	−0.04
C(52)–C(53)	1.48	C(39)–C(36)–C(34)	118	H(61)	+0.06
C(53)–C(60)	1.47	C(27)–C(30)–C(35)	123	H(62)	+0.06
C(54)–C(53)	1.38	C(30)–C(35)–C(36)	123	H(63)	+0.06
H(55)–C(51)	1.09	C(42)–C(34)–C(36)	118	H(64)	+0.06

H(56)–C(54)	1.09	C(30)–C(35)–H(37)	117	C(65)	−0.04
C(57)–C(52)	1.47	C(29)–C(33)–H(38)	117	C(66)	−0.04
C(58)–C(57)	1.36	C(35)–C(36)–C(39)	123	C(67)	−0.02
C(59)–C(58)	1.45	C(41)–C(40)–C(39)	119	C(68)	−0.04
C(60)–C(59)	1.36	C(36)–C(39)–C(40)	123	C(69)	−0.04
H(61)–C(57)	1.09	C(48)–C(41)–C(40)	118	C(70)	−0.02
H(62)–C(60)	1.09	C(42)–C(41)–C(40)	119	H(71)	+0.06
H(63)–C(59)	1.09	C(47)–C(48)–C(41)	124	H(72)	+0.06
H(64)–C(58)	1.09	C(48)–C(41)–C(42)	123	C(73)	−0.02
C(65)–C(70)	1.46	C(41)–C(42)–H(43)	120	C(74)	−0.02
C(66)–C(65)	1.47	C(36)–C(39)–H(44)	117	C(75)	−0.04
C(67)–C(66)	1.46	C(39)–C(40)–C(45)	123	C(76)	−0.04
C(68)–C(67)	1.37	C(41)–C(40)–C(45)	118	H(77)	+0.05
C(68)–C(69)	1.48	C(40)–C(45)–C(46)	124	H(78)	+0.06
C(69)–C(120)	1.46	C(45)–C(46)–C(47)	119	C(79)	−0.02
C(70)–C(69)	1.38	C(124)–C(46)–C(47)	118	C(80)	−0.04
H(71)–C(67)	1.09	C(46)–C(47)–C(48)	119	C(81)	−0.04
H(72)–C(70)	1.09	C(123)–C(47)–C(48)	123	C(82)	−0.02
C(73)–C(65)	1.38	C(47)–C(48)–H(49)	119	H(83)	+0.06
C(74)–C(66)	1.38	C(40)–C(45)–H(50)	117	H(84)	+0.06
C(75)–C(74)	1.46	C(3)–C(4)–C(51)	123	C(85)	−0.02
C(75)–C(76)	1.47	C(5)–C(4)–C(51)	118	C(86)	−0.03
C(76)–C(73)	1.46	C(53)–C(52)–C(51)	119	C(87)	−0.04
H(77)–C(74)	1.09	C(4)–C(51)–C(52)	123	C(88)	−0.02
H(78)–C(73)	1.09	C(60)–C(53)–C(52)	118	H(89)	+0.06
C(79)–C(76)	1.39	C(54)–C(53)–C(52)	119	H(90)	+0.06
C(80)–C(79)	1.45	C(59)–C(60)–C(53)	122	C(91)	−0.02
C(80)–C(81)	1.46	C(60)–C(53)–C(54)	123	C(92)	−0.02
C(81)–C(82)	1.45	C(4)–C(51)–H(55)	117	C(93)	−0.04
C(82)–C(75)	1.39	C(53)–C(54)–H(56)	120	C(94)	−0.04

H(83)–C(79)	1.09	C(51)–C(52)–C(57)	123	H(95)	+0.06
H(84)–C(82)	1.09	C(53)–C(52)–C(57)	118	H(96)	+0.06
C(85)–C(80)	1.40	C(52)–C(57)–C(58)	122	C(97)	−0.02
C(86)–C(85)	1.43	C(57)–C(58)–C(59)	121	C(98)	−0.04
C(86)–C(87)	1.45	C(58)–C(59)–C(60)	121	C(99)	−0.02
C(87)–C(88)	1.43	C(52)–C(57)–H(61)	118	C(100)	−0.04
C(88)–C(81)	1.40	C(59)–C(60)–H(62)	120	H(101)	+0.06
H(89)–C(88)	1.09	C(58)–C(59)–H(63)	118	H(102)	+0.06
H(90)–C(85)	1.09	C(57)–C(58)–H(64)	121	C(103)	−0.02
C(91)–C(86)	1.42	C(69)–C(70)–C(65)	123	C(104)	−0.04
C(92)–C(87)	1.42	C(70)–C(65)–C(66)	118	C(105)	−0.04
C(93)–C(92)	1.41	C(73)–C(65)–C(66)	119	C(106)	−0.02
C(93)–C(94)	1.45	C(65)–C(66)–C(67)	118	H(107)	+0.06
C(94)–C(91)	1.41	C(74)–C(66)–C(67)	123	H(108)	+0.06
H(95)–C(91)	1.09	C(69)–C(68)–C(67)	120	C(109)	−0.04
H(96)–C(92)	1.09	C(66)–C(67)–C(68)	123	C(110)	−0.06
C(97)–C(93)	1.44	C(120)–C(69)–C(68)	118	C(111)	−0.06
C(98)–C(97)	1.39	C(70)–C(69)–C(68)	120	C(112)	−0.04
C(98)–C(100)	1.46	C(119)–C(120)–C(69)	123	H(113)	+0.06
C(98)–C(106)	1.45	C(120)–C(69)–C(70)	123	H(114)	+0.06
C(99)–C(94)	1.44	C(66)–C(67)–H(71)	117	H(115)	+0.06
C(100)–C(99)	1.39	C(69)–C(70)–H(72)	120	H(116)	+0.06
H(101)–C(99)	1.09	C(70)–C(65)–C(73)	123	C(117)	−0.02
H(102)–C(97)	1.09	C(65)–C(66)–C(74)	119	C(118)	−0.04
C(103)–C(100)	1.45	C(76)–C(75)–C(74)	118	C(119)	−0.04
C(104)–C(103)	1.38	C(66)–C(74)–C(75)	123	C(120)	−0.02
C(104)–C(105)	1.47	C(73)–C(76)–C(75)	118	H(121)	+0.05
C(105)–C(112)	1.47	C(79)–C(76)–C(75)	119	H(122)	+0.05
C(106)–C(105)	1.38	C(65)–C(73)–C(76)	123	C(123)	−0.02
H(107)–C(106)	1.09	C(66)–C(74)–H(77)	120	C(124)	−0.02

H(108)–C(103)	1.09	C(65)–C(73)–H(78)	120	H(125)	+0.05
C(109)–C(104)	1.47	C(73)–C(76)–C(79)	123	H(126)	+0.05
C(110)–C(109)	1.36	C(81)–C(80)–C(79)	118		
C(111)–C(110)	1.45	C(76)–C(79)–C(80)	122		
C(112)–C(111)	1.36	C(82)–C(81)–C(80)	118		
H(113)–C(112)	1.09	C(88)–C(81)–C(80)	119		
H(114)–C(111)	1.09	C(75)–C(82)–C(81)	122		
H(115)–C(110)	1.09	C(74)–C(75)–C(82)	123		
H(116)–C(109)	1.09	C(76)–C(75)–C(82)	119		
C(117)–C(68)	1.46	C(76)–C(79)–H(83)	120		
C(118)–C(117)	1.38	C(75)–C(82)–H(84)	120		
C(118)–C(119)	1.49	C(79)–C(80)–C(85)	123		
C(118)–C(123)	1.44	C(81)–C(80)–C(85)	119		
C(119)–C(124)	1.44	C(87)–C(86)–C(85)	119		
C(120)–C(119)	1.38	C(80)–C(85)–C(86)	122		
H(121)–C(117)	1.09	C(88)–C(87)–C(86)	119		
H(122)–C(120)	1.09	C(92)–C(87)–C(86)	119		
C(123)–C(47)	1.42	C(81)–C(88)–C(87)	122		
C(124)–C(46)	1.42	C(82)–C(81)–C(88)	123		
H(125)–C(123)	1.09	C(81)–C(88)–H(89)	119		
H(126)–C(124)	1.09	C(80)–C(85)–H(90)	119		
		C(85)–C(86)–C(91)	123		
		C(87)–C(86)–C(91)	119		
		C(88)–C(87)–C(92)	123		
		C(94)–C(93)–C(92)	119		
		C(87)–C(92)–C(93)	122		
		C(91)–C(94)–C(93)	119		
		C(99)–C(94)–C(93)	118		
		C(86)–C(91)–C(94)	122		
		C(86)–C(91)–H(95)	119		
		C(87)–C(92)–H(96)	119		

		C(92)–C(93)–C(97)	123		
		C(94)–C(93)–C(97)	118		
		C(100)–C(98)–C(97)	119		
		C(106)–C(98)–C(97)	123		
		C(93)–C(97)–C(98)	122		
		C(99)–C(100)–C(98)	119		
		C(105)–C(106)–C(98)	123		
		C(103)–C(100)–C(98)	118		
		C(91)–C(94)–C(99)	123		
		C(94)–C(99)–C(100)	122		
		C(106)–C(98)–C(100)	118		
		C(94)–C(99)–H(101)	118		
		C(93)–C(97)–H(102)	118		
		C(99)–C(100)–C(103)	123		
		C(105)–C(104)–C(103)	119		
		C(100)–C(103)–C(104)	123		
		C(112)–C(105)–C(104)	118		
		C(106)–C(105)–C(104)	119		
		C(111)–C(112)–C(105)	122		
		C(112)–C(105)–C(106)	123		
		C(105)–C(106)–H(107)	120		
		C(100)–C(103)–H(108)	117		
		C(103)–C(104)–C(109)	123		
		C(105)–C(104)–C(109)	118		
		C(104)–C(109)–C(110)	122		
		C(109)–C(110)–C(111)	121		
		C(110)–C(111)–C(112)	121		
		C(111)–C(112)–H(113)	120		
		C(110)–C(111)–H(114)	118		
		C(109)–C(110)–H(115)	121		
		C(104)–C(109)–H(116)	118		

		C(67)–C(68)–C(117)	123		
		C(69)–C(68)–C(117)	118		
		C(119)–C(118)–C(117)	119		
		C(123)–C(118)–C(117)	123		
		C(68)–C(117)–C(118)	123		
		C(124)–C(119)–C(118)	118		
		C(47)–C(123)–C(118)	124		
		C(120)–C(119)–C(118)	119		
		C(46)–C(124)–C(119)	124		
		C(123)–C(118)–C(119)	118		
		C(124)–C(119)–C(120)	123		
		C(68)–C(117)–H(121)	117		
		C(119)–C(120)–H(122)	120		
		C(46)–C(47)–C(123)	118		
		C(45)–C(46)–C(124)	123		
		C(47)–C(123)–H(125)	118		
		C(46) –C(124) –H(126)	118		

TABLE 21.3 Total Energy (E_0), Maximal Charge on the Hydrogen Atom (q_{max}^{H+}), and Universal Factor of Acidity (pKa) of Molecules Dekacene and Eicosacene.

Molecules	E_0 **(kDg/mol)**	q_{max}^{H+}	**pKa**
Dekacene	−5,50,105	+0.06	33
Eicosacene	−10,69,853	+0.06	33

KEYWORDS

- **dekacene**
- **eicosacene**
- **molecular graphene model**
- **H-acids**
- **method MNDO**

REFERENCES

1. Novoselov, K. S., et al. Electric Field Effect in Atomically Thin Carbon Films, *Science* **2004**, *306*, 666, DOI:10.1126/science.1102896.
2. Shmidt, M. W.; Baldrosge, K. J.; Elbert, J. A.; Gordon, M. S.; Enseh, J. H.; Koseki, S.; Matsvnaga., N.; Nguyen, K. A.; Su, S. J., etal. *J. Comput. Chem.***1993**, *14*, 1347–1363.
3. Bode, B. M.; Gordon, M. S. *J. Mol. Graphics Mod.* **1998**, *16*, 133–138.
4. Babkin, V. A.; Fedunov, R. G.; Minsker, K. S., et al. *Oxid. Commun.* **2002**, 25 (1), 21–47.
5. Babkin, V. A.; Ignatov, A. V.; Ignatov, A. N.; Gulyukin, M. N.; Dmitriev, V. Yu.; Stoyanov, O. V.; Zaikov, G. E. *Quantum-chemical Calculation of Some Molecules of Triboratols*; "Vestnik" of Kazan State Technological University: Kazan, RU, 2013; Vol. 16, N2, pp 15–17.
6. Babkin, V. A.; Ignatov, A. V.; Stoyanov, O.V.; Zaikov, G. E. *Quantum-chemical Calculation of Some Monomers of Cationic Polymerisation with Small Cycles;* "Vestnik" of Kazan State Technological University: Kazan, Ru, 2013; Vol. 16, N 4, pp 21–22.
7. Babkin, V. A.; Trifonov, V. V.; Lebedev, N. G.; Dmitriev, V. Yu.; Andreev, D. S.; Stoyanov, O. V.; Zaikov, G. E. *Quantum-chemical Calculation of Tetracene and Pentacene by MNDO in Approximation of the Linear Molecular Model of Graphene*; "Vestnik" of Kazan State Technological University: Kazan, RU, 2013., Vol. 16, N 7, pp 16–18.

INDEX

B

C

D

G

H

P